PHYSICAL CHEMISTRY OF EXTRACTIVE METALLURGY

Edited by
V. Kudryk and Y. K. Rao

PHYSICAL CHEMISTRY OF EXTRACTIVE METALLURGY

Proceedings of an international symposium sponsored by the Physical Chemistry Committee and the Copper, Nickel, Cobalt and Precious Metals Committee of The Metallurgical Society of AIME held at the AIME Annual Meeting, New York, New York, February 24-28, 1985.

Edited by

Val Kudryk
ASARCO Incorporated
South Plainfield, New Jersey 07080

Y. Kris Rao
University of Washington
Seattle, Washington 98195

A Publication of   The Metallurgical Society of AIME

Library of Congress Cataloging in Publication Data
Main entry under title:

Physical chemistry of extractive metallurgy.

Held during the AIME Annual Meeting, Feb. 24-28,
1985, New York, N.Y.
Includes bibliographies and index.
1. Metallurgy--Congresses. 2. Chemistry,
Physical and theoretical--Congresses. I. Kudryk, Val.
II. Rao, Y. K. III. Metallurgical Society of AIME.
Physical Chemistry of Extractive Metallurgy Committee.
IV. Metallurgical Society of AIME. Cooper, Nickel and
Precious Metals Committee. V. AIME Meeting (1985 :
New York, N.Y.)
TN605.P49 1985   669'.94   84-29561
ISBN 0-89520-486-X

A Publication of The Metallurgical Society of AIME
420 Commonwealth Drive
Warrendale, Pennsylvania 15086
(412) 776-9000

The Metallurgical Society and American Institute of Mining, Metallurgical, and Petroleum Engineers are not responsible for statements or opinions and absolved of liability due to misuse of information contained in this publication.

© 1985 by American Institute of Mining, Metallurgical,
and Petroleum Engineers, Inc.
345 East 47th Street
New York, NY 10017

Printed in the United States of America.
Library of Congress Catalogue Number 84-29561
ISBN NUMBER 0-89520-486-X

Authorization to photocopy items for internal or personal use or the internal or personal use of specific clients, is granted by The Metallurgical Society of AIME for libraries and other users registered with the Copyright Clearance Center (CCC) Transactional Reporting Service, provided that the base fee of $3.00 per copy is paid directly to Copyright Clearance Center, 29 Congress Street, Salem, Massachusetts 01970.

Foreword

The Herbert H. Kellogg Symposium on the Physical Chemistry of Extractive Metallurgy was planned, organized and sponsored by the Physical Chemistry Committee and the Copper, Nickel, Cobalt and Precious Metals Committee of The Metallurgical Society of AIME. The proceedings, comprised of 33 papers authored by distinguished contributors, reflect the active interest in extractive metallurgy and the high esteem held for Professor H. H. Kellogg.

The Committees concluded that it was an appropriate time for a symposium to focus on current studies and directions for the future in the field of extractive metallurgy. It was also deemed fitting to name it the Herbert H. Kellogg Symposium in recognition of active participation in AIME Committees and for his contributions to this field. His numerous publications and awards attest to his prolific works of superior quality, as well as his early influence in utilizing thermodynamics for a better understanding of metallurgical reactions. His services as a teacher, scientist and consultant have been greatly appreciated as we look forward to a continuing source of enlightenment and accomplishments. The 1985 Annual Meeting in New York was an ideal locale for the Symposium since it is the home of Columbia University with which Professor Kellogg has had a long and outstanding association.

A wide array of topics were presented including studies in thermodynamics, kinetics and process development as related to extractive metallurgy. Subject matter ranged from basic research to practical applications in industry. Most of the papers have been included in this volume of the proceedings.

The editors would like to express their appreciation to the Session Chairmen for their participation and directing their respective sessions. Particular thanks are due to the authors not only for the time and effort in preparing the papers but also for sharing their knowledge which no doubt will have a significant bearing on future developments. It was especially gratifying to have several of Professor Kellogg's students present papers.

Much of the success for this symposium can be attributed to the diligent work performed by the TMS-AIME staff and we would like to thank Alexander Scott, Frederick Pettit, Elizabeth Luzar, Barbara Kamperman and Marilyn Zabel for making the task so much easier. The assistance of Lorraine Kaczenski at Asarco is gratefully acknowledged for communicating with the authors and assembling the papers for the volume.

Val Kudryk
ASARCO Incorporated
South Plainfield, New Jersey

Y. Kris Rao
University of Washington
Seattle, Washington

October 1984

ORGANIZATION OF
HERBERT H. KELLOGG SYMPOSIUM
PHYSICAL CHEMISTRY OF EXTRACTIVE METALLURGY

SYMPOSIUM CHAIRMEN

Val Kudryk
ASARCO Incorporated
South Plainfield, New Jersey

Y. Kris Rao
University of Washington
Seattle, Washington

Session Chairmen

P. Duby
Columbia University
New York, NY

D. R. Fosnacht
Inland Steel Company
East Chicago, IN

M. C. Jha
AMAX Extractive R&D, Inc.
Golden, CO

S. E. Khalafalla
Bureau of Mines
Minneapolis, MN

V. Kudryk
ASARCO Incorporated
South Plainfield, NJ

W. R. Opie
AMAX Base Metals, R&D, Inc.
Carteret, NJ

Y. K. Rao
University of Washington
Seattle, WA

N. J. Themelis
Columbia University
New York, NY

B. L. Tiwari
General Motors Research
Warren, MI

S. J. Warner
INCO Ltd.
Mississauga, Ont.

W. L. Worrell
University of Pennsylvania
Philadelphia, PA

S. Young
ARCO Metals Company
Tucson, AZ

Contents

Foreword V

THERMODYNAMICS AND KINETICS

Mathematical Correlation of Thermochemical
Properties for Molten Cu-Bi, Cu-Sb, Bi-Sb and Cu-Bi-Sb 3
    H. H. Kellogg, Y. H. Kim, T. Stapurewicz,
    D. Verdonik and G. Archer

Metal, Matte and Slag Solution Thermodynamics 23
    T.R.A. Davey and G. M. Willis

A Solid-State EMF Study of the Fe-Ni-S-O
Quarternary System 41
    K. Hsieh and Y. A. Chang

Estimation of the Standard Free Energies of
Formation of Tin Chlorides 49
    S. S. Guzman

Kinetics and Thermodynamics of Chalcopyrite
Oxidation 63
    P. C. Chaubal and H. Y. Sohn

Thermodynamics of Sulfation of Impurities in
Titaniferrous Slags 79
    K. Borowiec and T. Rosenquist

Viscosity Measurements of Industrial Lead
Blast Furnace Slags 97
    R. Altman, G. Stavropoulos, K. Parameswaran
    and R. P. Goel

The Effect of Iron Activity and Oxygen Pressure
on the Copper and Iron Transfer in Fayalite
Slag at $1200^\circ C$ 117
    T. Nakamura, B. Chan and J. M. Toguri

PROCESS DEVELOPMENT

Influence of Organic Additives on Electrolytic Tin
Deposition and Dissolution in Stannous Sulfate
Electrolyte 133
    R. Kammel, U. Landau and B. Szesny

Electrolytic Extraction of Magnesium from
Commercial Aluminum Alloy Scrap 147
    B. L. Tiwari, R. A. Sharma and B. S. Howie

Electrochemical Evaluation of Zinc Sulfate
Electrolyte Containing Cobalt, Antimony and
Organic Additives 165
    T. J. O'Keefe and M. W. Mateer

Pyrohydrolysis of Nickel Chloride Solution in a
30-inch Diameter Fluidized-Bed Reactor 179
    M. C. Jha, B. J. Sabacky and G. A. Meyer

Contribution to Chloride Hydrometallurgy 209
    R. Winand

Extraction of Silver Through Complexation in
the Vapor Phase 231
    J. P. Hager, M. C. Rupert and W. A. May

Recovery of Fluorite and Byproducts from the
Fish Creek Deposit, Eureka County, Nevada 251
    D. G. Foot, Jr., F. W. Benn, and J. L. Huiatt

The Influence of LiF and Bath Ratio on Properties
of Hall Cell Electrolytes 263
    S. Young and R. O. Loutfy

COPPER

Residence Time Predictions During
Fluid-Bed Roasting 277
    C. A. Natalie, J. P. Hager and T. Li

Rate Phenomena in the Outokumpu Flash
Smelting Reaction Shaft 289
    N. J. Themelis, J. K. Makinen and N. D. H. Munroe

Progress of Copper Sulfide Continuous Smelting 311
    T. Nagano

Toward a Basic Understanding of Injection
Phenomena in the Copper Converter 327
    J. K. Brimacombe, A. A. Bustos, D. Jorgensen and
    G. G. Richards

The Behavior of Arsenic, Antimony and Bismuth in the
Solidification and Electrolysis of Nickel-Oxygen-Bearing
Copper Anodes 353
    O. Forsen, E. Hettula and K. Lilius

Secondary Copper Smelting 379
    W. P. Opie, H. P. Rajcevic and W. D. Jones

Electrochemical Sensors Using Li_2SO_4 - Ag_2SO_4
Electrolytes for the Detection of SO_2 and/or SO_3 387
    Q. G. Liu and U. L. Worrell

Pyrometallurgical Refining of Chloride
Leach/Electrowin Copper 397
    R. R. Bhappu and W. G. Davenport

IRON AND STEEL

Carbothermic Reduction of Minerals in a
Plasma Environment 417
    J. J. Moore, M. M. Murawa and K. J. Reid

Direct Method to Prepare Low Carbon Ferrochrome 433
    S. E. Khalafalla and J. E Pahlman

Continuous Casting of Steel 457
    R. D. Pehlke

Characterization and Utilization of Iron-Bearing
Steel Plant Waste Materials 479
    D. R. Fosnacht

Author Index 493

Subject Index 495

Thermodynamics and Kinetics

MATHEMATICAL CORRELATION OF THERMOCHEMICAL PROPERTIES FOR MOLTEN Cu-Bi, Cu-Sb, Bi-Sb AND Cu-Bi-Sb

H.H. Kellogg, Y.H. Kim, T. Stapurewicz, D. Verdonik
and G. Archer

Henry Krumb School of Mines

Columbia University

New York, N.Y. 10027

Summary

The thermochemical properties of molten solutions in the binaries Cu-Bi, Cu-Sb and Bi-Sb, as well as the ternary Cu-Bi-Sb, have been correlated and described in mathematical form by use of the associated species concept combined with the three-suffix Margules equations to account for the non-ideal behavior of the solution species. Agreement of the correlations with experimental data is excellent (relative deviation of less than 2.5% in the values of the activity coefficients) in most cases. The correlations are applicable over the entire composition range of each system, and in the temperature range 700°-1250°C.

Introduction

The aim of the study is to show that the three-suffix Margules equations, combined where appropriate with the assumption of hypothetical associated solution species, are capable of accurate description of the thermochemical properties for a wide variety of binary solution systems, and that the method readily lends itself to description of ternary and high-order systems. Liquid solutions in the system Cu-Bi-Sb were chosen to illustrate the method because of their industrial importance, and because experimental data were available for the ternary and the three binaries.

The three-suffix Margules equations, and their application to systems with associated species, have been discussed and applied in several previous papers (1,2,3,4). For a solution of n species at constant temperature, the equations take the form:

$$\ln \gamma_i = 1/2 \sum_j (k_{ij} + k_{ji})N_j - 1/2 \sum_j \sum_p k_{jp} N_j N_p +$$

$$\sum_j (k_{ij} - k_{ji})(N_j/2 - N_i)N_j + \sum_j \sum_p (k_{jp} - k_{pj})N_j^2 N_p \quad \ldots \quad (1)$$

where the summations are carried out over the n species and $k_{ii} = k_{jj} = k_{pp} = 0$. Each pair of species, i and j, is characterized by two parameters, k_{ij} and k_{ji}, which are constants for a given temperature, and which describe the non-ideal behavior. The equations are consistent with Raoult's and Henry's laws and have built-in agreement with the Gibbs-Duhem relation. Thus, in a system of two species, if a given pair of k_{ij} and k_{ji} values correctly describe the known behavior of γ_i, it follows that the same values of k_{ij} and k_{ji} will yield the behavior of γ_j, which is consistent with the Gibbs-Duhem relation, without recourse to integration. By appropriate mathematical operations the equations of $\ln \gamma_i$ (equation 1) can be transformed to yield expressions for all other partial and integral solution properties.

The temperature dependence of k_{ij} should, in general, take the form:

$$k_{ij} = C_{ij}/T + D_{ij} \quad \ldots \ldots (2)$$

where C_{ij} and D_{ij} are constants independent of temperature, provided the heat capacity change on mixing is zero or very small. It can be readily shown that the C_{ij} values determine the partial and integral enthalpies, and that the D_{ij} values determine the partial and integral excess entropies of the system. In previous studies (1,2,3,4) the simplifying assumption was made that all D_{ij} values were zero (i.e., that the excess entropies were zero), an assumption often referred to as the "regular solution assumption". In this study the assumption proved inadequate, and the full form of equation (2) was employed in most cases.

The application of the Margules equations and associated species concept, employed in this study, are essentially empirical -- i.e., the values of C_{ij}, D_{ij} and the association constants are determined by fitting experimental data to the equations. Despite this limitation, the thermodynamic consistency and flexibility of the equations make it possible to describe virtually any non-ideal behavior, over the full range of composition. For this reason, and because the method can be extended to multi-component systems, the authors believe that it should be more widely applied to the complex solutions (alloys, slags and mattes) of industrial importance.

The Binary Cu-Sb

Of the three binaries considered here, that for Cu-Sb shows the most marked departures from ideal behavior, and considerable asymmetry between the behavior of Cu and Sb. The selected data of Hultgren et al (5) can be used to calculate the excess stability ($St^{xs} = -2RT d\ln \gamma_1 / dX_2^2$) for the liquid solution at 1190 K, and the resulting plot of St^{xs} vs X_{Cu} displays a sharp maximum of 33 Kcal/mol at $X_{Cu} \cong 0.75$. Darken (6) pointed out that, when excess stability peaks are found for the liquid behavior, a stable solid compound of near to the same composition as the peak usually exists at lower temperature in the same system. Such is the case here, where the solid β- phase, approximating to the composition Cu_3Sb, is a major intermediate phase at temperatures below the liquidus.

It can be readily shown that the Margules equations applied to a two-species (Cu and Sb) system are mathematically incapable of yielding a peak in excess stability. Accordingly, and following the practice of Larrain and Kellogg (2), the description of this system was formulated in terms of three species*, Cu, Cu_3Sb and Sb, assumed to be in homogeneous equilibrium ($3\underline{Cu} + \underline{Sb} = \underline{Cu_3Sb}$), with the non-ideal behavior of each species described by the Margules equations. As has been shown in previous studies (1,2,3), the assumption of the partial association to form $\underline{Cu_3Sb}$ provides the mathematical form required to describe the rapid change in activity coefficients that is responsible for a peak in excess stability.

Table I gives the values of the Margules parameters and the association constant for formation of $\underline{Cu_3Sb}$, as determined by fitting the system of equations to available data. For this purpose we relied mostly on the selected data of Hultgren et al (5) for γ_{Cu} and integral heat of mixing at 1190 K, but some compromise was made in selection of parameters to yield improved agreement with the experimental data of Azakami and Yazawa (7).

The resulting correlation, summarized in Table I, accurately reproduces Hultgren's selected data for γ_{Cu} and γ_{Sb} at 1190 K, as illustrated in Figure 1. At 1190 K the correlation of Table I reproduces Hultgren's values with a relative deviation of ±1.4% in γ_{Cu} and ±1.8% in γ_{Sb}; these values compare with the relative uncertainty of ±4.5% that Hultgren assigns to his selected data. If Hultgren's data are extrapolated to higher temperature (assuming that his values for $\Delta \bar{H}_i$ and $\Delta \bar{S}_i^{xs}$ are independent of temperature), the correlation of Table I continues to show excellent agreement (for example, at 1523 K: the relative deviation of the correlation from Hultgren's values is ±1.3% in γ_{Cu} and ±2.6% in γ_{Sb}).

---

*Hypothetical solution species are underlined throughout this paper to distinguish them from the components, Cu (1) and Sb (1) in this case.

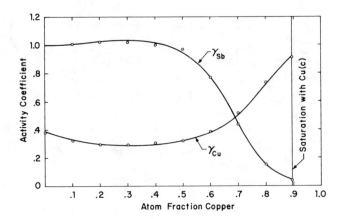

Fig. 1. Activity coefficients in molten Cu-Sb at 1190 K. Points are the selected values of Hultgren et al (5). Lines predicted by correlation (Table I).

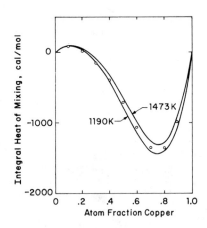

Fig. 2. Integral heat of mixing for Cu-Sb melts. Points are from Hultgren et al (5) for 1190 K. Lines predicted by correlation (Table I) for 1190 and 1473 K. The study of Cu-Sb melts by E. Hayer, K. Komarek and R. Castanet (Z. Metallkde. 68 (10), 1977, pp. 688-96) which gives experimental evidence for a temperature dependence of ΔH similar to that predicted here, was not available to the authors until after this paper was written.

Table I

Parameters for the Binary Correlation Cu-Sb

Species: $\underline{Cu}$, $\underline{Sb}$, $\underline{Cu_3Sb}$

Equilibrium: $3\underline{Cu} + \underline{Sb} = \underline{Cu_3Sb}$ .... (a)

$$\ln K_a = 3787.7/T + 0.4360$$

Margules Parameters: $k_{ij} = C_{ij}/T + D_{ij}$

i\j	Cu	Sb	Cu$_3$Sb
Cu	0.0/T + 0.0	845.9/T−1.6506	3233.6/T−0.1237
Sb	−1931.5/T−1.3263	0.0/T+0.0	−5577.8/T+1.1030
Cu$_3$Sb	−3684.0/T+2.2737	−92.9/T+.0009	0.0/T+0.0

The correlation shows only fair agreement with the data of Azakami and Yazawa (7) at 1473 K. Their data yield lower values of γ_{Sb} than the correlation, and Hultgren (5) has also pointed out the discrepancy between his selected data and that of Azakami and Yazawa. The vapor pressure method used by Azakami and Yazawa is subject to uncertainties because of the polymerization of antimony vapor, and probably becomes less accurate at low antimony concentration. This may account for the fact that the maximum discrepancy between the correlation and their data occurs at the lowest concentration of antimony (X_{Sb}=0.2) studied by them. At infinite dilution (X_{Sb}=0), the correlation of Table I predicts $\gamma°_{Sb}$ = .025 at 1473 K, compared to the value of .013 obtained by Azakami and Yazawa by extrapolation of their measurements from X_{Sb}=0.2 to zero. We believe the correlation value is to be preferred, both because of the possible inaccuracy in vapor-pressure measurements at X_{Sb}=0.2, and the considerable uncertainty in extrapolation of γ_{Sb} from X_{Sb}=0.2 to zero concentration.

Enthalpy Effects for Cu-Sb

Figure 2 compares the integral heat of mixing for the system Cu-Sb, predicted from the correlation of Table I, with the selected values reported by Hultgren et al (5). The correlation is in excellent agreement, with a standard deviation of only ± 48 cal/mol at 1190K, compared to Hultgren's estimated uncertainty of ±550 cal/mol. The correlation faithfully follows the complex shape of the integral heat behavior.

Of particular interest, however, is that the correlation predicts a small but significant temperature dependence for ΔH, whereas one normally assumes in alloy systems that ΔH is nearly independent of temperature (insofar as ΔC_p for the mixing process is usually quite small). The temperature dependence of ΔH, shown in Figure 2 is inherent in the assumptions that underlie our correlating method, and may be understood by the following considerations: If X mols of Cu(l) and (1−X) mols of Sb(l) are mixed at constant temperature to form a liquid containing n_{Cu} mols of $\underline{Cu}$,

n_{Sb} mols of Sb and n_{Cu_3Sb} mols of Cu$_3$Sb, and each of these solution species behaves non-ideally, then the integral heat on mixing will be given by:

$$\Delta H = n_{Cu}\Delta \bar{H}_{Cu} + n_{Sb}\Delta \bar{H}_{Sb} + n_{Cu_3Sb}\Delta \bar{H}_{Cu_3Sb} + n_{Cu_3Sb}\Delta H° \quad \ldots \quad (3)$$

where $\Delta \bar{H}_i$ is the partial molar heat of species i, and $\Delta H°$ is the molar heat of formation of species Cu$_3$Sb from Cu(l) and Sb(l). The values of $\Delta \bar{H}_i$ and $\Delta H°$ in equation (3) will be independent of temperature, provided that the Margules constants and equilibrium constant for formation of Cu$_3$Sb obey the simple temperature dependence shown in Table I. However, for any given overall solution composition, the values of n_i in equation (3) will be temperature dependent because the degree of association to form Cu$_3$Sb will depend on temperature. It may be concluded, in general, that the assumption of an intermediate associated species will result in some temperature dependence for the heat of mixing, the magnitude of which will depend on the degree of association and the temperature dependence of the association constant.

This conclusion suggests the need for precise measurements of integral heat of mixing as a function of temperature both for systems where associated species are likely and those where they are not. If it can be shown by experiment that a significant temperature dependence of ΔH is characteristic of systems such as Cu-Sb, this would provide evidence consistent with the associated species concept.

The Binary Bi-Sb

Despite the simple phase diagram for this system - complete miscibility in both solid and liquid - the activity behavior is complex, showing both positive and negative deviations from ideal behavior and a rapid change in the activity coefficient of both species near to $X_{Bi}=0.5$. The selected data of Hultgren et al (5) indicate a weak peak in excess stability at 1200 K (St^{xs} = 9.6 Kcal/mol) at the composition near to $X_{Bi}=0.55$. To correlate the data for this system we found it necessary to assume an intermediate associated species, BiSb, even though the phase diagram showed no evidence of compound formation at lower temperature. In this respect the system is similar in behavior to the system Cu-Fe, where Kellogg (1) was able to correlate the thermodynamic properties of the liquid by assumption of the associated species CuFe, despite the absence of a solid phase of this composition at lower temperature.

Table II gives the parameters of the correlation, which were determined by fitting the selected data of Hultgren et al (5) to the system of equations. Figure 3 compares the correlation results with those of Hultgren at 1200 K, and the agreement is excellent, except for the limiting activity coefficient for antimony; the relative discrepancy for the composition range of $0.1 \leq X_{Bi} \leq 0.9$ is ±2.2% for both γ_{Bi} and γ_{Sb}, compared to Hultgren's estimated uncertainty of ±5% for his values. The deviations between the correlation and Hultgren's values of γ_i remain near to ±2% for other temperatures between 1200 and 1523 K (calculated on the assumption that Hultgren's values of $\Delta \bar{H}_i$ and $\Delta \bar{S}_i^{xs}$ are independent of temperature).

Table II

Parameters for the Binary Correlation Bi-Sb

Species: Bi, Sb, BiSb

Equilibrium: Bi + Sb = BiSb .... (b)

$$\ln K_b = -443.5/T - 0.7330$$

Margules Parameters: $k_{ij} = C_{ij}/T + D_{ij}$

i\j	Bi	Sb	BiSb
Bi	0.0/T + 0.0	249.5/T - 0.8559	-1035.7/T - 4.1519
Sb	-153.7/T + 0.8121	0.0/T + 0.0	-803.9/T - 2.3321
BiSb	159.6/T - 0.5620	386.4/T - 0.3509	0.0/T + 0.0

Since the measurements on which Hultgren's selected values are based do not extend below X_{Sb} = 0.1, we attribute the poor agreement of the values for $\gamma°_{Sb}$ to inaccurate extrapolation to X_{Sb} = 0. by Hultgren and co-workers. On Figure 3 the unlikely extrapolation that he makes to evaluate $\gamma°_{Sb}$ is indicated by the dashed line.

Table III compares values of the integral heat of solution calculated from the correlation of Table II with those reported by Hultgren et al (5). At 1200 K the correlated values agree with Hultgren's values with a mean discrepancy of ± 18 cal/mol, compared to the uncertainty of ± 25 cal/mol that Hultgren assigns to his values. The correlated values of ΔH show only a slight temperature dependence as shown by the values for 1200 and 1523 K

Table III

Integral Heat of Solution for Bi-Sb Liquid Alloys

X_{Bi}	(cal/gm mol) Hultgren (5)	Correlation Values	
	1200 K	1200 K	1523 K
0.1	50	60	61
0.2	88	104	106
0.3	114	126	128
0.4	130	126	126
0.5	134±25	110	110
0.6	127	95	96
0.7	110	87	89
0.8	83	72	75
0.9	46	42	45

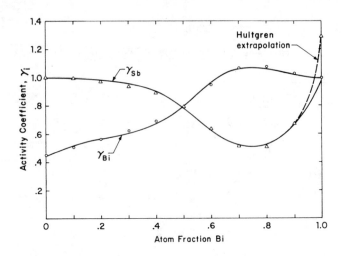

Fig. 3 Activity coefficients in molten Bi-Sb at 1200 K. Points are the selected values of Hultgren et al (5). Solid lines predicted by correlation (Table II). Dashed line: extrapolation by Hultgren et al (5).

in Table III. The very much smaller temperature dependency, compared to that exhibited by the system Cu-Sb (see Figure 2), results from the much smaller heat of association for BiSb ($\Delta H° = 881.$ cal/mol) compared to Cu_3Sb ($\Delta H° = -7527$ cal/mol) which, in turn, yields very little change with temperature in the degree of association of the Bi-Sb system. For example, the correlation of Table II predicts that the mol fraction of species BiSb is 0.227 at 1200 K and 0.229 at 1523 K for the overall composition $X_{Bi} = 0.5$.

The Binary Cu-Bi

This simple eutectic system shows moderate positive deviations from ideal behavior for both components, and no indication of a maximum in excess stability. There appears to be no justification for assuming the existence of an intermediate associated species, so that the Margules equations are applied, in this case, to the two simple species Cu and Bi, assumed to be identical in properties to Cu(l) and Bi(l). Only four adjustable parameters (C_{Cu-Bi}, D_{Cu-Bi}, C_{Bi-Cu}, D_{Bi-Cu}) are available to describe both the composition and temperature dependence of all thermodynamic properties for this system.

In determining the values of these four parameters, we first employed only the selected data of Hultgren et al (5) (data for γ_{Cu} and integral heat of mixing at 1200 K). The correlation obtained in this manner showed satisfactory agreement with all the Hultgren data, but was in poor agreement with the precise data of Nathans and Leider (8) on the composition of the copper liquidus*. In the belief that these liquidus data were more reliable than the e.m.f. data employed by Hultgren et al (5), we adjusted our choice of parameters to improve the fit to the liquidus, with some sacrifice of agreement with Hultgren's data for γ_{Cu} in the dilute region, as discussed later.

The final choice of parameters (listed in Table IV) yields a good fit to the liquidus data as illustrated in Figure 4; the root-mean square deviation of the correlation from 15 smoothed points (700°-1050°C) of Nathans and Leider (8) is ± 1.5 atom % Cu, compared to their reported experimental uncertainty of ± 1.4 atom % Cu. The agreement with the more recent liquidus data of Taskinen and Niemela (9) is equally good, but the experimental scatter in their 23 data points (728-1040°C) results in a root-mean-square deviation of ±2.3 atom % Cu compared to our correlation.

For this two-species system, the Margules expression for the integral heat of mixing takes the simple form:

$$\Delta H = (C_{Cu-Bi} X_{Bi}^2 X_{Cu} + C_{Bi-Cu} X_{Cu}^2 X_{Bi}) R \quad \ldots \ldots \ldots \ldots (4)$$

Figure 5 compares the heat of mixing, calculated with the values of C_{ij} from Table IV, with the values reported by Hultgren et al (5) and by Taskinen and Niemela (9). The agreement is good, both as to the shape of the curve (slightly asymmetric, with a maximum on the copper-rich side) and the magnitude of the values. The correlation values are, on average, 128 cal/mol less endothermic than Hultgren's values, well within his estimated uncertainty of ±200 cal/mol. Both sets of literature values for heat of mixing are calculated from e.m.f. measurements over relatively short temperature intervals (70°-100°C) which leads to considerable uncertainty in calculation of ΔH.

Figure 6 compares the values of γ_{Cu} and γ_{Bi} at 1200 K obtained from the correlation, with the values reported by Hultgren et al (5). The

Table IV

Parameters for the Binary Correlation Cu-Bi

Species: Cu, Bi

Margules Parameters: $k_{ij} = C_{ij}/T + D_{ij}$

i\j	Cu	Bi
Cu	0.0/T + 0.0	1750./T - 0.4594
Bi	3050./T - 1.0510	0.0/T + 0.0

---

*To calculate the composition of the copper liquidus from the activity of supercooled Cu(l) predicted by the correlation, we used the following free energy equation for the fusion of pure copper, based on the data reported in reference (10).

$\Delta G° = 311.7 + .00157\ T^2 - 4.20\ T \ln T + 27.9355\ T$, cal/mol (973-1358 K)

11

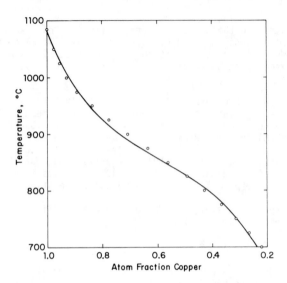

Fig. 4 Copper liquidus for system Cu-Bi. Points are the smoothed data of Nathans and Leider (8). Curve predicted by correlation (Table IV)

agreement is good except for γ_{Cu} when copper is dilute. In the range $0.2 \leqslant X_{Cu} \leqslant 0.78$, the correlation deviates from Hultgren's values by ±1.2% in the value of γ_{Cu} and ±1.7% in the value of γ_{Bi} and these are identical to the uncertainty that Hultgren assigns to his values. Taskinen and Niemela (9) determined γ_{Cu} at eight compositions (X_{Cu} = 0.55 − 0.76) at temperatures that included 1200 K and the correlation deviates from their measurements by ±1.5%.

The limiting value, $\gamma°_{Bi}$, for dilute solution of bismuth in copper is of practical importance in copper smelting and refining. At 1200°C (1473 K) the correlation of Table IV yields $\gamma°_{Bi}$ = 2.77. This agrees very well with the value 2.7 obtained by Azakami and Yazawa (7) from their vapor pressure measurements, and the value 2.75 obtained by Taskinen and Niemela (9) by extrapolation of their e.m.f. measurements from lower temperature; it is significantly larger than the value 2.17 obtained by Arac and Geiger (11) from their vapor pressure measurements.

The Ternary Cu-Bi-Sb

To extend the binary correlations discussed above to the ternary, one must incorporate all of the binary species and binary parameters into the ternary correlation without change. This will assure that the ternary correlation will extrapolate to the proper binary behavior, in the limit where one of the components approaches zero concentration. Additionally, the ternary correlation may involve new ternary species and new two-species Margules parameters that were not present in the binaries.

Proceeding from the simplest to the more complex, we first assumed that no new species or Margules parameters were required to describe the ternary. The five species from the binaries (<u>Cu</u>, <u>Bi</u>, <u>Sb</u>, <u>Cu$_3$Sb</u>, <u>BiSb</u>),

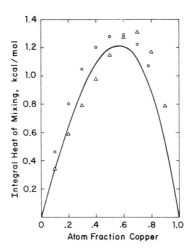

Fig. 5 Integral heat of mixing for Cu-Bi melts. Curve predicted by correlation (Table IV). Points: o Hultgren et al (5), △ Taskinen and Niemela (9).

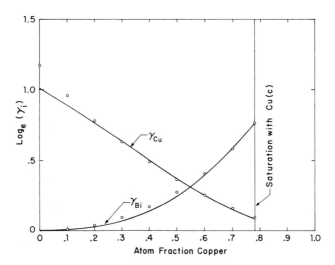

Fig. 6 Activity coefficients in molten Cu-Bi at 1200 K. Points are the selected values of Hultgren et al (5). Lines predicted by correlation (Table IV).

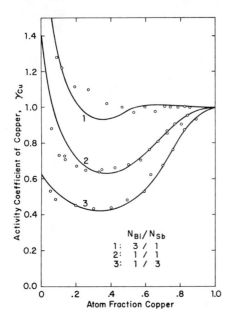 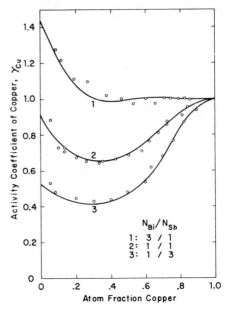

Fig. 7 Activity coefficient of copper in Cu-Bi-Sb melt at 1215 K. Lines predicted by correlation A (see text). Points are the experimental values of Lomov and Krestovnikov (12-14).

Fig. 8 Activity coefficient of copper in Cu-Bi-Sb melt at 1215 K. Lines predicted by correlation B (see text). Points are the experimental values of Lomov and Krestovnikov (12-14).

along with the two equilibria ((a) and (b)) describing the formation of the complex species, and the 14 expressions ($k_{ij} = C_{ij}/T + D_{ij}$) for the Margules parameters were gathered from Table I, II and IV, and used to predict the ternary behavior. This ternary description, called correlation A, is compared in Figure 7 with values of γ_{Cu} at 1215 K determined from e.m.f. measurements by Lomov and Krestovnikov (12-14) for the ternary system, at three different atom ratios, Bi/Sb. The correlation correctly predicts the general trends of measurements, with a relative error in γ_{Cu} ranging from 3.1% to 7.8% for the different Bi/Sb ratios.

Correlation A might be considered an adequate description of the ternary behavior. However, it can be substantially improved by assigning finite values to those Margules parameters amongst the five species that were not involved in the binaries, namely, interactions between Bi and Cu_3Sb, between Cu and BiSb, and between Cu_3Sb and BiSb. These interactions give rise to six additional pairs of C_{ij} and D_{ij} values, and the values of these were determined such as to yield a best fit to the experimental data of Lomov and Krestovnikov (12-14), and are reported in Table V. Because of the limited range of temperature (1115-1215 K) studied by Lomov and Krestovnikov, and the scatter of their data, we were unable to determine separate values of C_{ij} and D_{ij} which were statistically significant; as a result the "regular solution assumption" (i.e., $D_{ij}=0.0$) has been adopted for the values reported in Table V.

Table V

Additional Margules Parameters for the Ternary Cu-Bi-Sb for use in Correlation B

i-j	$k_{ij} = C_{ij}/T + D_{ij}$
Cu - BiSb	$-2264/T + 0.0$
BiSb - Cu	$-2395/T + 0.0$
Bi - Cu$_3$Sb	$147/T + 0.0$
Cu$_3$Sb - Bi	$1237/T + 0.0$
Cu$_3$Sb - BiSb	$-1436/T + 0.0$
BiSb - Cu$_3$Sb	$679/T + 0.0$

Combination of parameters of Table V with those of Tables I, II and IV constitutes correlation B for the ternary. The ternary behavior of γ_{Cu} at 1215 K, according to correlation B, is compared to the experimental ternary data in Figure 8. The relative error between the correlation and the experimental data has been reduced to 2.4-2.8% for the different Bi/Sb ratios and this is within the estimated uncertainty of 3-5% for the measurements. From here on, correlation B will be adopted as the final correlation for the ternary.

With the parameters of correlation B, all other partial and integral thermodynamic properties of the ternary melts can be readily calculated (see Appendix). As examples, Figures 9, 10, 11 show the contours of equal activity coefficient for Cu, Bi and Sb on ternary composition plots at 1200°C (1473 K), and Figures 12 and 13 give detailed information of the same kind for γ_{Bi} and γ_{Sb} at 1250°C (1523 K) for melts rich in copper ($X_{Cu} \geqslant$ 0.95). The accuracy of the correlation can only be ascertained in the regions of temperature and composition where measurements exist, and these comparisons have already been presented throughout the text. The internal consistency of the Margules equations with Gibbs-Duhem relation, Raoult's and Henry's laws, and the good agreement with the known binaries and ternary data lead the authors to believe that correlation B should be generally valid for all compositions in the temperature range 700° - 1250°C (973 -1523 K). In the temperature range where most measurements have been made (700-950°C, 973-1223 K) the accuracy is probably better than ±5% in the value of γ_i; at higher temperatures, ±10% in the value of γ_i is probably more realistic.

For those who prefer to use the Wagner interaction coefficient method for calculation of activity coefficients, Table VI lists the interaction coefficients and limiting activity coefficients predicted by correlation B for dilute solutions of bismuth and antimony in copper. Only first-order coefficients have been listed, and these should be fully adequate for solutions containing less than 1 atom % Bi and/or Sb. For more concentrated solutions it is recommended that Figures 9-13 be used, or that correlation B be solved for the particular compositions and temperatures of interest.

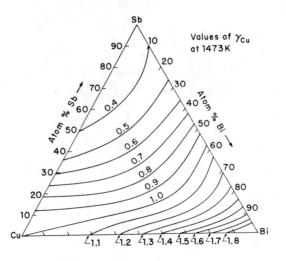

Fig. 9 Iso-activity coefficient contours for copper in Cu-Bi-Sb melts at 1473 K. All lines predicted by correlation B.

Table VI

Parameters for use of Wagner Interaction Coefficients*

$$\ln \gamma_i = \ln \gamma^°_i + \varepsilon^i_i X_i + \varepsilon^j_i X_j$$

$$\varepsilon^{Sb}_{Sb} = 19734/T - 4.668$$

$$\varepsilon^{Bi}_{Bi} = -8621/T + 3.249$$

$$\varepsilon^{Bi}_{Sb} = \varepsilon^{Sb}_{Bi} = 4086/T + 0.274$$

$$\ln \gamma^°_{Sb} = -5506/T + 0.063$$

$$\ln \gamma^°_{Bi} = 3050/T - 1.051$$

---

*These parameters were derived for the temperature range 1423-1523 K. For solutions with less than 1 atom % Bi and/or less than 1 atom % Sb, these Wagner parameters will reproduce the values calculated from the correlations with deviations less than 0.2% in the value of γ_{Sb} or γ_{Bi}.

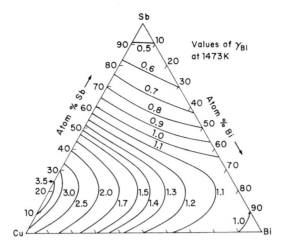

Fig. 10 Iso-activity coefficient contours for bismuth in Cu-Bi-Sb melts at 1473 K. All lines predicted by correlation B.

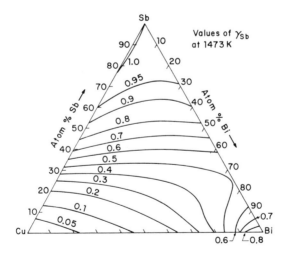

Fig. 11 Iso-activity coefficient contours for antimony in Cu-Bi-Sb melts at 1473 K. All lines predicted by correlation B.

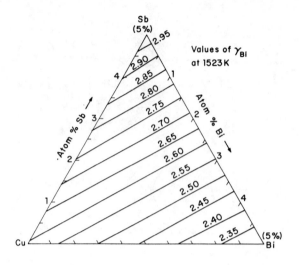

Fig. 12 Iso-activity coefficient contours for bismuth for dilute solutions of Bi and Sb in copper at 1523 K. All lines predicted by correlation B.

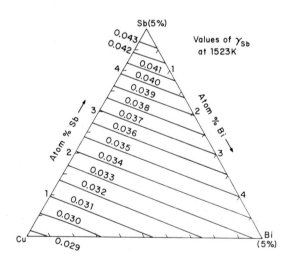

Fig. 13 Iso-activity coefficient contours for antimony for dilute solutions of Bi and Sb in copper at 1523 K. All lines predicted by correlation B.

References

1. H.H. Kellogg, "Thermochemical Modelling of Molten Sulfides", pp 49-68 in *Physical Chemistry in Metallurgy*, R.M. Fisher, R.A. Oriani and E.T. Turkdogan, ed; U.S. Steel Co. Research Lab., Monroville, Pa., 1976.

2. J.M. Larrain and H.H. Kellogg, "Use of Chemical Species for Correlation of Solution Properties", pp. 130-144 in *Calculation of Phase Diagrams and Thermochemistry of Alloy Phases*, Y.A. Chang and J.F. Smith, ed; TMS-AIME, Warrendale, Pa., 1979.

3. R.P. Goel, H.H. Kellogg and J.M. Larrain, "Mathematical Description of the Thermodynamic Properties of the System Fe-O and Fe-O-SiO_2", *Metall Trans.*, 11B(1) (1980) pp. 107-117.

4. J.M. Larrain and S.L. Lee, "Thermodynamic Properties of Copper-Nickel-Sulfur Melts", *Can. Met. Quart.*, 19 (1980) pp. 183-190.

5. R. Hultgren et al, Selected Values of the Thermodynamic Properties of Binary Alloys, Am. Soc. Metals, Menlo Park, Ohio, 1973.

6. L.S. Darken, "Thermodynamics of Binary Metallic Solutions", *Trans. Met. Soc. AIME*, 239 (1967) pp 80-89.

7. T. Azakami and A. Yazawa, "Activity of Bismuth and Antimony in Liquid Copper Base Alloys", *J. Min. Met. Inst. Japan*, 83 (1967) pp. 666-672.

8. M.W. Nathans and M. Leider, "Studies on Bismuth Alloys. I. Liquidus curves of the Bismuth-Copper, Bismuth-Silver, and Bismuth-Gold Systems", *J. Phys. Chem.*, 66 (1962) pp. 2012-15.

9. P. Taskinen and J. Niemela, "Thermodynamics and Liquidus Equilibria in Copper-Bismuth Alloys at 700-1100°C", *Scand. J. Met.*, 10 (1981) pp 195-200.

10. L.B. Pankratz, Thermodynamic Properties of Elements and Oxides, U.S. Bur. Mines, Bull 672, 1982.

11. S. Arac and G.H. Geiger, "Thermodynamic Behavior of Bismuth in Copper Pyrometallurgy: Molten Matte, White Metal and Blister Copper Phases", *Metall. Trans.* 12B(3) (1981) pp. 569-578.

12. A.L. Lomov and A.N. Krestovnikov, "Investigation of the Thermodynamic Properties of the Ternary Metal System Bi-Cu-Sb", *Doklady Akad. Nauk USSR*, 156(6) (1964) pp. 1389-1390.

13. A.L. Lomov and A.N. Krestovnikov, "Investigation of the Thermodynamic Properties of the Ternary Metallic System Bi-Cu-Sb along the line $N_{Bi}:N_{Sb}$ = 1 by the Electromotive Force Method", *J. Appl. Chem. USSR* (English Trans.) 38 (1965) pp. 179-182.

14. A.L. Lomov and A.N. Krestovnikov, "A Potentiometric Study of the Thermodynamics of the Ternary System Bi-Cu-Sb along the $N_{Bi}:N_{Sb}$ = 3/1 Section", *Russ. J. Phys. Chem.* (English Trans.) 38 (1964), pp.1441-1443.

Appendix

Solving the Ternary Correlation

With all of the parameters of correlation B known from Table I, II, IV and V, solving the correlation for any desired temperature and solution composition is readily accomplished with the aid of a computer. A general method is outlined below:

Given conditions: Temperature = T

Composition (atom fraction): X_{Cu}, X_{Bi}, X_{Sb}

Answers desired: $a_{Cu} = \gamma_{Cu} X_{Cu}$, $a_{Bi} = \gamma_{Bi} X_{Bi}$, $a_{Sb} = \gamma_{Sb} X_{Sb}$

Unknowns (13 in total):

Mole fraction of species: N_{Cu}, N_{Bi}, N_{Sb}, N_{BiSb}, N_{Cu_3Sb}

Activity coeff. of species: f_{Cu}, f_{Bi}, f_{Sb}, f_{BiSb}, f_{Cu_3Sb}

Activity coeff. of components: γ_{Cu}, γ_{Bi}, γ_{Sb}

Available equations (13 in total):

1) $N_{Cu} + N_{Bi} + N_{Sb} + N_{BiSb} + N_{Cu_3Sb} = 1.0$

2) $X_{Bi} = (N_{Bi} + N_{BiSb})/(1 + N_{BiSb} + 3N_{Cu_3Sb})$

3) $X_{Sb} = (N_{Sb} + N_{Cu_3Sb})/(1 + N_{BiSb} + 3N_{Cu_3Sb})$

4) $K_a = (N_{Cu_3Sb} \cdot f_{Cu_3Sb})/(N_{Cu}^3 \cdot f_{Cu}^3 \cdot N_{Sb} \cdot f_{Sb}) = f_a(T)$

5) $K_b = (N_{BiSb} \cdot f_{BiSb})/(N_{Bi} \cdot f_{Bi} \cdot N_{Sb} \cdot f_{Sb}) = f_b(T)$

6) - 10) $\ln f_i$ = (Margules function (equation 1 of text) of T, N_i, C_{ij}, D_{ij}; One equation for each species)

11) $\gamma_{Cu} = (N_{Cu}/X_{Cu}) \cdot f_{Cu}$

12) $\gamma_{Bi} = (N_{Bi}/X_{Bi}) \cdot f_{Bi}$

13) $\gamma_{Sb} = (N_{Sb}/X_{Sb}) \cdot f_{Sb}$

Procedure to solve

1) Guess values for $N_{\underline{BiSb}}$ and $N_{\underline{Cu_3Sb}}$
2) Calculate $N_{\underline{Bi}}$ from eq. (2)
3) Calculate $N_{\underline{Sb}}$ from eq. (3)
4) Calculate $N_{\underline{Cu}}$ from eq. (1)
5) Calculate five f_i values from eq. (6)-(10)
6) Calculate apparent values of K_a and K_b (designated as K_a^* and K_b^*) by substitution of N_i and f_i values into equations (4) and (5).
7) Iterate guessed values of $N_{\underline{BiSb}}$ and $N_{\underline{Cu_3Sb}}$ by use of the following relations*

$$(N_{\underline{BiSb}})_{new} = (N_{\underline{BiSb}})_{old} \cdot (K_b/K_b^*)^{1/n}$$

$$(N_{\underline{Cu_3Sb}})_{new} = (N_{\underline{Cu_3Sb}})_{old} \cdot (K_a/K_a^*)^{1/n}$$

8) Repeat iteration until $K_a^*=K_a$ and $K_b^*=K_b$ to the desired precision - say 1 part in 10,000. The final sets of N_i and f_i values will then agree with all 13 equations.
9) Calculate γ_{Cu}, γ_{Bi}, γ_{Sb} from equations (11), (12) and (13).

The authors can supply a simple FORTRAN program, designed to make such calculations, to those who may be interested.

---

* These are not rigorous equations, but they will adjust the guesses in the right direction, and they do result in a convergent solution provided n is properly chosen. For most solutions to the Cu-Bi-Sb system n=3 works well, for some compositions near to $X_{Cu}=1$, it is necessary to use n=10 in order to achieve a convergent solution.

METAL, MATTE AND SLAG SOLUTION THERMODYNAMICS

By T.R.A.Davey   and   G.M.Willis

    Consultant,   Senior Associate,
    Metafor,   Chem. Engineering Dept.,
    5 Rhodes Drive,   University of Melbourne,
    Glen Waverley,   Carlton,
    Victoria 3150   Victoria 3052,
    Australia.   Australia.

Abstract:

A simple model for metal solution behaviour may be derived, on the basis of dilute solution solubilities (where log concentration of solute varies linearly with reciprocal temperature) and not-so-dilute region solubilities (where the log of the alpha-function at saturation frequently varies linearly with reciprocal temperature) and assuming activity to vary according to the quadratic formalism in between.

Matte binary systems usually exhibit regular solution behaviour when the constituent species are chosen appropriately.

In many recent papers, metal solubilities in slags are expressed as the sum of several different and supposedly independant molecular species (e.g. oxides, sulfides, chlorides). Such a view is untenable because of the now proven ionic nature of slags. Some possible reasons for the great effect of sulfur on copper solubilities in slags are discussed.

Introduction

Kellogg's recent contributions to considering solutions as if they contained associations of solutes is interesting, and seems to be fruitful, whether or not the mathematical models correspond to some physical reality (1). He joins Chipman and Darken, of this country, Lumsden and Richardson of England, and Wagner and Oelsen of Germany, as one of the foremost contributors to our understanding and appreciation of the importance of solutions to practical metallurgy, both in smelting and refining processes. It therefore seems appropriate today to offer a minor contribution to this subject, more with the object of stimulating further work along certain lines, than with the intention of presenting a great mass of fully digested material.

Lumsden (2) introduced many of us to the principle of Occam's razor - that one should choose the simpler of alternative hypotheses - and to the advantages of plotting thermodynamic functions in various ways, many of them new to practising metallurgists. He preferred mathematical expressions with terms which, in general, correspond to a physical reality, or to which a physical meaning can be ascribed. Richardson showed the significance of the ionic nature of metallurgical melts, especially of slags, and this has been overlooked by many authors over the last decade. Because slags are ionic, it is usually misleading to consider molecular compounds as being in solution in slags, since metal ions which are common to two or more types of compound cannot have <u>separate</u> and <u>independant</u> solubilities as the different compounds.

Metal Solutions

In the 1939 edition of Metals Handbook, Anderson & Rodda (3) used results obtained by their colleagues in the New Jersey Zinc Co's Research Division to determine the monotectic point in the lead-zinc system. This must be one of the earliest applications of thermodynamics in non-ferrous extraction metallurgy. The necessary extrapolation was based on the now well-known fact that the plot of log concentration against reciprocal temperature is linear. The reference given for this method of plotting was to a paper on aluminium alloys by Fink & Freche (4), which was given at the AIME Fall Meeting in New York - in October 1934. This paper concluded with a half-page bibliography on thermodynamics of solutions, and Fink was the Chairman at the session at which the original experimental work was presented. Now, some fifty years after these pioneering applications, the role of thermodynamics in smoothing experimental results in new ways is far from exhausted.

Although it should now be a routine matter to determine eutectic and monotectic points in the manner of Anderson & Rodda, as recently as 1967 Davey (5) showed in this way that the Fe-Sn eutectic composition was considerably lower than the value generally accepted at that time. Kleppa (6) showed how accurate solubility data in the dilute solution region can be used to determine heats and entropies of solution in binary systems, and hence obtain solute activities. Davey (7) has shown that, if the solubility data are accurate enough, extrapolation of such activity data to very high temperatures may be justified.

Figure 1 shows a simple binary metallic system, with a liquidus curve MLE extending from the eutectic point, E, to the melting point of the solute, M; L is the limit of dilute solution behaviour, and at temperatures or compositions above this, Henry's Law no longer applies. When the arithmetic coordinates of Figure 1 are replaced by the plot of Figure 2, showing logarithm of solute concentration against reciprocal temperature, the curve MLE takes on a different configuration: the portion LE is now linear, which aids greatly in smoothing the solubility data. Until now, however, there has been no such method available for smoothing the data for the section ML. As shown in Figure 3, depicting the solubilities of copper in lead, and copper in bismuth, it may be possible to draw the solubility curve with reasonable certainty through the experimental points, if the solubility measurements are reasonably concordant, and the scatter is not too great. Even so, without very sophisticated thermodynamic models, it is usually not possible to find a simple and serviceable equation for such a solubility curve, so that its position could be fixed algebraically, without hand-plotting and subjective judgement.

We have discovered an extremely simple function that provides a linear relationship for the section ML of the solubility curve, which completely and elegantly overcomes this difficulty: the log of the <u>alpha-function</u> at saturation is linear against reciprocal temperature. This relationship is illustrated, for the two systems of Figure 3 (Cu-Pb and Cu-Bi) in Figure 4. The plots show clearly that there is no systematic departure of solubilities from the linear plot, but only random scatter,

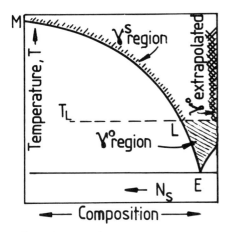 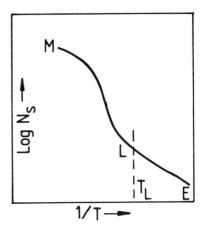

Fig. 1 - Typical eutectic phase diagram, plotted arithmetically. γ^0 = activity coefft. of solute in dilute soln.; γ^S = activity coefft. of solute in satd. soln.; L = limit of dilute soln. region above which Henry's law no longer holds; M = melting point of solute.

Fig. 2 - Solubility curve corresponding to Fig. 1, but arithmetic coordinates are replaced by log concn. of solute, and reciprocal of absolute temperature.

in the case of Cu-Bi. In the case of Cu-Pb, at the highest temperatures the slight apparent upward trend of deviations from linearity is probably due to experimental difficulties in this region. Figures 5 - 8 show this method of plotting for several other systems, illustrating how useful it can be when the scatter is great.

Not only can this technique be used to smooth solubility data, but by an extension of Kleppa's methods, we can use it to obtain an estimate of activities in the non-dilute solution regions, and indeed over the whole range of temperature and composition, in the event that Darken's quadratic formalism is assumed to be applicable - that is, that there is no large value for the excess stability function in the concentrated solutions.

Determination of Activities

At saturation, the solute in solution has an activity equal to that in the pure solid solute (or sometimes to a solid solution whose component's activities can be estimated with reasonable accuracy). Referred to the hypothetical pure super-cooled liquid solute as standard state, this activity is given by:

$$\log a_s = \Delta H_f (1 - 1/F)/RT , \qquad [1]$$

where ΔS_f ($= \Delta H_f/T$) is the entropy of fusion of the solute, ΔH_f is the enthalpy of fusion of the solute, F is the temperature of fusion of the solute, R is the gas constant, a is the activity of the solute at temperature T K, and Δc_p has been neglected.

(To include the Δc_p correction, let:
$$\Delta c_p = c_{p(solid)} - c_{p(liquid)} = X + YT .$$

Then:
$$\log a_s = \Delta H_f/RF - \Delta H_f/RT + (X/R)[1 - F/T + \ln(F/T)]$$
$$+ (Y/R)[F - F^2/2T - T/2]$$

where ln is natural logarithm.)

Alternatively:
$$\log a_s = A_f + B_f/T \, , \qquad [2]$$

where $A_f = \Delta S_f/R$, and $B_f = -\Delta H_f/R$.

In the dilute solution region:
$$\log N_s = A_s + B_s/t \, , \qquad [3]$$

where subscript "s" refers to the saturated solution, and thus:
$$\log \gamma^s = \log a_s - \log N_s \, , \qquad [4]$$
$$= A_f - A_s + (B_f - B_s)/T \, . \qquad [5]$$

If we assume that Henry's law applies to this region, then:
$$\gamma = \gamma^s = \gamma^o \, ,$$

where superscripts "s" and "o" refer to the saturated and infinitely dilute regions, respectively, and:
$$\log \gamma^o = A_o + B_o/T \, , \qquad [6]$$

where $A_o = A_f - A_s$, and $B_o = B_f - B_s$.

Relation [6] was first noted by Kleppa (6), and enables the calculation of γ^o to be made from accurate determinations of solubility in the dilute, low temperature region, if the heat and temperature of fusion of the solute are known, and Δc_p is small enough to be neglected; however, the activities are not defined at concentrations outside the dilute range.

The Alpha-Function at Saturation

The logarithm of the alpha-function plotted against $1/T$ almost invariably exhibits a straight line relationship, as shown for example by Figure 4:
$$\log \alpha_s = \log \gamma^s/(1 - N_s)^2 = (A' + B'/T) \, . \qquad [7]$$

Figures 5-8 show what a sensitive means for smoothing solubility data is afforded by the alpha-function for the saturated solutions, and this technique provides for the first time an excellent means of smoothing data, comparable with the use of $\log N_s$ against $1/T$ at low temperatures. Such a plot readily shows whether individual points should be discarded, and the remainder can be fitted with the straight line of best fit.

Substituting for γ^s in equations [5] and [7]. we obtain:
$$\log N_s = A_f + B_f/T - (A' + B'/T)(1 - N_s)^2 \, , \qquad [8]$$

and smoothed values of N_s for any temperature may be obtained by evaluating equation [8].

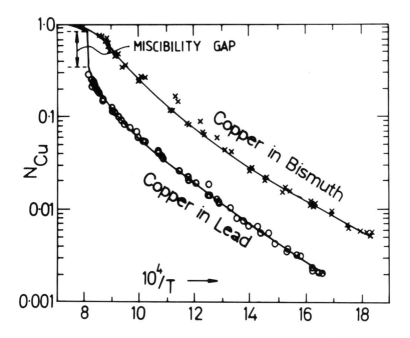

Fig. 3 - Solubilities of copper in lead or bismuth.

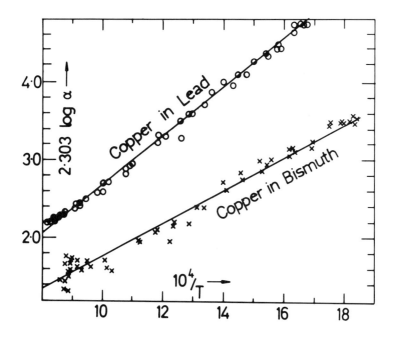

Fig. 4 - solubilities of copper in lead or bismuth: the same data as in Fig 3.

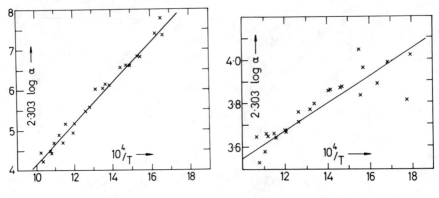

Fig. 5 - Al (solute) in Lead.   Fig. 6 - Al (solute) in Bismuth.

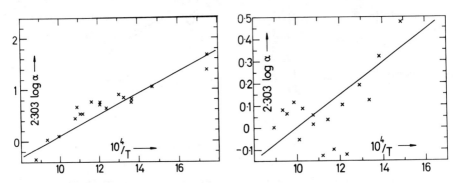

Fig. 7 - Ge (solute) in Indium.   Fig. 8 - Ge (solute) in Zinc.

Concentrations Intermediate Between Dilute and Saturated

Equation [6] gives the activity coefficient of solute in dilute solutions (obtained from the slope and intercept of the solubility curve in the low-temperature straight region), and equation [7] gives the activity coefficient of solute in saturated solutions (obtained from the slope and intercept of the alpha-function plot for saturated solutions, using equation [8] to obtain smoothed values of N_s for any temperature. Both equations may be extrapolated to higher temperatures, well beyond the experimental data, when required.

In order to obtain activities for concentrations between the infinitely dilute, and the saturated, a further relation must be introduced. Darken's quadratic formalism (8,9) has been shown to provide an excellent correlation of activity with concentration variation, at constant temperature, for many systems. It may therefore be introduced to define the activity coefficient at intermediate concentrations in this model. At any constant temperature:

$$\log \gamma = X(1 - N)^2 + Y \ . \quad [9]$$

When $N = 0$,

$$\log \gamma^o = X + Y, \quad [10]$$

and when $N = N_s$,

$$\log \gamma^s = X(1 - N_s)^2 + Y \ .$$

From [6] and [10]:
$$X + Y = A_o + B_o/T , \qquad [12]$$
and from [7] and [11]:
$$nX + Y = n(A' + B'/T) , \qquad [13]$$
$$\text{where } n = (1 - N_s)^2 .$$

Solving [12] and [13] for X and Y:
$$X = A_1 + B_1/T , \qquad [14]$$
$$\text{and } Y = A_2 = B_2/T , \qquad [15]$$
$$\text{where } A_1 = (nA' - A_o)/(n-1),$$
$$B_1 = (nB' - B_o)/(n-1),$$
$$A_2 = n(A_o - A')/(n-1),$$
$$B_2 = n(B_o - B')/(n-1).$$

Substituting [14] and [15] in [9]:
$$\log \gamma = (A_1 + B_1/T)(1-N)^2 + (A_2 + B_2/T) . \qquad [16]$$

Equation [16] defines the activity coefficient of the solute at temperature T and concentration N mole fraction, from infinite dilution up to saturation, for temperatures from the freezing point of the solvent up to the limits of extrapolation. The activity of the solvent is given by:
$$\log \gamma_{solvent} = (A_1 + B_1/T)N^2 , \qquad [17]$$
where N is still the concentration of the solute.

We have arrived at a four-parameter equation, [16], describing one side of a binary system, the parameters having been derived entirely from solubility data, plus the entropy and temperature of fusion of the solute. Writing out in full the evaluations of the A and B parameters used in equations [16] and [17]:
$$A_1 = (nA' - \Delta S_f/R + A_s)/(n-1) , \qquad [18]$$
$$B_1 = (nB' + \Delta H_f/R + B_s)/(n-1) , \qquad [19]$$
$$A_2 = n(A_s + A' - \Delta S_f/R)/(1-n) , \qquad [20]$$
$$B_2 = n(B_s + B' + \Delta H_f/R)/(1-n) . \qquad [21]$$
$$n = (1 - N_s)^2$$

A_1, A_2, B_1, and B_2 are independent of composition, but not of temperature, since they are functions of n, and hence of N_s, which varies with temperature. The other A's and B's are independent of both temperature and composition. It will be remembered that A_s and B_s are obtained from the plot of log N_s against 1/T in the dilute region, whereas A' and B' are obtained from the plot of log alpha against 1/T for more concentrated ranges. Alternatively, they can be obtained, without plotting, from a regression calculation.

In the above derivations, it has been assumed that the saturated solutions are in equilibrium with the pure solid solute. When the solid is not pure, but a solid solution, equation [2] above becomes:

$$\log a_s = A_f + B_f/T + \log N'',$$

where N'' = concentration of solute in the solid phase in equilibrium with the liquid solution.

Frequently N'' can be expressed as a linear function of reciprocal temperature:
$$\log N'' = A'' + B''/T,$$

and then equation [2] becomes:
$$\log a_s = A_f + A'' + (B_f + B'')/T. \quad [2a]$$

In this case, the solid solutions can be allowed for by replacing, in equations [18] - [21]:
$$\Delta S_f \text{ by } (\Delta S_f + RA''),$$
and
$$\Delta H_f \text{ by } (\Delta H_f + RB'').$$

Comparison With Some Similar Solution Models

In an *ideal* solution, the activity is defined at all temperatures and compositions without any parameters requiring empirical determination, since $\gamma = 1$, or $\log \gamma = 0$.

In a *regular* solution, one activity determined for a single temperature and composition suffices to fix all activities, since $\log \gamma = (\beta/RT)(1-N)^2$; the alpha function, $\alpha = (\log \gamma)/(1-N)^2 = B/T$, where B is the empirical constant.

A *semi-regular* solution (6) is completely defined by two pieces of empirical data, and the entropy of mixing is no longer ideal:
$$\log \gamma = (A + B/T)(1-N)^2.$$

A simpler form of four-parameter equation than the one presented here has previously been employed (9), with all A's and B's independent of both temperature and pressure:
$$\log \gamma = (A_0 + B_0/T) + (A_1 + B_1/T)(1-N)^2. \quad [22]$$

In principle only four solubility points are required to determine the empirical constants, the A's and B's in equation [22], but of course normally they would be determined by a least squares method from more than four solubility determinations.

This equation [22] is implicit in Turkdogan & Darken (9), and Lumsden's four-parameter equation also reduces to it when Lumsden's $1/T^2$ term is much smaller than the others, as it usually is (2,11). This equation [22] requires that $\log \gamma$ varies as $1/T$ throughout all concentration ranges, whereas our equation [16] requires this only in the infinitely dilute range, and is therefore more flexible.

In principle our equation [16] can be derived from only three solubility determinations - two in the dilute (Henry's law) region, and one in the higher temperature, higher concentration region. However, like equation [22], it will normally be determined from many points by a least squares technique.

Equation [16] was applied to the copper in lead and copper in bismuth systems (12,13) and found to agree exactly with the (graphically presented) experimentally determined activities. The experimental solubility points for these and other systems (as depicted in Figs. 3-8) were taken from all primary sources referred to by Hultgren et al. (14).

Mixtures of Molten Sulfides

Pseudo-binary Sulfides

Most experimental determinations have been carried out on supposedly binary sulfide melts, which were in fact pseudo-binaries. Table 1 gives a selection of typical results from the literature. Temkin's rule (15) for mixing is commonly applied to the results; it was originally derived for molten salt mixtures, assumed to be separate and distinct cation and anion lattices. This nomenclature has drifted into the description of molten sulfides, which however are semi-conductors, and not electrolytes. In this case, it would be better to refer to "metal", and either "non-metal" or "sulfur" lattices. In a matte on the FeS-Cu_2S join, it is then assumed that the Cu and Fe mix on the metal lattice, while the sulfur lattice or matrix remains unchanged. Mixing of metal atoms on this Temkin model is generally indicated by writing the species as FeS and $CuS_{\frac{1}{2}}$.

Unless the sulfide melts have similar structures (e.g. FeS, CoS, NiS), this formulation presents problems. It is generally assumed that the liquid sulfides have structures similar to those of the solids at high temperature, and therefore Temkin's model would not be expected to apply to a system such as Cu_2S-FeS, where the terminal solids have quite different structures. Nor can it be assumed that similar sulfide formulae imply similar structures - e.g. FeS and PbS. Thus mixing cannot be assumed to be simply that of metal atoms on their own sub-lattice.

Nevertheless, as Table 1 shows clearly, the sub-lattices in the liquid are apparently much more accommodating than might be expected on structural grounds alone, as the behaviour is quite simple: they are mostly not far from ideal, usually being regular solutions with a small regularity constant, and the only exceptions above are for Cu_2S-Ni_3S_2 - if confirmed - and for Ag_2S-Sb_2S_3, which must be due to the stability of the composition corresponding to the solid compound $Ag_3Sb_2S_3$. Intermediate compounds (sulfosalts) of this type are uncommon at elevated temperatures, in pseudo-binary sulfide systems; mostly the phase diagrams exhibit eutectics - with substantial miscibility in both solid and liquid phases - or, rarely, miscibility gaps.

Metal-rich Sulfide Solutions

It is generally recognised that mattes in practice contain metal well in excess of the pseudo-binary melt. A small part of the deficiency can be due to the presence of oxygen, as is the case in copper mattes.

However, many binary metal-sulfur systems (such as Cu-S, Sn-S, Ag-S) contain a miscibility gap between liquid metal and liquid sulfide, the latter containing a substantial amount of metal in solution at smelting temperatures. For low-melting metal (e.g. bismuth or lead) systems which do not possess a miscibility gap, there is a continuous range of liquid from metal to sulfide - i.e., an infinite solubility of metal in the liquid sulfide. The same applies in ternary and higher order systems, and many cases are known where multi-component mattes possess a high solubility for metal above the stoichiometric equivalent of the sulfur. Figure 9 shows the ability of Cu-Pb-Fe mattes to dissolve considerable amounts of lead metal, which is an unfortunate feature of lead-copper metallurgy.

TABLE 1

Activities in Pseudo-binary Sulfide Melts

System	Properties	Reference
$FeS-CuS_{\frac{1}{2}}$	Almost ideal	Koh & Yazawa(16)
	Temkin mixing	Bale & Toguri(17)
FeS-PbS	Ideal solution	Eric & Timucin(18)
	" "	Willis(19)
FeS-SnS	Regular solution	Davey & Joffre(20)
$PbS-AgS_{\frac{1}{2}}$	Ideal	Lumsden(21)
$CuS_{\frac{1}{2}}-NaS_{\frac{1}{2}}$	Regular, with some ordering	Richardson & Antill(22)
	Regular	Lumsden(23)
$PbS-CuS_{\frac{1}{2}}$	Regular	Lumsden(24)
	"	Davey(25)
	Negative deviations	Eric & Timucin(18)
BUT: $Cu_2S-Ni_3S_2$	Almost ideal	Koh & Itagaki(26)
$CuS_{0.5}-NiS_{0.67}$	Strong positive deviations from ideality	" (26)
$Ag_2S-Sb_2S_3$	Negative deviations from ideality; high excess stability function	" (26)
$FeS-Ni_3S_2$	Close to ideal (calcd.)	Chuang & Chang (27)

Ternary Metal-sulfur Solutions

Larrain & Kellogg (1) and Chang & colleagues (27,29) have extended their associated solution models to ternary systems such as Cu-Ni-S (29,30,31) and Fe-Ni-S (27,29). Most experimental work has been nominally on the binary joins already discussed, and comparatively little information is available on the more metal-rich melts. It is often assumed that the activities of the metal sulfides (e.g. of FeS or Cu_2S in copper mattes) are not greatly affected by departures from the pseudo-binary compositions, despite the known fact that, in the binary Fe-S system, the activity

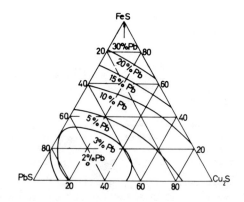

Fig. 9 - Lead metal solubilities in copper-iron-lead mattes. (28)

of FeS drops to about 0.7 at metal saturation, as was shown by Bale & Toguri (17). The pioneering results of Krivsky & Schuhmann(32) show increasing divergence between the activities of FeS in the pseudo-binary and in the metal-saturated mattes with increasing fraction of iron, expressed as (Fe/[Fe + ½Cu]), until there is almost a factor of 2 between them when this fraction is 0.573. Figure 8, based on the results of Bale & Toguri (17), shows that there are real changes in activity as the sulfur/metal ratio changes. Calculations based on the activities in the pseudo-binaries can therefore be misleading. In the Cu-S system, the activity of Cu_2S changes very little with the composition of the sulfide, and it appears to have been assumed that this is true for the Cu-Fe-S melt also, although, as already mentioned, the activity of FeS changes substantially.

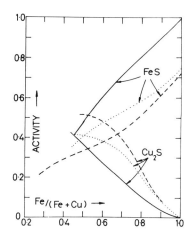

Fig. 10 - Activities in Cu-Fe-S mattes at 1473 K, from ref.(17).

———— Cu_2S pseudo-binary;

········ N_S = 0.42;

— — — metal saturated mattes.

Calculations of copper losses in slags are often made using data from the Cu_2S-FeS binary, which is unnecessary, since the activities of Cu and Fe in the melt are available from ternary Gibbs-Duhem integration of p_{S_2} measurements, such as those of Bale & Toguri (17), or Koh & Yazawa (16), on the Cu-Fe-S system.

Instead of:
$$[FeS]_{matte} + (Cu_2O)_{slag} = (FeO)_{slag} + [Cu_2S]_{matte}$$
one can use:
$$[Fe]_{matte} + (Cu_2O)_{slag} = (FeO)_{slag} + 2[Cu]_{matte} .$$

In the latter, a_{Fe}/a_{FeO} gives the p_{O_2} necessary to calculate a_{Cu_2O}, (or, better, $CuO_{\frac{1}{2}}$). Unfortunately, according to unpublished work by Willis (19), the differences in activities due to changing metal/sulfur ratios do not appear to be sufficient to explain the so-called "sulfide" solubility of copper in slags.

Metals and Sulfur in Slags

There is no doubt that iron silicate slags (generally saturated with silica for experimental convenience) dissolve more copper when equilibrated with a sulfur-bearing melt than with a sulfur-free metal, when the copper content is clearly established by the equilibrium:

$$2Cu + \tfrac{1}{2}O_2 = 2Cu^+ + O^{2-} . \qquad [23]$$

Toguri & Santander (33) explicitly assume that "N_O is approximately constant", and hence also the activity of O^{2-}. Intuitively, one ascribes this constancy to buffering due to the various polymeric equilibria involving O^{2-} ions. Taylor & Jeffes (34) found the behaviour of Cu_2O to be unaffected by changes in iron/silica ratios from 1.66 to 4.62, hence the activity of O^{2-} ions must be constant over a large range of slag compositions. For iron silicate slags, saturated in silica, Richardson & Billington (35) showed the Fe^{3+}/Fe^{2+} ratio to be proportional to $p_{O_2}^{1/4}$, in accordance with:

$$2Fe^{2+} + \tfrac{1}{2}O_2 = 2Fe^{3+} + O^{2-} \ . \qquad [24]$$

Again, $a_{O^{2-}}$ appears to be constant, because – in their opinion – of the saturation with SiO_2. It appears to be generally true that variations in the activity of oxygen ions have not been found in equilibria of this sort – perhaps because they have not been looked for.

The failure of [23] to describe (%Cu) as a function of p_{O_2} in the presence of matte must be ascribed to some effect of sulfur in the slag. The simplest way to describe this is to say that sulfur lowers the activity coefficient of copper in the slag, thus increasing the copper content above that in the sulfur-free slags (36). Lumsden (37) pointed out that the activity of FeS and Cu_2S in oxygen-bearing mattes should be closer to that calculated by Flood's method than that by simple Temkin mixing. Rosenqvist (38) has shown recently that the depression of a_{FeS} and enhancement of a_{Cu_2S} are not inconsistent with the results in the literature. Sehnalek & Imris (39) assumed that the <u>activities</u> (not the activity coefficients) of the O^{2-} and S^{2-} ions are constant, and then calculate independently the solubilities of Cu_2O and Cu_2S in the slag. Davey (40), in discussion of this paper, pointed out that, if copper is present in ionic form in the slags, there is no way of distinguishing between "oxide copper" and "sulfide copper", and the effect of sulfur on the solubility of copper should be expressed by way of interaction coefficients. This manner of proceeding was elaborated at the Bombay Conference on Chemical Metallurgy (41). In the treatment which follows we are replacing the $N_{S_\frac{1}{2}}$ and N_{O} used at Bombay, for calculation of the interactions, by $p_{S_2}^{\frac{1}{2}}$ and $p_{O_2}^{\frac{1}{2}}$ respectively, as they are more suitable terms because they are in principle measurable. As N_O cannot be measured in any known way, in Bombay it was suggested that $N_{FeO_{1.5}}$ should be used in its place.

Whatever the mechanism for the influence of sulfur on the solubility of copper in the slag, the following relations must hold, and (whereas other formulations may be valid for interpolation only) should be valid generally, if the interaction coefficients can be accurately determined:

$$a_{Cu_2S} = k_1 \cdot a_{Cu}^2 \cdot a_S = a_{Cu_{\frac{1}{2}}S}^2$$

$$a_{Cu_2O} = k_2 \cdot a_{Cu}^2 \cdot a_O = a_{Cu_{\frac{1}{2}}O}^2$$

$$a_{Cu} = \gamma_{Cu} \cdot N_{Cu} \ ; \quad a_S = \gamma_S \cdot N_S \ ; \quad a_O = \gamma_O \cdot N_O$$

$$\log \gamma_{Cu} = \log \gamma_{Cu}^o + \varepsilon_1 \cdot N_{Cu} + \varepsilon_2 \cdot p_{S_2}^{\frac{1}{2}} + \varepsilon_3 \cdot p_{O_2}^{\frac{1}{2}}$$

$$\log \gamma_O = \log \gamma_O^o + \varepsilon_4 \cdot p_{O_2}^{\frac{1}{2}} + \varepsilon_3 \cdot N_{Cu} + \varepsilon_5 \cdot p_{S_2}^{\frac{1}{2}}$$

$$\log \gamma_S = \log \gamma_S^o + \varepsilon_6 \cdot p_{S_2}^{\frac{1}{2}} + \varepsilon_2 \cdot N_{Cu} + \varepsilon_5 \cdot p_{O_2}^{\frac{1}{2}}$$

Shimpo et al. recently (42) put forward a similar view, ascribing changes in solubility to changes in activity coefficients of the various neutral species in the slag. It must be remembered that the activity coefficients deduced depend upon the species assumed, and the forms of the equations depend upon the particular statistical-mechanical relation assumed between the thermodynamic properties and the overall composition. Jalkanen (43) assumed quite elaborate complexes, such as $Fe(CuS)_2$, in the slag, in order to explain the effect of slag sulfur on dissolved copper.

Two other possible explanations should be examined. One is that high-S slags could be structurally different from S-free slags, so that the sites into which S enters are different in number or character, so that the solubilities in the one type of slag cannot be calculated from those in the other type.

A more specific reason can be suggested: that sulfur may dissolve in the slag by replacing O^{2-} ions:

$$[S]_{matte} + (O^{2-})_{slag} = (S^{2-})_{slag} + [O]_{matte} \quad [25]$$
$$\updownarrow \qquad \qquad \qquad \qquad \qquad \qquad \updownarrow$$
$$\tfrac{1}{2}S_{2(g)} \qquad \qquad \qquad \qquad \qquad \tfrac{1}{2}O_{2(g)}$$

Then high-S slags would have a lower activity of oxygen ions than S-free slags, and so would dissolve more Cu^+ at a given p_{O_2} and a_{Cu}. This, of course, contradicts the assumption of constant a_{O_2} as required by equations [23] and [24]. However, if iron oxide-silica slags are buffered with respect to O^{2-} when oxides such as $CuO_{\frac{1}{2}}$ are added, it does not follow that they are buffered with respect to [25]. The number of exchangeable oxygen ions in silicate slags appears to be limited. Richardson (44) believed that the exchangeable oxygens are those not bonded to silicon – the so-called "free" O^{2-} ions. The dissolution of metal oxides does not necessarily involve these exchangeable oxygen ions in any way. Similar considerations apply, of course, to other metals, where the general reaction is of the form:

$$M + \tfrac{1}{2}O_2 = M^{2-} + O^{2-} \ .$$

The activity of FeS in the matte is the primary influence on the sulfur content of the slag. Thus, with slags of approximately constant FeO content, and hence constant Fe^{2+} concentration (42), high a_{FeS} in the matte must give high S in slag.

Equation [25] has been written to include O_2 and S_2 in the gas phase, and will be recognised as the fundamental equation on which the concept of sulfur capacities is based. Unfortunately, sulfur capacities of iron silicate slags do not appear to have been measured at temperatures around 1200-1300°C. If solutions in slags are dilute enough for the slag capacity C_s to be unaffected, equilibria involving p_{S_2} and p_{O_2} can be written in terms of (%S), since:

$$C_s = (\%S)(p_{O_2}/p_{S_2})^{\tfrac{1}{2}} \ .$$

Despite considerable elaboration of the theory of slags recently, little attention has been given to the fundamental aspects of sulfur capacities, and to the role of sulfur in relation to metal solubilities in slags. We would urge that experimental work should be undertaken in this field, as it could provide results of great practical value, as well as theoretical significance such as we have discussed above.

Conclusions

We have presented a novel method for smoothing solubility data in metal systems, based on the observation that the logarithm of the alpha-function at saturation is frequently a linear function of reciprocal temperature. Combining this with the fact that in dilute solution the logarithm of concentration of a solute is always a linear function of reciprocal temperature, and the assumption that Darken's quadratic formalism applies, we have derived an equation for the activity coefficient of a solute in a binary syatem requiring at least three solubility determinations at different temperatures: two in the dilute region, and one in the concentrated region; the equation requires that $\log \gamma$ varies linearly with reciprocal temperature only in the infinitely dilute region, but not necessarily elsewhere.

We have shown that nearly all binary sulfide solutions close to the pseudo-binary composition behave regularly, with a rather small regularity constant, so that they are not very far from ideal; when they contain excess metal, over the stoichiometric equivalent of the sulfur present (as is the case for mattes in most practical applications), the activities of the sulfide species may be considerably different from those in the pseudo-binaries.

We have argued that it is incorrect to ascribe separate and independent solubilities to metal oxide and metal sulfide compounds dissolved in slags, because the metals almost certainly are in ionic form, and a correct formulation of the activity and solubility relationships must recognise this, and allow for interactions between the various species present. We have speculated on the reason for the very great effect of sulfur on the solubility of copper in slags, and proposed lines of investigation which could elucidate this, and so might be fruitful in suggesting ways to reduce the losses in practice.

References

1. J.M.Larrain & H.H.Kellogg,"Use of Chemical Species for Correlation of Solution Properties," pp.130-144 in Calculation of Phase Diagrams and Thermochemistry of Alloy Phases, Y.A.Chang & J.F.Smith, eds.; AIME, New York,NY,1980.

2. John Lumsden, Thermodynamics of Metals, p.384; Institute of Metals, London,1952.

3. E.A.Anderson & T.L.Rodda, "Constitution of Zinc-Lead Alloys", p.1748-9 in Metals Handbook, ASM, Cleveland, 1939.

4. W.L.Fink & H.R.Freche,"Correlation of Equilibrium relations in Binary Aluminium Alloys of High Purity," Trans. AIME, 111 (1934) pp. 304-317.

5. T.R.A.Davey,"The Tin-Iron System: 1 - The Phase Diagram Liquidus Boundaries," Trans. Instn. Mining & Metallurgy, 76 (1967) pp.C66-7.

6. O.J.Kleppa & J.A.Weil,"The Solubility of Copper in Liquid Lead below $950°$.J. Am. Chem. Soc., 73 (1951) pp. 4848-50.

7. T.R.A.Davey, V.Ramachandran & J.White,"Determination of Activities from Solubility Data in Binary Systems by Use of Solution Models," p.207-215 in Advances in Extractive Metallurgy, ed. M.J.Jones; Instn. Mining & Metallurgy, London,1977.

8. L.S.Darken,"Thermodynamics of Binary Metallic Solutions - Part I," Trans. TMS-AIME, 242 (1967) pp. 80-89.

9. E.T.Turkdogan & L.S.Darken, "Thermodynamics of Binary Metallic Solutions - Part II," Trans. TMS- AIME, 242 (1968) pp. 1997-2005.

10. T.R.A.Davey & J.V.Happ,"Solubility of Copper in Liquid Tin", Trans. Instn. Mining & Metallurgy, 78 (1969) pp.C108-110.

11. T.R.A.Davey,"The Tin-Iron System", Trans. Instn. Mining & Metallurgy, 76 (1967) pp. C278-81.

12. E.Schürmann & A.Kaune,"Kalorimetrie und Thermodynamik der Kupfer-Blei-Legierungen", Z.Metallkunde, 56 (1965) pp.453-61.

13. W.Oelsen, E.Schürmann & D.Buchholz,"Kalorimetrie und Thermodynamik der Kupfer-Wismut-Legierungen", Archiv. Eisenhüttenwesen, 32 (1961) pp.39-46.

14. R.Hultgren et al., Selected Values of Thermodynamic properties of Metal and Alloys, p.963; Wiley, New York, 1963.

15. M.Temkin, "Mixtures of Fused Salts as Ionic Solutions," Acta Physicochim. U.R.S.S.,20 (1945) pp.411-420.

16. J.Koh & A.Yazawa,"Thermodynamic Properties of the Cu-S, Fe-S and Cu-Fe-S Systems," Bull. Res. Inst. Min. Dress. & Metall. Tohoku Univ., 38(Part 2),(1982) pp.107-117.

17. C.W.Bale & J.M.Toguri,"Thermodynamics of the Cu-S, Fe-S and Cu-Fe-S Systems," Can. Met. Quart.," 14 (1976) pp.1-14.

18. H.Eric & M.Timucin,"Activities in Cu_2S-FeS-PbS Melts at $1200^{\circ}C$," Met. Trans.B, 12B (1981) pp.493-500.

19. G.M.Willis, unpublished work.

20. T.R.A.Davey & J.E.Joffre,"Vapour Pressures and Activities of SnS in Tin-Iron Mattes," Trans. Instn.Min. & Metall., 82C0. T.R.A.Davey & J.E.Joffre,"Vapour Pressures and Activities of SnS in Tin-Iron Mattes," Trans. Instn.Min. & Metall., 82C (1973) pp.C145-150.

21. John Lumsden, Thermodynamics of Molten Salt Mixtures, p.259; Academic Press, Lon & NY, 1966.

22. F.D.Richardson & J.Antill,"Thermodynamic Properties of Cuprous Sulphide and its Mixtures with Sodium Sulphide", Trans. Faraday Soc., 51 (1955) pp.22-33.

23. John Lumsden, p.261 in ref. 21.

24. John Lumsden, p.260 in ref. 21.

25. T.R.A.Davey,"Phase Systems Concerned with the Copper Drossing of Lead", *Trans. Instn. Min. & Metall.*, 75 (1963) pp.553-620. Discussion on above, pp.771-2.

26. J.Koh & K.Itagaki,"Measurements of Thermodynamic Quantities for Molten $Ag_2S-Sb_2S_3$ and $Cu_2S-Ni_3S_2$ Systems by Quantitative Thermodynamic Analysis," *Trans. Jap. Inst. Met.*, 25 (1984) pp.367-373.

27. Y.-Y.Chuang & Y.A.Chang,"A Thermodynamic Analysis of Ternary Cu-Ni-S and Fe-Ni-S," pp.73-79 in *Advances in Sulfide Smelting*, Vol.1, H.Y.Sohn, D.B.George & A.D.Zunkel, eds.; AIME, New York, 1983.

28. T.R.A.Davey,"The Physical Chemistry of Lead Refining," pp.477-507 in *Lead-Zinc-Tin '80*, J.M.Cigan, T.S.Mackey & T.J.O'Keefe, eds.; AIME New York, 1980.

29. Y.A.Chang & R.C.Sharma,"Application of an Associated Solution Model to the Metal-Sulfur Melts and the Calculation of Metal-Sulfur Phase Diagrams", pp. 145-174 in ref.1.

30. Y.-Y.Chuang & Y.A.Chang,"Extension of the Associated Solution Model to Ternary Metal-Sulfur Melts: Cu-Ni-S," *Met. Trans.B*, 13B (1982) pp.379-385.

31. J.M.Larrain & S.L.Lee,"Thermodynamic Properties of Copper-Nickel-Sulfur Melts," *Can. Met. Quart.*, 19 (1980) pp.183-190.

32. W.A.Krivsky & R.Schuhmann Jr.,"Thermodynamics of the Cu-Fe-S System at Matte Smelting Temperatures," *Trans. Met. Soc. AIME*, 209 (1957) pp.981-988.

33. J.M.Toguri & N.H.Santander,"The Solubility of Copper in Fayalite Slags at $1300^{\circ}C$," *Can. Met. Quart.*, 8 (1969) pp.167-171.

34. J.R.Taylor & J.H.E.Jeffes,"Activity of Cuprous Oxide in Iron Silicate Slags of Various Compositions," *Trans. Instn, Min & Metall.*,84C (1975) pp.C17-24.

35. F.D.Richardson & J.C.Billington,"Copper and Silver in Silicate Slags," *Trans. Inst. Min. & Metall.*, 65 (1955-6) pp.273-297.

36. A.Geveci & T.Rosenqvist,"Equilibrium Relations between Liquid Copper, Iron-Copper Matte and Iron Silicate Slag at $1200^{\circ}C$," Trans. Instn. Min. & Metall.," 82C (1973) pp.C193-201.

37. John Lumsden,"Thermodynamics of Iron-Copper-Sulphur-Oxygen Melts," pp.155-169 in *Metal-Slag-Gas Reactions*, Z.A.Foroulis & W.W.Smeltzer, eds.; The Electrochemical Soc., Princeton, 1975.

38. T.Rosenqvist, "Thermodynamics of Copper Smelting," pp.239-255 in ref. 27.

39. F.Sehnalek & I.Imris,"Slags from Continuous Copper Production," pp.39-62 in *Advances in Extractive Metallurgy and Refining*, M.J.Jones ed.; Instn. min. & Metall., London, 1972.

40. T.R.A.Davey, Discussion of ref.39, ibid., p.64.

41. T.R.A.Davey,"Developments in Direct Smelting of Base Metals - a Critical Review," Paper No.6 in Symposium: Advances in Chemical Metallurgy, Bombay, Jan'79, C.V.Sundaram ed.; Indian Inst. Metals, Bombay, 1979.

42. R.Shimpo, S.Goto, O.Ogawa & I.Isakura,"A Study of the Equilibrium between Copper Matte and Slag," Paper No.6 presented at CIM, Annual Conference of MetallA Study of the Equilibrium between Copper Matte and Slag," Paper No.6 presented at CIM, Annual Conference of Metallurgists, Quebec, August 1984: Copper'84, Vol.2: Copper Pyrometallurgy," A.Ismay, ed.; CIM, 1984.

43. H.Jalkanen,"Copper and Sulfur Solubilities in Silica Saturated Iron Silicate Slags from Copper Matte," Scand.J.Metall. 10 (1981) pp.177-184.

44. F.D.Richardson, Physical Chemistry of Melts in Metallurgy, Vol.2, p.300; Academic Press, London & NY, 1974.

A SOLID-STATE EMF STUDY OF THE Fe-Ni-S-O QUARTERNARY SYSTEM

Ker-Chang Hsieh
Y. Austin Chang

Department of Metallurgical and Mineral Engineering
University of Wisconsin-Madison
1509 University Avenue
Madison, WI 53705

Summary

Four three-phase equilibria and one two-phase equilibrium in the Fe-Ni-S-O quaternary system in the vicinity of 1023 K and p_{SO_2} = 1 atm were studied using the following emf cell

$$Pt, O_2(air) | ZrO_2 \cdot CaO | P_{SO_2} = 1 \text{ atm, mixture of phases, Au}$$

The four three-phase equilibria are sp + (Fe,Ni)S + Ni_3S_2, sp + Ni_3S_2 + NiO, sp + NiO + $NiSO_4$ and Fe_2O_3 + sp + (Fe,Ni)SO_4. The two-phase equilibrium is sp + (Fe,Ni)S. The two-phase equilibrium was measured by using mixtures containing a major portion of one phase and a minor portion of the other. The results are presented in terms of oxygen potential as a function of $y_{Ni} = n_{Ni}/(n_{Ni}+n_{Fe})$ at constant temperature and p_{SO_2} = 1 atm. The stability diagram was also calculated using appropriate thermodynamic models for all the pertinent phases. The calculated diagram is in accord with the measured ones.

Introduction

The quaternary system Fe-Ni-S-O is of technological interest in the fields of nickel extraction and hot corrosion of structural alloys containing iron and nickel. A knowledge of the thermochemistry of this system would be useful to the process development of nickel extraction and at the same time to the kinetics of simultaneous oxidation-sulfidation. In the present study, an emf method employing a calcia-stabilized zirconia electrolyte is used to measure the oxygen potentials of several three phase equilibria* and one two-phase equlibrium.* The measurements are carried out at constant temperatures with a fixed pressure of SO_2 = 1 atm. According to the phase rule, the number of degrees of freedom, F, of a quaternary system is

$$F = 6 - P$$

where P is the number of phases.

When we keep T = constant and p_{SO_2} = constant = 1 atm, the number of degrees of freedom reduces to

$$F = 4 - P$$

Accordingly, when we have a three-phase equilibrium P=4 with the fourth phase being the gaseous phase. The system is invariant and has a unique value for the oxygen potential. This value may be measured using an appropriate emf cell. For two-phase mixtures, the system is univariant and an additional constraint must be imposed. In the present study, the composition for one of the co-existing phases is fixed. This is done by using a two-phase mixture containing a major portion of one phase and a minor portion of the second phase. In this manner, the composition of the major phase changes so slightly during the equilibration that its composition may be taken to the constant.

Fig. 1 shows a schematic diagram of Fe-Ni-S-O at 1023 K and p_{SO_2} = 1 atm. The stable phases under these conditions are (Fe,Ni)S (δ), (Fe,Ni)$_3$S$_2$ (β), NiS$_2$, NiO (ϵ), (Fe,Ni)SO$_4$ (ξ), Fe$_3$O$_4$-NiFe$_2$O$_4$ (sp), Fe$_2$O$_3$ and Fe$_2$(SO$_4$)$_3$. In the present study, the oxygen potentials of δ +sp, δ + β+sp, β+ϵ+sp, ϵ+ξ+sp and sp+ξ+Fe$_2$O$_3$ are measured in the vicinity of 1023 K and p_{SO_2} = 1 atm. The results are shown in Fig. 2 as a function of $y_{Ni} = n_{Ni}/(n_{Fe}+n_{Ni})$ where n_{Fe} and n_{Ni} are the moles of Fe and Ni. The numbers 5, 4, 3, 2 and 1 in both Figs. 1 and 2 indicate the different three-phase equilibria. These two diagrams specify the phase relationships and the thermodynamics of this system at 1023 K and p_{SO_2} = 1 atm.

---

*In the present study, the three-phase and two-phase equilibria refer to only the condensed phases.

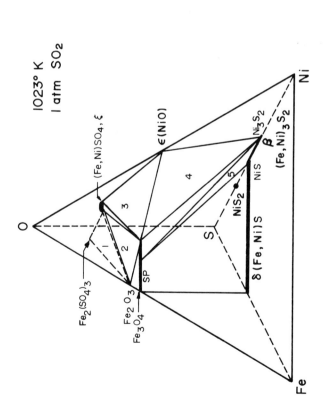

Fig. 1 A schematic representation of Fe-Ni-S-O at 1023 K and P_{SO_2} = 1 atm. Numbers indicate coexistence of three solidphases.

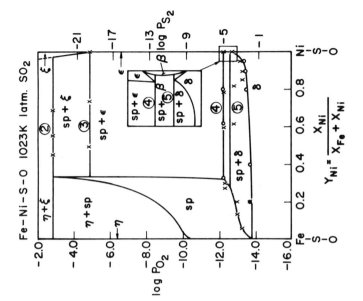

Fig. 2 Fe-Ni-S-O stability diagram at 1 atm. of SO_2 and 1023 K. Numbers as in Fig. 1.

Experimental Method

The emf cell used in the present study is given below,

$$\text{Pt}, O_2(\text{air}) \mid ZrO_2\cdot CaO \mid \text{mixture of phases, Au} \\ SO_2, S_2, O_2 \text{ with } p_T = 1 \text{ atm}$$

A schematic diagram of the emf cell is shown in Fig. 3. A ceramic tube, placed inside the electrolyte, was used to press the sample and to minimize free space available. This would minimize any disturbance of the equilibrium condition of the gases over the condensed-phase mixtures.

A schematic diagram for the experimental set-up is shown in Fig. 4. Prior to each experimental run, the electrolyte tube was evacuated with a mechanical pump and then filled with dry $SO_2(g)$ by passage through silica gel. An atmosphere of SO_2 was maintained in the cell by by-passing this gas to the exhaust as shown in Fig. 3. This was done in order to prevent the minor species such as $S_2(g)$, $O_2(g)$ and $SO_3(g)$ which were formed at the gas sample interface from being swept away with the gas flowing through the cell. The flowing through SO_2 gas would disturb the equilibrium condition around the sample. The exhausted SO_2 gas was absorbed by passage through a column of glass beads by a sodium hydroxide solution.

Attainment of equilibrium condition during the measurement is essential if the data obtained have any meaning. The design of the apparatus as mentioned above was made to be assured that equilibrium condition could be achieved readily. In order to be certain this was the case, the following procedure and criteria were adopted (1,2).

(i) The cell resistance was estimated and it must be at least four orders of magnitude smaller than the internal resistance of the instrument. The Keithley model 616 digital volt meter used has an internal resistance of 10^{14} ohm.

(ii) The emf vaues at the temperature of interest must be stable for a period of at least one hour. After the measurement is complete at a fixed temperature T', the sample was held at a higher and then a lower temperature for some time; after which, the sample was brought back to the original temperature T' and repeated measurements were carried out. A typical emf/temperature cycle is shown in Fig. 5. Only these measurements that are independent of prior history of the sample were regarded as reliable.

(iii) The cells were polarized with an opposite potential using a 1.5 V dry cell for 15 seconds; after which, the time for return to the original emf values was measured. A short time of 10-15 minutes for such recovery was an indication of satisfactory cell performance, i.e. the cell was reversible. Fig. 6 shows a typical result.

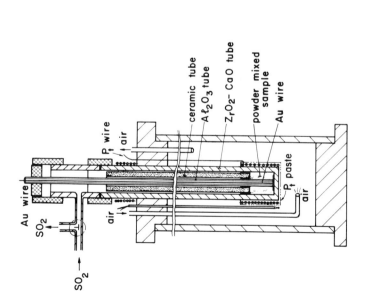

Fig. 3 Cell design

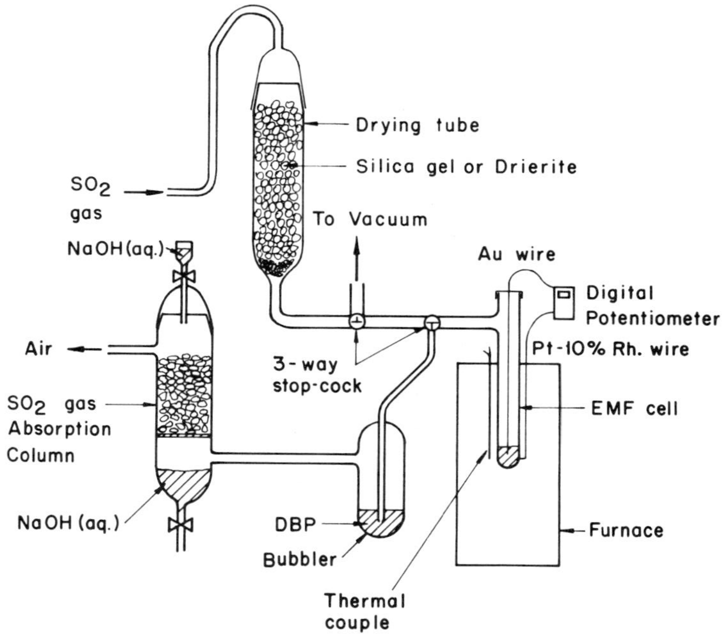

Fig. 4 Experimental apparatus

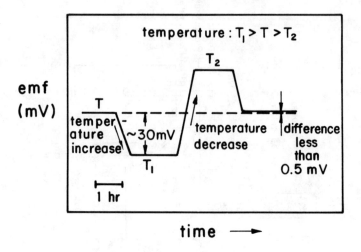

Fig. 5 A typical temperature/emf cycle for checking the reproductivity.

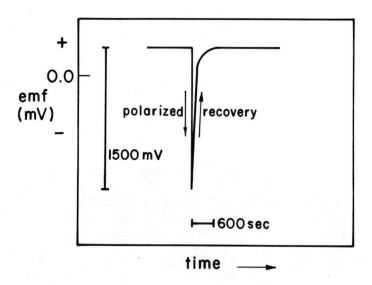

Fig. 6 Emf vs. time during polarization.

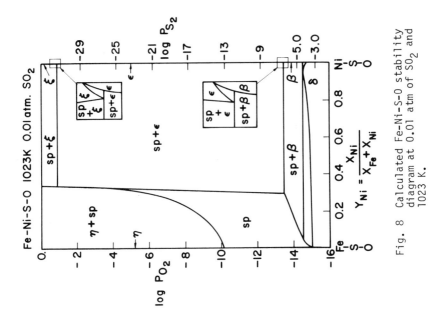

Fig. 8 Calculated Fe-Ni-S-O stability diagram at 0.01 atm of SO_2 and 1023 K.

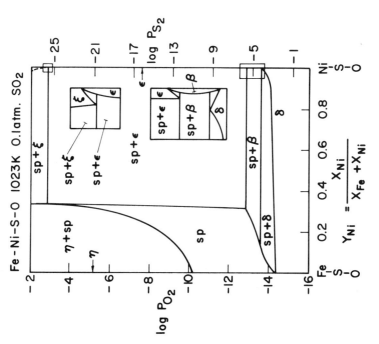

Fig. 7 Calculated Fe-Ni-S-O stability diagram at 0.1 atm of SO_2 and 1023 K.

Results

The experimental results at 1023 K at p_{SO_2} = 1 atm. are given in Fig. 2. The data of Klubnes and Rosenqvist (3) obtained using the same method and ours are in good agreement. The solid lines are calculated from thermodynamic properties of the pertinent phases. The calculated values of the oxygen potentials agree with the experimental data within the uncertainties of the measurements. A detailed description of the models used and the method of calculation are given elsewhere (4). It suffices to state that the monosulfide (Fe,Ni)S, δ, the $(Fe,Ni)_3S_2$ phase, β, the monoxide (Fe,Ni)O, ϵ and the sulfate $(Fe,Ni)SO_4$, ξ, were treated as pseudobinaries of the compound phases. Either the quasi-regular or quasi-subregular models were used. For the spinel phase, a structural model (5,6) was used.

Two other stability diagrams corresponding to p_{SO_2} = 0.1 and 0.01 atm were also calculated as shown in Figs. 7 and 8. We are currently measuring the oxygen potentials of the various equilibria of Fe-Ni-S-O at p_{SO_2} = 0.01 atm. Preliminary results show that the calculated values are in agreement with the experimental data.

Acknowledgement

The authors wish to thank the National Science Foundation for financial support through Grant Nos. NSF-DMR-78-04066 and NSF-DMR-83-10529. They are also grateful to Professor T. Rosenqvist of Norwegian Institute of Technology for much help concerning the experimental method used and for providing their data not yet published. Professor Rosenqvist was a Brittingham Professor at the University during Spring 1983.

References

1. A. W. Espelund and H. Jynge, "The System Zn-Fe-S-O. Equilibria between Sphalerite or Wurtzite, Pyrrhotite, Spinel, ZnO and a Gas Phase", Scandanavian Journal of Metallurgy, 6 (1977), pp. 256-262.

2. Terkel Rosenqvist and Audun Hofseth, "Phase Relations and Thermodynamics of the Copper-Iron-Sulphur-Oxygen System at 700-1000°C", Scandanavian Journal of Metallurgy, 9 (1980), pp. 129-138.

3. S. Klubnes and Terkel Rosenqvist: unpublished research, University of Trondheim, 1979.

4. Ker-Chang Hsieh, "The Thermodynamics and Phase Equilibria of the Fe-Ni-S-O System", PhD Dissertation, University of Wisconsin-Madison, Madison, WI 1983.

5. Hermann Schmalzried and J. D. Tretjakow, "Disorder in Ferrite", Berichte der Bunsengesellschaft fur Physikalische Chemie, 70 (1966), pp. 180-189.

6. A. D. Pelton, Hermann Schmalzried and J. Sticker, "Computer-Assisted Analysis and Calculation of Phase Diagrams of the Fe-Cr-O, Fe-Ni-O and Cr-Ni-O Systems", Journal of Physics and Chemistry of Solids, 40 (1979), pp. 1103-1122.

ESTIMATION OF THE STANDARD FREE ENERGIES
OF FORMATION OF TIN CHLORIDES

Salustio S. Guzman*

Henry Krumb School of Mines
Columbia University
New York, New York 10027
U.S.A.

The standard free energies of formation of tin chlorides were estimated with reviewed and estimated basic thermodynamic data, and measured reversible potentials for the disproportionation of tin dichloride into metallic tin and tin tetrachloride. Values for C_p°, S°, $H^\circ - H_{298}^\circ$, $-(G^\circ - H_{298}^\circ)/T$, ΔH_f°, ΔG_f°, and log Kp are given in tabular form, these estimations are compared with estimations reported in the literature.

* Present Address: Alcan International Limited, Kingston Laboratories, P.O. Box 8400, Kingston, Ontario K7L 4Z4, Canada.

Introduction

Although a number of compilations of thermodynamic data have appeared over the last decades which make available a large amount of thermodynamic data of the expanding knowledge of high temperature chemistry, the metallurgist seeking to make use of equilibrium calculations will encounter systems where one or more thermodynamic properties he requires for his problem are unknown.

The estimation of thermodynamic properties is rapid and sufficiently accurate for many purposes, as most of the principles which are applied have been well established for considerable time.

A thermodynamic analysis of tin chloride volatilization (1) was made so as to foresee some of the problems that may arise in a proposed new chlorination - electrowinning process for tin extraction from ores and concentrates (2).

The present paper reports estimations made of some basic thermodynamic data for tin chlorides, measured reversible potentials for the disproportionation of tin dichloride into metallic tin and tin tetrachloride in the range 260-350°C and, estimated standard free energies of formation of tin chlorides which are compared with estimations reported in the literature.

Thermodynamic Data on Tin Chlorides

There is very limited information in the literature on the standard free energies of formation of tin chlorides, the only sources constitute the estimations made by Kellogg (3) and Kubaschewski (4). The Kellogg data is in the range 298-2000°K for both chlorides, the Kubaschewski data for $SnCl_{4(g)}$ is in the range 500-1200°K and for $SnCl_{2(\ell)}$ from 520-925°K ($SnCl_{2(c,g)}$ phases are not covered).

In the usual sources of thermodynamic data, there is no reported value for the heat capacity of $SnCl_{2(c,\ell,g)}$. Furthermore, there is conflicting data on the standard enthalpy of formation ($\Delta H°_{298}$) and standard entropy at 298°K ($S°_{298}$) for $SnCl_{2(c)}$. On the other hand, the thermodynamic information of $SnCl_{4(g)}$ in the literature is more reliable and consistent.

The thermodynamic data for $SnCl_{2(c,\ell,g)}$, $SnCl_{4(g)}$, $Sn_{(c,\ell)}$ and Cl_2 used in this investigation are given as follows:

Tin dichloride

Table I. Basic Thermodynamic Data of $SnCl_{2(c,\ell,g)}$

		Reference
$SnCl_{2(c)}$		
$\Delta H°_{f298}$	= to be estimated	
$S°_{298}$	= 30.5 (±1), (cal/°K mole)	estimated
$C°_p$	= 13.81 + 14.4 × 10^{-3}T (±0.1), (cal/°K mole)	estimated
$\Delta H°_{fus}$	= 3,470 (cal/mole)	4
T_{fus}	= 520.15°K	4
$SnCl_{2(\ell)}$		
$C°_p$	= 25.9 (±0.5), (cal/°K mole)	estimated
$\Delta H°_{vap}$	= 19,500, (cal/mole)	4
T_{vap}	= 925.15°K	4
$SnCl_{2(g)}$		
$C°_p$	= 13.864 + 2.819 × 10^{-5}T − 0.3988 × $10^5/T^2$ (±0.1), (cal/°K mole)	estimated

Tin Tetrachloride

Table II. Basic Thermodynamic Data of $SnCl_{4(g)}$

		Reference
$SnCl_{4(g)}$		
$\Delta H°_{f298}$	= −116,900, (cal/mole)	4
$S°_{298}$	= 88.5 (±1), (cal/°K mole)	estimated
$C°_p$	= 25.57 + 0.20 × 10^{-3}T − 1.87 × $10^5/T^2$ (±0.1), (cal/°K mole)	5

Tin

Table III. High Temperature Data of $Sn_{(c,\ell)}$. Reference 6.

T, °K	Condensed Phase (cal/°K g-atom)			(cal/g-atom)
	C_p	$S_T - S_{st}$	$-\dfrac{G_T - H_{st}}{T}$	$H_T - H_{st}$
298.15(β)	6.450	0.000	12.236	0
400	6.891	1.958	12.497	679
500	7.323	3.542	12.998	1390
505.06(β)	7.345	3.615	13.023	1428
505.06(ℓ)	7.100	6.941	13.023	3108
600	6.875	8.138	14.101	3764
700	6.817	9.193	15.065	4448
800	6.800	10.101	15.927	5128
900	6.800	10.902	16.685	5808
1000	6.800	11.618	17.367	6488
1200	(6.800)	12.858	18.554	7848

	Reference
S°_{298} = 12.236 (±0.1), (cal/°K mole)	6
ΔH_{fus} = 1,680 (±50), (cal/mole)	6
T_{fus} = 505.06°K	6

Chlorine

Table IV. High Temperature Data of $Cl_{2(g)}$. Reference 6

	Gas, $1/2\ Cl_{2(g,\ ideal)}$			
	(cal/°K g-atom)			(cal/g-atom)
T, °K	C_p	$S_T - S_{st}$	$-\dfrac{G_T - H_{st}}{T}$	$H_T - H_{st}$
298.15	4.057	0.000	26.646	0
400	4.218	1.217	26.807	422
500	4.309	2.169	27.116	849
600	4.367	2.960	27.467	1283
700	4.405	3.636	27.822	1722
800	4.433	4.226	28.167	2164
900	4.455	4.750	28.498	2608
1000	4.472	5.220	28.811	3054
1200	4.498	6.038	29.390	3952

Free Energies For The Cell Reaction

Reversible potentials were measured for the disproportionation of tin dichloride into metallic tin and tin tetrachloride, the apparatus and experimental procedure are described in another paper (7). The results are summarized in Table V.

Table V. Tin Dichloride Decomposition Potentials. Reference 7

$$2SnCl_{2(\ell)} = Sn_{(\ell)} + SnCl_{4(g)} \qquad (1)$$

Temperature		E° (volts)	ΔG° (cal/mole)
°C	°K		
260	533.15	0.563	25,970
300	573.15	0.542	25,000
350	623.15	0.513	23,660
accuracy		±0.004	±185

Correlation of Data

The standard enthalpy of formation at 298°K for the cell reaction (reaction 1) may be obtained from the experimental data, $\Delta G°$, by the so-called third law method. If the heat capacity of each substance in the reaction is expressed as an empirical equation.

$$C_p° = a + bT + c/T^2 \qquad (2)$$

the change in heat capacity for the reaction, $\Delta C_p°$, may be expressed as:

$$\Delta C_p° = \Delta a + \Delta bT + \Delta c/T^2 \qquad (3)$$

The value of $\Delta H°$ and $\Delta S°$ at any temperature, T, may be found from the relations:

$$\Delta H_T° = \int \Delta C_p° \, dT = \Delta aT + \frac{\Delta bT^2}{2} - \frac{\Delta c}{T} + I_H \qquad (4)$$

$$\Delta S_T° = \int \frac{\Delta C_p°}{T} \, dT = \Delta a \ln T + \Delta bT - \frac{\Delta c}{2T^2} + I_S \qquad (5)$$

where I_H and I_S are integration constants which can be evaluated by substitution of known values of $\Delta H°$ and $\Delta S°$ into these equations.

The basic definition of free energy leads directly to the relation

$$\Delta G° = \Delta H° - T\Delta S° \tag{6}$$

where $\Delta H°$ and $\Delta S°$ are, respectively, the standard molar enthalpy and entropy changes for the reaction.

Combination of (4), (5) and (6) results in the equation:

$$\Delta G° = I_H + (\Delta a - I_S)T - \frac{\Delta b T^2}{2} - \frac{\Delta c}{2T} - \Delta a T \ln T \tag{7}$$

The combination of the empirical constants $(\Delta a - I_S)$ is often denoted simply by I,

$$I = \Delta a - I_S \tag{8}$$

If empirical heat capacity equations are used to describe the temperature dependence of $\Delta H°$ and $\Delta S°$, then Σ' function may be used for correlation and extrapolation of the experimental data. The definition of this function is:

$$\Sigma' = \Delta G° + \frac{\Delta b T^2}{2} + \frac{\Delta c}{2T} + \Delta a T \ln T \tag{9}$$

Comparisons of equations (9) and (7) shows that,

$$\Sigma' = I_H + IT \tag{10}$$

it follows that a plot of Σ' vs T will yield a straight line of slope I and intercept I_H.

Making the third law correlation to compute $\Delta H°_{298}$ for the cell reaction, we may write the following reactions:

$$2SnCl_{2(c)} = Sn_{(c)} + SnCl_{4(g)} \quad (298.15 - 505.06°K) \quad (11)$$

$$2SnCl_{2(c)} = Sn_{(\ell)} + SnCl_{4(g)} \quad (505.06 - 520.15°K) \quad (12)$$

$$2SnCl_{2(\ell)} = Sn_{(\ell)} + SnCl_{4(g)} \quad (520.15 - 925.15°K) \quad (13)$$

The application of equation (5) to reactions (11), (12) and (13) with the corresponding data for heat capacities and entropy for each temperature range yield: $I_S = 160.591$ at $520.15°K$, and from equation (8) $I = -182.362$.

Σ' and I_H can be computed for each experimental value of the free energy of the cell reaction with equations (9) and (10) respectively:

T,°K	Σ'	I_H
533.15	-46,378	50,848
573.15	-53,676	50,845
623.15	-63,002	50,637
		$\overline{I_H} = 50,777$

The lack of any significant trend in I_H with T suggest that the estimated values for $S°$ and $C_p°$ of $SnCl_{2(\ell)}$ and $SnCl_{4(g)}$ are consistent.

The application of equation (4) to reactions (13), (12) and (11) with the corresponding data for the heat capacities and enthalpy for each temperature range yield: $\Delta H°_{298} = 46,452$ (cal/mole) for reaction (11).

The standard free energy of the cell reaction and the standard free energy of formation of $SnCl_{4(g)}$ can now be computed from the known and estimated thermochemical properties of $SnCl_{4(g)}$, $SnCl_{2(c,\ell,g)}$ and, the data for $Sn_{(c,\ell)}$ and Cl_2 given in Tables III and IV by the equation:

$$\Delta G°_T = \Delta H°_{298} + T\Delta fef_T \quad (14)$$

Results are shown in Tables VI and VII.

Table VI. Estimated Thermochemical Properties of $SnCl_{4(g)}$ At Elevated Temperatures

T,°K	$C_p^°$	$S°$	$-(G°-H_{298}°)/T$	$H°-H_{298}°$	$\Delta H_f°$	$\Delta G_f°$	$\log K_p$
	(cal/°K mole)				(kcal/mole)		
298.15	23.526	88.500	88.500	0	-116.900	-107.860	79.057
400	24.481	95.567	89.438	2.452	-116.815	-104.785	57.247
500	24.922	101.082	91.234	4.924	-116.762	-101.786	44.487
505.06	24.938	101.334	91.334	5.050	-116.762	-101.635	43.976
505.06	24.938	101.334	91.334	5.050	-118.442	-101.635	43.976
600	25.171	105.650	93.267	7.430	-118.366	- 98.479	35.868
700	25.328	109.543	95.321	9.955	-118.281	- 95.178	29.713
800	25.438	112.933	97.315	12.494	-118.190	- 91.876	25.097
900	25.519	115.934	99.220	15.042	-118.098	- 88.589	21.511
1000	25.583	118.626	101.029	17.597	-118.007	- 85.318	18.645
1100	25.635	121.067	102.741	20.158	-117.910	- 82.077	16.306
1200	25.680	123.299	104.362	22.724	-117.832	- 78.798	14.350

Table VII. Estimated Standard Free Energy Of The Cell Reaction

$2SnCl_{2(c,\ell,g)} = Sn_{(c,\ell)} + SnCl_{4(g)}$	
T, °K	$\Delta G°$ (kcal/mole)
298.15	34.605
400	30.661
500	26.988
505.06	26.808
520.15	26.228
600	24.312
700	22.222
800	20.402
900	18.833
925.15	18.480
1000	20.566
1100	23.354
1200	26.033

The standard free energy of formation of tin dichloride is now calculated by combining the standard free energy of the cell reaction and standard free energy of formation of $SnCl_{4(g)}$, and the results are given in Table VIII.

Table VIII. Estimated Thermochemical Properties Of $SnCl_{2(c,\ell,g)}$ At Elevated Temperatures.

T, °K	C_p°	S°	$-(G^\circ-H_{298}^\circ)/T$	$H^\circ-H_{298}^\circ$	ΔH_f°	ΔG_f°	log K_p
	(cal/°K mole)			(kcal/mole)			
298.15	18.103	30.500	30.500	0	− 81.676	− 71.232	52.210
400	19.570	36.025	31.229	1.918	− 81.280	− 67.723	36.999
500	21.010	40.547	32.651	3.947	− 80.816	− 64.387	28.141
505.06	21.083	40.758	32.732	4.054	− 80.792	− 64.221	27.788
505.06	21.083	40.758	32.732	4.054	− 82.472	− 64.221	27.788
520.15	21.300	41.382	32.974	4.374	− 82.387	− 63.682	26.755
520.15	25.900	48.055	32.974	7.844	− 78.917	− 63.682	26.755
600	25.900	51.754	35.234	9.912	− 78.094	− 61.395	22.361
700	25.900	55.746	37.886	12.502	− 77.066	− 58.700	18.325
800	25.900	59.205	40.340	15.092	− 76.040	− 56.139	15.335
900	25.900	62.255	42.609	17.682	− 75.018	− 53.711	13.042
925.15	25.900	62.969	43.152	18.333	− 74.761	− 53.129	12.549
925.15	13.843	84.047	43.152	37.833	− 55.261	− 53.129	12.549
1000	13.852	85.124	46.255	38.869	− 55.402	− 52.942	11.569
1100	13.862	86.445	49.849	40.255	− 55.589	− 52.716	10.473
1200	13.870	87.652	52.950	41.642	− 55.786	− 52.415	9.545

The standard heat of formation for $SnCl_2$,

$$\Delta H_{f298}^\circ = -81,676 \text{ cal/mole}$$

was calculated by combining the standard heat of formation of $SnCl_{4(g)}$ with the standard heat of formation of the cell reaction. This estimated value is within the range of estimates for the standard heat of formation of $SnCl_2$ reported in the literature, as seen in Table IX.

Table IX. Standard Heat Of Formation Of $SnCl_2$

ΔH°_{f298} (cal/mole)	Reference	
-81,676	This investigation	
-77,700	NBS 270-3	, (9)
-81,100	Wicks and Block	, (10)
-79,100	Kubaschewski	, (4)
-83,600	Rossini	, (11)
-80,800	Budgen and Shelton	, (12)
-78,360	Vasil'ev, et al.	, (13)

The estimated standard free energies of formation, ΔG°, for $SnCl_2$ and $SnCl_4$ from its elements are plotted as a function of temperature, Figures 1 and 2, along with the estimations made by Kellogg and Kubaschewski.

Kubaschewski's and Kellogg's estimates show a greater deviation from my estimates at lower temperatures, this may be due to their selected values of the standard enthalpy of formation and standard entropy at 298°K for $SnCl_{2(c)}$ and $SnCl_{4(g)}$. The deviations at high temperatures may be attributed to the differences in the estimated heat capacities and phase changes data.

Acknowledgements

The author wishes to thank Professor H.H. Kellogg for criticizing the manuscript. The work reported herein was part of a doctoral dissertation (2).

References

(1) S.S. Guzman and H.H. Kellogg, "Cassiterite Chlorination with Tin Tetrachloride", paper presented at the 111th AIME Annual Meeting, Dallas, Texas, February 16, 1982, unpublished.

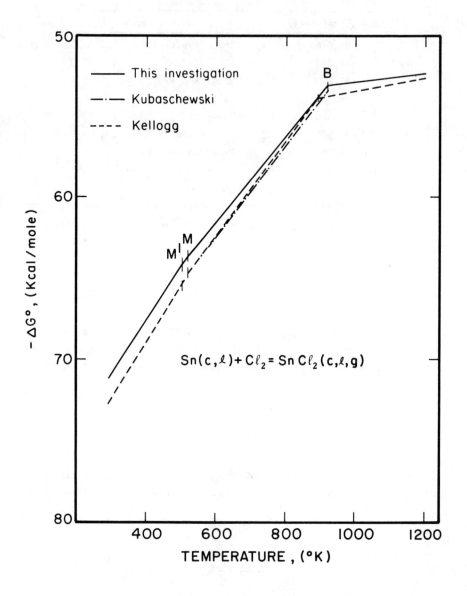

Figure 1. Estimated standard free energy of formation of $SnCl_2(c,\ell,g)$ from 298 to 1200°K.

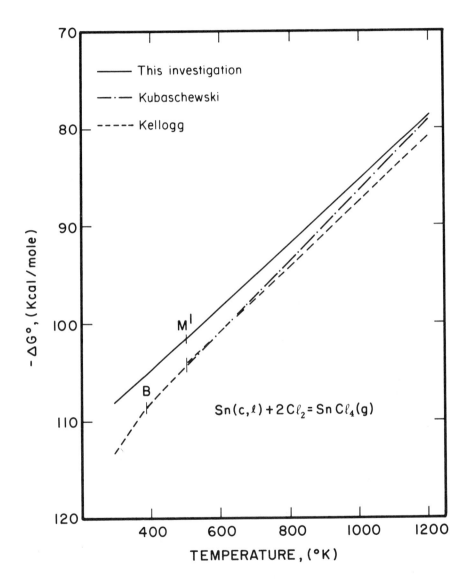

Figure 2. Estimated standard free energy of formation of $SnCl_{4(g)}$ from 298 to 1200°K.

(2) S.S. Guzman, Study Of A Potential Process For The Recovery Of Tin By Chlorination And Electrolysis, doctoral thesis, Columbia University, New York, N.Y., 1982.

(3) H.H. Kellogg, "Thermodynamic Relationships in Chlorine Metallurgy", Journal of Metals, 188 (6) (1950), pp. 862-872.

(4) O. Kubaschewski and C.B. Alcock, Metallurgical Thermochemistry, 5th ed., Pergamon Press Inc., New York, 1979, and 4th ed., 1967.

(5) K.K. Kelley, Contributions to the Data of Theoretical Metallurgy - XIII High Temperature Heat Content, U.S. Bur. of Mines, Bull. 584, 1960.

(6) R. Hultgren, et al., Selected Values of the Thermodynamic Properties of the Elements, American Society for Metals, Ohio, 1973.

(7) S.S. Guzman and P.F. Duby, in Chlorine Electrometallurgy, pp. 115-129, P.D. Parker, ed., AIME, New York, N.Y., 1982.

(8) H.H. Kellogg, Chemical Equilibria in Metallurgy, Henry Krumb School of Mines, Columbia University, New York, unpublished.

(9) D.D. Wagman, et al., Technical Note 270-3, Selected Values of Chemical Thermodynamic Properties, U.S. Nat. Bur. of Stand., 1968.

(10) C.E. Wicks and F.E. Block, Thermodynamic Properties of 65 Elements - Their Oxides, Halides, Carbides and Nitrides, U.S. Bur. of Mines, Bull. 605, 1963.

(11) F.D. Rossini, et al., Selected Values of Chemical Thermodynamic Properties, U.S. Nat. Bur. Stand., Bull. 500, 1962.

(12) W.G. Budgen and R.A.J. Shelton, "Thermodynamic Properties of $PbCl_2$, $SnCl_2$, and CuCl from E.M.F. Measurements on Galvanic Cells With Solid Electrolyte", Institution of Mining and Metallurgy, Transactions, Sect. C, 79 (9) 1970, pp. C215-C220.

(13) V.P. Vasil'ev, et al., "Standard Enthalpies of Formation of Cristalline Tin (II) Chlorides", Russian Journal of Inorganic Chemistry, (English translation), 18 (6) 1973, pp. 774-776.

KINETICS AND THERMODYNAMICS OF CHALCOPYRITE OXIDATION

P. C. Chaubal and H. Y. Sohn

Department of Metallurgy and Metallurgical Engineering
University of Utah
Salt Lake City, Utah 84112-1183

Abstract

The overall kinetics of oxidation of single particles of chalcopyrite have been studied for temperatures up to 1273 K. The experiments were conducted in a stationary bed reactor using a non-isothermal technique. Any mass and heat transfer effects on the kinetics of oxidation were eliminated by using a high-enough gas flowrate. Up to temperatures of 873 K, pore-blocking kinetics were applicable, whereas above 873 K, power-law kinetics gave a satisfactory fit of the data. The activation energy was 215 kJ/mol below 754 K and 23.7 kJ/mol above 1,060 K.

Predominance area diagrams were constructed at various temperatures and used in conjunction with X-ray analyses of partially oxidized samples to determine the intermediate phases formed during the reaction. This analysis also provides a justification for the kinetic models used.

The results indicate that the reaction rate is first order with respect to oxygen concentration and inversely proportional to the square of the particle size throughout the ranges of temperature studied.

Introduction

The oxidation of chalcopyrite has received wide attention in the past, due to the development of roasting as a unit process in the recovery of copper from chalcopyrite ore. Knowledge of the rate of oxidation and its dependence on oxygen pressures, temperatures, and particle size are of considerable importance in ensuring that desired phases occur as final products and that industrially feasible rates of roasting are achieved (1).

Most of the previous work at understanding the mechanism and kinetics of chalcopyrite oxidation was limited to roasting temperatures of practical interest -- below 973 K. At temperatures around 773 K, cupric sulfate and ferric oxide were the final roasting products, whereas, at around 923 K, cupric and cuprous oxides were observed (2-5). Bumazhnov and Lenchev (3) extended their work to higher temperatures and reported ferrite formation at temperatures above 973 K. Yazawa (6) and Rosenqvist (7) have analyzed the process thermodynamically by constructing predominance area diagrams and reported that the observed phases agree with their analysis.

Leung (8) has suggested that oxidation in commercial roasters occurs through two paths -- the direct oxidation of chalcopyrite to Fe_3O_4 and by oxidation of pyrrhotite resulting from the decomposition of chalcopyrite. Other workers (2,4) believe that bornite forms first followed by magnetite formation. Razouk et al. (9) suggest that, below 673 K, covellite forms first and is then oxidized to oxide and sulfate. At higher temperatures, $CuSO_4$ results from the interaction between CuO and $Fe_2(SO_4)_3$. Above 873 K, both copper and iron sulfate decompose, and the resulting oxides react above 1173 K to form ferrites. Thornhill and Pidgeon (10) conducted a micrographic study of the roasting process at 923 K and concluded that the formation of digenite and covellite was the key and accounted for the selective oxidation of the iron component of chalcopyrite.

Few workers have attempted a basic kinetic study of the oxidation process. Using particles in the size range 50-250 microns, Bumazhnov and Lenchev (3) found that nucleation and growth kinetics described their data, whereas Agarwal and Gupta (11) used a shrinking core model to describe the oxidation of pelletized agglomerates. The results are summarized in Table I.

Table I. Reported Data on Kinetics of Oxidation of Chalcopyrite

Temp. Range	Activation Energy kJ/mole	Kinetic Model	Source
673-789 K	176	Nucleation and growth	Bumazhnov and Lenchev (3)
789-973 K	112	Nucleation and growth	Bumazhnov and Lenchev (3)
> 973 K	40.0	Nucleation and growth	Bumazhnov and Lenchev (3)
773-973 K	49.9	Shrinking Core	Agarwal and Gupta (11)

The studies mentioned above used a fluid bed reactor to carry out the tests. The elimination of mass transfer effects cannot be considered to be complete in such reactors. Hence the data cannot be used as is to study the kinetics in reactors where fluid dynamics are different from those in the fluid bed reactor.

For example, in the flash smelting process the description of the initial heat up to ignition requires a knowledge of the intrinsic kinetics of single-particle oxidation. The data can be combined with equations describing mass and heat transfer effects to describe the ignition process.

In this study the oxidation of chalcopyrite was carried out in the absence of heat and mass transfer effects. The effects of oxygen pressure and particle size have also been studied using a non-isothermal technique. X-ray phase analysis combined with predominance area diagrams has been used to determine the intermediate phases occurring in the oxidation process.

Theoretical Considerations

The intrinsic kinetics of reaction between a solid and a gas may be represented by the following general equation:

$$\frac{dx}{dt} = k f_1(p_A) f_2(x) f_3(d) \tag{1}$$

x: fraction reacted

d: average particle size

p_A: partial pressure of gaseous reactant

The function $f_1(p_A)$ accounts for the dependence of the reaction rate on the gaseous reactant concentration, $f_2(x)$ describes the dependence on fraction reacted, and $f_3(d)$ describes the dependence on the particle size. "k" is the reaction rate constant and accounts for the temperature dependence on the reaction rate as follows:

$$k = k_o \exp(-E/RT) \tag{2}$$

where E is the activation energy of the process.

When Eq. (1) is integrated under isothermal conditions, the following expression results:

$$\int_0^x \frac{dx}{f_2(x)} \equiv g(x) = \int_0^t f_1(p_A) f_3(d) dt \tag{3}$$

The function $g(x)$ normally depends on the geometric change occurring in the solid with progress of reaction. Various forms encountered in gas-solid systems are listed below:

a) the shrinking-core scheme (topochemical reaction)

$$g(x) = 1 - (1 - x)^{1/F_g} \qquad F_g = 1, 2, \text{ or } 3 \qquad (4)$$

b) the power-law kinetics

$$g(x) = (1 - x)^{-m} - 1 \qquad m > 0 \qquad (5)$$

c) nucleation and growth kinetics

$$g(x) = [-\ln(1 - x)]^{1/n} \qquad (6)$$

d) pore-blocking model

$$g(x) = \exp(\tfrac{x}{\lambda} - 1) \qquad (7)$$

The pore-blocking kinetics are not normally encountered. However, in porous solids, the formation of product larger in volume than the reactant, can cause partial blockage of the pores. Evans (12) has shown that the above expression (Eq. (7)) can be obtained by considering the rate of blockage to be dependent on the number of blocked pores and the rate of reaction to be dependent on the number of open pores.

Determination of the appropriate form of g(x) must generally be done experimentally. The actual experimental measurements may be carried out under isothermal or nonisothermal conditions. In the former, experiments have to be conducted at various temperatures, whereas in the latter, in principle, one experiment is sufficient. The justification for the use of the nonisothermal technique is given in the next section. Whichever technique is employed, the direct determination of intrinsic kinetics requires elimination of external heat and mass transfer with a high-enough gas flow rate.

In the nonisothermal technique, the solid sample is brought in contact with the reactant gas, and, once the gas flow is started, the temperature of the reactants is raised linearly. So,

$$T = T_0 + at \qquad (8)$$

therefore, $dT = a\,dt$ \hfill (9)

Equations (2) and (9) may be substituted in Equation (1) to obtain, after some rearrangement, the following:

$$\ln\left[\frac{dx}{dT} \cdot g'(x)\right] = \ln\left[\frac{k_0 f_1(p_A) f_3(d)}{a}\right] - \frac{E}{RT} \qquad (10)$$

Therefore, a plot of the lefthand side vs. 1/T should be linear. The slope (and hence activation energy) and the pre-exponential factor k_0 can then be obtained once the appropriate form of g(x) is determined.

Experimental

The aim of this study was to determine the intrinsic kinetics of single-particle oxidation in the absence of mass and heat transfer effects. In order to choose an appropriate technique, a number of factors had to be considered, as discussed below.

Firstly, conventional kinetic studies involve heating the particle up to a predetermined temperature in an inert atmosphere; once steady state is reached, the reactant gases are allowed to react with the solid and the progress of the reaction (in this case the amount of sulfur burnt) determined. In some preliminary tests, chalcopyrite particles were suddenly introduced into a heated inert atmosphere furnace and the weight change measured. At temperatures above 673 K, significant losses due to sulfur vaporization were observed. Therefore if isothermal tests were conducted, significant sulfur losses would occur during the heating-up period, and at steady state the sample would be comprised of chalcopyrite, bornite, and other decomposition products depending on the temperature and heating time.

The use of fluidized bed reactors to carry out isothermal tests as proposed by some workers suffers from the drawbacks of finite initial heat-up time to temperature, and, more importantly, the elimination of heat and mass transfer effects is difficult. The use of a higher flow rate to eliminate these effects will result in elutriation of particles and will render the method ineffective.

Sulfide ores are known to undergo "ignition" (13,14). Although some workers (15) have attempted to determine an ignition temperature, this is not an intrinsic property of the ore but, more correctly, a phenomenon based on mass and heat transfer effect in addition to reaction kinetics. Fluidized reactors will likely be subject to the ore igniting (14). Therefore, conditions have to be maintained (in nonisothermal studies) where particle temperatures increase at a predetermined linear rate and not show a sharp change as is associated with ignition.

Due to the above considerations, a nonisothermal, stationary-bed reactor was used in this study. Figure 1 shows the experimental set-up. The apparatus consists of a silica reaction shaft with a porous plug. Chalcopyrite particles suspended in mineral wool were placed over this porous plug. The furnace is heated by a Eurotherm constant-rate heating source on which different heating rates can be set. Temperature was controlled and monitored with two Pt-Pt 10% Rh thermocouples inserted in the reaction bed.

The reactor is operated in two modes -- gas flow from the bottom and from the top (Figure 1 shows the top-flow mode). In either case, the offgas is cooled and analyzed for sulfur dioxide content using a Beckman model 865 SO_2 infrared analyzer. The reaction rate is thus monitored by measuring the SO_2 strength of the offgas.

When the oxidant gases are introduced from below, the offgas is collected through a central pipe in the porous plug. A high-pressure, low-flowrate stream of N_2 is introduced as shown in Figure 1 to reduce any mixing effects in the upper part of the reaction tube. The tube through which this small N_2 flow is introduced is used to introduce the oxidation gas flow in the second mode. The central pipe is sealed off, and gases are taken out from below. This mode is more cumbersome to operate since a simple arrangement cannot be used to direct the hot offgases to the SO_2 analyzer. However, when high flow rates are used, the first mode results in a fluidized bed, and hence the latter mode must be used.

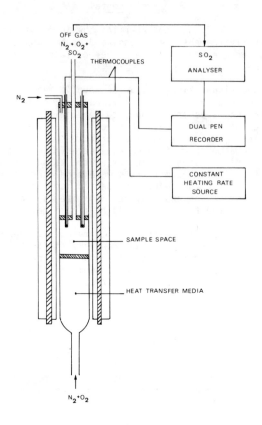

Figure 1. Experimental apparatus showing the set-up for gas flow upward.

In the experiments, the particles were suspended in mineral wool. The presence of the wool can lead to temperature gradients along the width of the reaction tube. To determine the presence of this gradient a number of thermocouples were introduced at various positions in the reaction bed. The bed was composed of mineral wool alone. The reaction tube was heated at a linear rate up to 1173 K. An air flow rate of 21 cm/sec at room temperature was used. Heating rates between 2-5 K/min were investigated. At 5 K/min, the maximum difference between the outermost and innermost thermocouple was not more than 5 K. Therefore uniformity of temperature along the bed could be assured.

To ensure that no mixing occurred between the SO_2 evolved and the gases in the upper zone of the reaction tube, a small SO_2 stream was suddenly introduced along with the in-coming air. The SO_2 analyzer indicated a step response indicating plug flow behavior of the reactor. If mixing is found to occur, it can be accounted for as shown by Fukunaka et al. (16).

Transvaal chalcopyrite was used in the experiments. Using a Leco sulfur analyzer, the sulfur content of the ore was found to be 34.9% S. X-ray analysis showed chalcopyrite to be the only phase present.

Chalcopyrite particles weighing about 0.5 grams were carefully sprinkled in mineral wool kept above the porous plug. The suspension of particles in mineral wool had to be done so that agglomeration was avoided. At the end of the run, if agglomerates were observed, the run was scrapped. To study the kinetics up to 923 K, a heating rate of 2 K/min and the gas flow in the reactor was in the upward direction. Flow rates of 21, 25, and 28 cm/sec at room temperature were investigated using 55% O_2 in the gas and 63.5 µ (200x270 mesh) particles. No change in the oxidation rate was observed. Therefore a flow rate of 21 cm/sec was considered sufficient to eliminate external heat and mass transfer effects. It should be pointed out that, when the sample was heated at a rate of 5 K/min and the gas flow rate was 21 cm/sec, the ore particles ignited.

To extend the measurement to higher temperature (1173 K), a higher heating rate of 3.5 K/min was used. Flow rates of 25, 28, and 31 cm/sec were tested under conditions mentioned above. No change was observed with flow rate of 28 and 31 cm/sec and hence 28 cm/sec was used in all experiments. Gas was blown in from the top in this series since at these flow rates some fluidization had been observed in the previous experiments (gas flow from the bottom).

Experiments were conducted using oxygen partial pressures of 15.2, 30.4, and 45.6 kPa and particle sizes of average diameter 127 µ (100x150 mesh), 89.5 µ (15x200 mesh), and 61.5 µ (200x270 mesh). The atmospheric pressure at Salt Lake City is 86.1 kPa (0.85 atm); hence the gas compositions used were 17.6%, 35.3%, and 52.9% O_2.

At the end of the runs the samples were analyzed with X-ray (Cu K_α radiation 33.5 KV, 15 mA) for presence of compounds arising due to the possible reaction between the particles and mineral wool. No such interaction was found, and hence it was assumed that the wool was essentially an inert material in the system. The reaction was also stopped at various stages and the samples subjected to X-ray analysis to determine the phases present.

Results and Discussion

Figure 2 shows a plot of conversion rate vs. temperature for various particles at a heating rate of 3.5°C/min and oxygen partial pressure of 0.15 atm. Plots are shown for various oxygen partial pressures. The plots are typical of the results obtained in the experimental runs. At lower heating rates (2°C/min), the first peak was more spread out and the conversion levelled off at temperature around 900 K so that the second peak was not observed. In order to analyze the conversion curves, a computer program was written. The Fortran code allows various kinetic models and reaction exponents to be tested. Simultaneous plotting of the results according to Eq. (12) allowed for visual judgement of the linear response. The goodness of fit was determined by computing the coefficient of correlation.

The first set of experiments were conducted at a heating rate of 2 K/min and spanned the temperature range of room temperature to 950 K. It was found that the pore-blocking model (Eq. (8)) fitted the data best. Although the value of the parameter λ can vary with temperature (17), it was found that a single value of λ = 0.07 adequately described the data. The data is plotted in Figure 3 using this value of λ. The plots in Figure 3 are for an average particle diameter of 127 microns. Since activation energy was not dependent on particle size, plots for other particle sizes are not shown.

In all calculations, the activation energy was first obtained by using the least square fit, then the straight line was drawn using an average slope

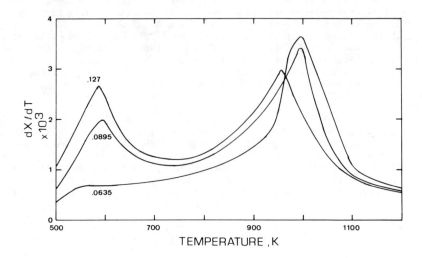

Figure 2. Plot of conversion rates against particle temperature. Average particle size in millimeters is shown alongside curves. p_{O_2} = 0.15 atm, heating rate = 3.5 K/min.

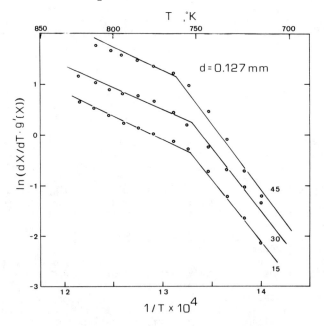

Figure 3. Low-temperature conversion data plotted according to Eq. (10) for 100x150 mesh particles.

(-E/R). The intercepts thus obtained were used for further calculations. Figure 3 show that two distinct kinetic regimes exist. Below 754 K, the activation energy (E) was found to be 216 kJ and above 754 K, E = 71.4 kJ/mole.

The second set of experiments was conducted at a higher heating rate (3.5 K/min). The best fit of the data was obtained using power-law kinetics (Eq. (6)) with m = 0.75. The use of this kinetic model gave a poor fit at temperatures below 873 K since pore-blocking kinetics are more appropriate in that temperature region. Data for the temperature range 873 K to 1223 K is plotted in Figure 4 according to Eq. (10). In Figure 4, two distinct kinetic regimes are observed.

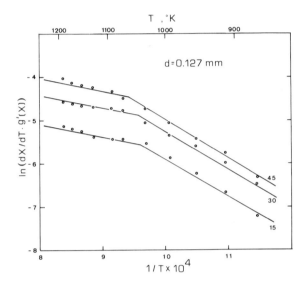

Figure 4. High-temperature conversion data plotted according to Eq. (10) for 100x150 mesh particles.

Between 923 K and 1600 K the activation energy was found to be the same as that obtained in the temperature range 754-923 K in the lower heating rate experiments, i.e., 71.4 kJ/mole, whereas above 1000 K the activation energy was found to be 23.7 kJ/mole. Bumazhnov and Lenchev (3) found a similar trend in that they found three kinetics regions, although they used nucleation and growth kinetics (Erofeev's equation (23)) to describe their data.

In order to understand the changes occurring during oxidation of chalcopyrite, stability diagrams for the Cu-Fe-S-O system were drawn at 700, 900, and 1100 K. Using data from various sources (6,7,13,18-22) and including solid solutions, the predominance area diagrams were developed and are shown in Figures 5-7. These diagrams are similar to those drawn by Yazawa (6) at 1100 K. The $p_{\Sigma O}$ = 0.1 atm and p_{SO_2} = 0.1 atm lines are also plotted on these figures to help understand the progress of the reaction.

At low temperatures, around 700 K, Figure 5 shows that the stable phases are the sulfates -- $CuSO_4$, $Fe_2(SO_4)_3$. Although not readily visible from this figure, a glance at the Fe-S-O diagram (6) shows that $FeSO_4$ can also be obtained at higher oxygen potentials. The formation of sulfates which would have a larger volume than the original chalcopyrite justifies the use of the pore-blocking model in the low-temperature region. As temperature increases (900 K), Figure 6 shows that the stable phases would most probably be iron oxide, and the copper would exist as the sulfate or basic sulfate. At still higher temperatures (1100 K), the oxides -- Cu_2O and Fe_2O_3 -- and the mixed oxides -- delafossite and ferrite -- will be the stable phases.

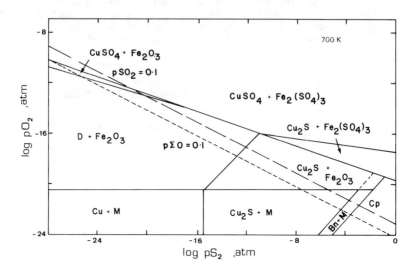

Figure 5. Cu-Fe-S-O predominance area diagram at 700 K. Bn ≡ Bornite, Cp ≡ Chalcopyrite, M ≡ Magnetite, D ≡ Delafossite.

The shift away from pore-blocking kinetics in the high-temperature range is consistent with the formation of different solid phases. The intermediate range 754-1060 K is a transition region where kinetic control shifts from pore-blocking to power-law kinetics. A small region between 823-923 K exists where both models only approximately describe the data. It is therefore suggested, based on the kinetic plots, that pore-blocking kinetics be used up to 873 K and power-law kinetics above 873 K.

Figures 5-7 show that the formation of bornite is necessarily the first step in oxidation of chalcopyrite. At low temperatures bornite will form as a result of the oxidation process, whereas at higher temperatures, chalcopyrite must decompose first to bornite and then undergo further oxidation. Using X-ray analysis, Bumazhnov and Lenchev (3) detected the presence of Fe_7S_8 at moderate temperatures (873-973 K) at the beginning of the reaction. This composition can be considered to fall in the "FeS" solid solution range and hence bears out the above analysis.

In order to confirm the analysis of the stability diagrams, X-ray analysis of samples collected at various stages was conducted. The results are shown in Table II. At temperatures below 773 K, $CuO \cdot CuSO_4$ and $CuSO_4$ were detected in minor amounts and Fe_2O_3 was found to be the major phase. Minor amounts of $FeSO_4$ were also observed but no ferric sulfate was observed. As seen from the stability diagrams, Fe_2O_3 will form prior to the formation of sulfate. Since temperature is increased continuously, Fe_2O_3 becomes more stable and hence no significant amount of sulfate may be detected. In isothermal tests, the sample is maintained at a particular temperature, and at

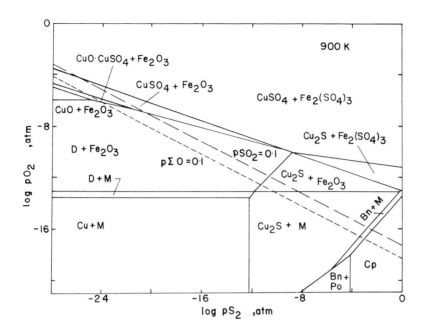

Figure 6. Cu-Fe-S-O predominance area diagram at 900 K. Sp ≡ Spinel ($CuFe_2O_4$).

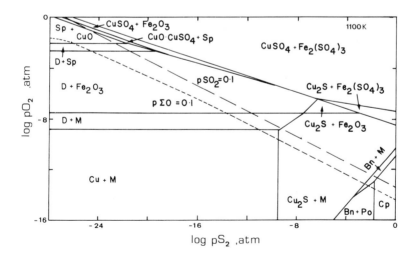

Figure 7. Cu-Fe-S-O predominance area diagram at 1,100 K.

Table II. X-Ray Analysis Results

Sample No.	Conditions	Phases Present	
III-1-a	2 K/min heating rate up to 723 K, quenched. atm: N_2, 21 cm/sec	major: minor:	$CuFeS_2$ none
III-1-b	2 K/min heating rate up to 723 K, quenched. atm: N_2 + 17.8% O_2, 21 cm/sec	major: minor:	$CuFeS_2$ Fe_2O_3, smaller amounts $CuSO_4$. Cu_5FeS_4 and $FeSO_4$ in smaller amounts.
III-1-c	2 K/min heating rate up to 783 K, quenched. atm: N_2 + 17.8% O_2, 21 cm/sec	major: minor:	Fe_2O_3, $CuFeS_2$ Cu_5FeS_4, $CuO \cdot CuSO_4$; in smaller amounts $CuSO_4$
III-1-d	2 K/min heating rate up to 833 K, quenched. atm: N_2 + 17.8% O_2, 21 cm/sec	major: minor:	Fe_2O, $CuFeS_2$ Cu_5FeS_4, some $CuO \cdot CuSO_4$, $CuFeO_2$ and $CuSO_4$ is also observed.
III-1-e	2 K/min heating rate up to 773 K and then N_2 + 17.8% O_2 up to 833 K, 21 cm/sec	major: minor:	Cu_5FeS_4, Fe_2O_3 $CuFeS_2$, $FeCu_2O_4$ in small amounts $CuO \cdot CuSO_4$ and possibly $FeSO_4$
III-1-f	same as III-1-e except heated to 873 K	major: minor:	Fe_2O_3, CuO $CuO \cdot CuSO_4$, Fe_2CuO_4, CuO, some Cu_5FeS_4, very small amount of $CuFeS_2$
II-2-c	2 K/min heating rate up to 923 K, N_2 + 17.8% O_2	major: minor:	Fe_2O_3 Cu_5FeSO_4, $CuSO_4$, $CuO \cdot CuSO_4$ in very small amounts $FeCu_2O_4$
III-1-g	5 K/min heating rate up to 1173 K, N_2 and 17.8% O_2, 30.69 cm/sec	major: minor:	Fe_2O_3, CuO Fe_2CuO_4, in small amounts Cu_2O

low temperatures ferrous and ferric sulfate are stable and can be detected by X-ray analysis. As temperature increases, the amount of bornite and iron oxide also increases. At temperatures above 873 K, the amount of bornite starts to decrease as compared with the iron oxide phase. At 1173 K the major phases detected are the individual oxides -- Cu_2O and Fe_2O_3 -- and some mixed oxides. Between 873-1173 K, delafossite and copper ferrite predominate.

When samples were heated up to 723 K in N_2 and then oxidized, a substantial amount of bornite was detected. Peaks slightly shifted from the stoichiometric FeS were also observed. This probably corresponds to the FeS solid solution which is produced by the decomposition of chalcopyrite.

The analysis described above shows that at temperatures around 700 K copper and iron sulfates are the stable phases. Above 1100 K the oxides Cu_2O and Fe_2O_3 are most stable. In the intermediate temperatures, mixed oxides are to be expected. This correlates well with the analysis of the stability diagrams also.

In order to study the effect of oxygen pressure on the reaction rate, the intercepts from plots in Figures 3 and 4 and similar plots for other particle sizes were plotted against the oxygen partial pressure. This is shown in Figures 8 and 9. The linear relationship indicates a first-order dependence on the oxygen partial pressure. Hence,

$$f_1(p_{O_2}) = p_{O_2} \tag{11}$$

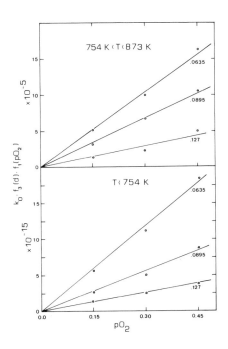

Figure 8. Plot showing the oxidation reaction to be first order with respect to oxygen partial pressure; low-temperature data.

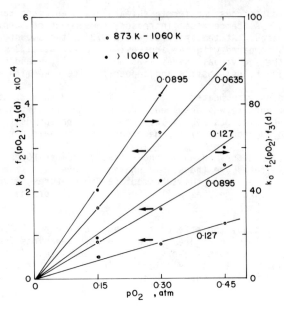

Figure 9. Plot showing the oxidation reaction to be first order with respect to oxygen partial pressure, high-temperature data.

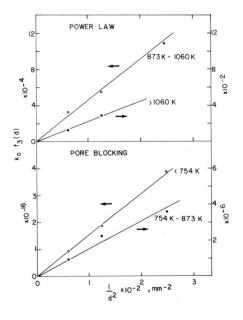

Figure 10. Plot showing that the reaction rate constant is inversely proportional to the square of the average particle diameter.

The intercept $k_0 f_2(p_{O_2}) \cdot f_3(d)$ was then divided by p_{O_2} and the average for each p_{O_2} was plotted as a function of $1/d^2$ -- d being the average particle diameter. This is shown in Figure 10. The linear relationship indicates that

$$f_3(d) = 1/d^2 \tag{12}$$

Table III summarizes the results of this study by listing the kinetic parameters that have been elucidated from the data.

Table III. Kinetics Parameters for Oxidation of Chalcopyrite

Temp. Range	Kinetic Model	K_0 cm^2-kPa^{-1}	E kJ/mole
< 754 K	Pore-blocking $\lambda = 0.07$	1.46×10^{10}	215
754-873 K	Pore-blocking $\lambda = 0.07$	1.56	71.4
873 K-1000 K	Power-law $m = 0.75$	4.59×10^{-2}	71.4
> 1060 K	Power-law $m = 0.75$	2.21×10^{-4}	23.7

Conclusions

In this study the intrinsic kinetics of oxidation of chalcopyrite have been determined. The study has identified three kinetics regions, below 754 K, 754-1060 K, and above 1060 K. The reaction is found to be first order with respect to the oxygen concentration. The data can be satisfactorily described by the pore-blocking kinetics and the power-law kinetics, below and above 873 K, respectively. Good agreement between X-ray phase analysis and the analysis of predominance area diagrams has been obtained providing a justification for the kinetic models utilized.

References

1. H. H. Kellogg, Trans. AIME, 206 (1956), 1105.

2. E. V. Margulis and V. D. Ponomarev, Zh. Prikl. Khim. 35 (1962), 970.

3. F. T. Bumazhnov and A. Lenchev, Metallurgiya (Sofia), 8-9 (1972), 32.

4. A. Lenchev, Rudy Met. Niezlaz, 21-9 (1976), 334.

5. C. Maurel, Bull. Soc. Franc. Min. Crist, 87-3 (1964), 377.

6. A. Yazawa, Met. Trans. B, 10B (1979), 307.

7. T. Rosenqvist, Met. Trans. B, 9B (1978), 337.

8. L. S. Leung, Met. Trans. B, 6B (1975), 341.

9. R. I. Razouk, G. A. Kolta, and R. Sh. Mikhail, J. Appl. Chem. (London), 15 (1965).

10. P. G. Thornhill and L. M. Pidgeon, Trans. AIME, 209 (1957), 989.

11. G. S. Agarwal and S. K. Gupta, Chem. Era., 12-3 (1976), 108.

12. U. R. Evans, The Corrosion and Oxidation of Metals: Scientific Principles and Practical Applications, Edward Arnold Ltd., London, 1960.

13. F. Habashi, Chalcopyrite, Its Chemistry and Metallurgy, McGraw Hill, New York, New York, 1978.

14. F. R. A. Jorgensen and E. R. Segnit, Proc. Australas. Inst. Min. Metall., 261 (1977), 34.

15. V. Volky and E. Sergievskaya, Theory of Metallurgical Process, Mir Publishers, Moscow, 1971.

16. Y. Fukunaka, T. Monta, Z. Asaki, and Y. Kondo, Met. Trans. B, 7B (1976), 307.

17. S. Won, Ph.D. Thesis, University of Utah, Salt Lake City, Utah, 1980.

18. K. T. Jacob, K. Fitzner, and C. B. Alcock, Met. Trans. B, 8B (1977), 451.

19. J. Paul Pemsler and C. Wagner, Met. Trans. B, 6B (1975), 311.

20. JANAF Thermochemical Tables, H. Stull and D. Prophet, ed., 2nd ed., U.S. Dept. of Commerce, NSRDS-NB737.

21. T. R. Ingraham, Trans. AIME, 233 (1965), 359.

22. U.S.G.S. Bulletin, R. A. Robie and D. R. Waldbaum, eds., 1970, 534-81.

23. H. Y. Sohn, Metallurgical Treatises, J. K. Tien and J. F. Elliott, eds., The Metallurgical Society of AIME, 1981.

THERMODYNAMICS OF SULFATION OF IMPURITIES IN TITANIFEROUS SLAGS

Krzyszof Borowiec and Terkel Rosenqvist

Division of Metallurgy
Norwegian Institute of Technology
N-7034 Trondheim-NTH, Norway.

Summary

With the aim of determining the thermodynamic conditions for sulfation of impurity elements, in particular magnesium, in titaniferous slags the equilibrium gas composition for the sulfation reaction:

$$MO + SO_2 + 1/2\ O_2 = MSO_4$$

has been studied by an EMF technique with a solid ZrO_2 + CaO electrolyte. Here M denotes the impurity elements: Mg, Ca and Mn, and the following oxide systems have been studied: TiO_2-MgO-Fe_2O_3, TiO_2-MnO-Mn_2O_3, and TiO_2-CaO-SiO_2, all in the temperature range 700-1000°C. Reproducible EMF values were obtained by adding small amounts of Na_2SO_4 to the oxide-sulfate mixture. From the observed results the Gibbs energy of formation of the phases Mg_2TiO_4, $MgTiO_3$, $MgTi_2O_5$, $MnTiO_3$, $CaTiO_3$, and $CaTiSiO_5$, as well as the activity of $MgTiO_5$ in the solid solution with Fe_2TiO_5 have been calculated. The results show that, from a thermodynamic viewpoint, sulfation of these impurities in titaniferous slags is favorable at around and below 800°C.

Introduction.

In the reduction smelting of ilmenite ores a metallic iron melt and a titaniferous slag are obtained. This slag may serve as raw material for the production of $TiCl_4$ and TiO_2, and represents an alternative to natural rutile. For the production of $TiCl_4$, in particular by fluid bed chlorination, various impurity elements may cause considerable complications. This applies in particular to magnesia and lime, which form liquid chlorides at the chlorination temperature, and thus destroy the fluidization. Removal of these impuirities from the slag, therefore, represents a challenging task.

Among the different chemical reactions which might be considered for the removal of impurities, conversion to water soluble sulfates appears to be one of the most promising, in particular for magnesium and calcium, iron and titanium being not easily sulfated. According to a patent from the U.S.Bureau of Mines (1) up to 60 % of the MgO content of titaniferous slag may be sulfated by treatment with sulfur trioxide in the temperature range 600-1100°C. In a later paper from the same laboratory (2) up to 90 % magnesia removal was obtained by treatment with a $SO_2 + O_2$ mixture and under the addition of Na_2SO_4. For other impurities even better removal was obtained. It was emphasized, however, that in order to obtain a good magnesia removal the iron oxide content of the slag should be below 5%.

In the smelting of ilmenite concentrate from the Sogndal area in SW Norway a slag that contains about 9 % MgO is obtained. Attempts in the authors' laboratory to sulfate this slag with a mixture of SO_2 and air in the temperature range 700-1000°C, and without any addition, gave magnesia removal of only about 40 %, the highest values being obtained for slags of low iron content. This rather poor result may be caused either by unfavorable thermodynamics or unfavorable kinetics. The present investigation had as its purpose to establish the thermodynamics, i.e., the equilibrium gas composition for sulfation as function of slag composition and temperature.

Industrial titanferous slags usually contain, after oxidation, a rutile phase of low impurity content, a M_3O_5 (pseudobrookite) phase, where impurities such as iron, magnesium and manganese are concentrated, and a silicate phase that contains lime as well as smaller amounts of magnesium, manganese, alumina. etc. In the present investigation the less complex systems TiO_2-MgO-Fe_2O_3, TiO_2-MnO-Mn_2O_3, and TiO_2-CaO-SiO_2 were studied. Measurements were based on a EMF technique where the oxygen potential for coexistence of oxide and sulfate phases under a total gas pressure of one atm. was determined, in the way previously reported by Skeaff and Espelund for simple oxide-sulfate systems (3).

Experimental

Materials

The oxide samples were prepared by intimate mixing of the pure, dehydrated oxides in calculated amounts, and were fired in air at 1200°C. The phase combinations of the products were ascertained by X-ray diffraction analyses.

For the TiO_2-MgO-Fe_2O_3 system the composition of the various samples is shown on Fig. 1, which also gives the phase relations at 850°C as recently reported by the present authors (4). For higher temperatures essentially the same phase diagram applies, whereas at temperatures below 800°C a miscibility gap may possibly exist for low magnesia contents in the $MgTi_2O_5$-Fe_2TiO_5 solid solution, where the rutile phase coexists with an iron-rich Fe_2O_3-$MgTiO_3$ solid solution. For the present study this miscibility gap is of minor importance.

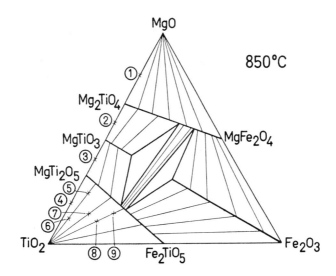

Fig. 1   The TiO_2-MgO-Fe_2O_3 system at 850°C.

Within the TiO_2-M_3O_5 two-phase region samples 5 and 6, (Fig. 1) have closely the same Mg/Fe ratio but different TiO_2 contents. The same is true of samples 8 and 9. These duplicates, therefore, have the same M_3O_5 phase composition, and should give the same gas equilibrium for the sulfation reaction.

For the TiO_2-MnO-Mn_2O_3 system Mn_3O_4 was first prepared by firing of $MnSO_4$, and was then mixed with TiO_2 corresponding to a molar ratio Mn/Ti $\approx$ 0.77. On firing in air at 1200°C this reacted to give rutile and $MnTiO_3$, where the latter may take some Mn_2O_3 into solid solution. For the TiO_2-CaO-SiO_2 system samples were prepared within the TiO_2-$CaTiO_3$ two-phase as well as within the TiO_2-$CaTiSiO_5$-SiO_2 three-phase regions.

The oxide samples thus prepared were mixed with anhydrous sulfates of the corresponding impurity metal. Usually duplicate runs were made with different amounts of added sulfates. The composition of the various mixtures is given in Tables I and II.

Table I. Samples for MgO-TiO_2-Fe_2O_3-$MgSO_4$ equilibrium.

Sample No.	Oxide composition (mole %)			wt. $MgSO_4$ / wt. oxide
	TiO_2	MgO	Fe_2O_3	
1	20	80	–	0.428
2	42	58	–	0.667
3	59	41	–	0.428
3a	59	41	–	1.000
4	80	20	–	0.571
4a	80	20	–	1.428
4b	80	20	–	1.000
5	70	25	5	0.428
6	85.5	12	2.5	0.571
7	75	15	10	1.000
7a	75	15	10	0.667
8	73	12	15	0.500
9	65	15	20	1.000

Table II. Samples for $MnO-Mn_2O_3-TiO_2-MnSO_4$ and $CaO-SiO_2-TiO_2-CaSO_4$ equilibria.

Sample No.	Oxide composition (mole %)				wt. MSO_4 / wt. oxide
	TiO_2	Mn_3O_4	CaO	SiO_2	
10	70	18	–	–	1.000
10a	70	18	–	–	0.428
11	73	–	27	–	0.250
11a	73	–	27	–	1.000
12	40	–	15	45	0.500
12a	40	–	15	45	1.000

Procedure.

The experimental method was essentially similar to that described previously from this laboratory (3, 5), and is shown in Fig 2. About 1 gram of the oxide-sulfate mixture was placed inside the ZrO_2-CaO tube, together with a platinum lead pressed against the bottom of the tube. The tube communicated with a slow flow of sulfur dioxide which maintained a total gas pressure ($SO_2+SO_3+O_2$) of one atmosphere inside the tube. The reference electrode was a piece of platinum gauze pressed against the outside bottom of the tube, and was flushed with a slow flow of oxygen of atmospheric pressure. A platinum lead, which also served as one leg of a Pt-Pt/Rh thermocouple was welded to the gauze. The oxygen potential inside the reaction mixture was calculated from the observed EMF by the Nernst Equation:

$$\log p_{O_2} \text{ (atm)} = \frac{-4FE}{2.303 \, RT}$$

For most of the temperature and composition range studied, SO_2 is the major gas species, and the SO_2 pressure is closely equal to one atmosphere. At high temperatures, however, the partial pressures of SO_3 and O_2 become significant, causing the partial pressure of SO_2 to be less than the total gas pressure. In these cases the partial pressure of SO_2 was calculated from the known equilibrium constant for the reaction:

$$2 \, SO_2 + O_2 = 2 \, SO_3$$

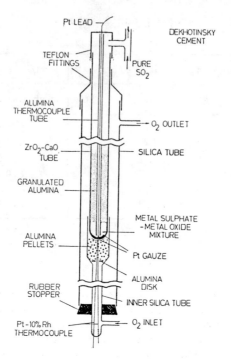

Fig. 2 Cell design, after Skeaff and Espelund (3).

in combination with $p_{SO_2} + p_{SO_3} + p_{O_2} = 1$ atm, in the way described previously (5). The equilibrium gas expression: $\log p_{O_2} + 2 \log p_{SO_2}$ which apply for the general sulfating reaction:

$$2 MO + 2 SO_2 + O_2 = 2 MSO_4$$

could then be calculated.

At high temperatures some of the SO_3 and oxygen that are formed from the oxide-sulfate mixture may diffuse out of the zirconia tube thus causing sulfate decomposition and upsetting the gas equilibrium in the reaction mixture. In order to minimize this diffusion the reaction mixture was covered with a layer of granulated alumina, see Fig 2. Nevertheless the measurements were limited to temperatures where the SO_3 and O_2

pressures were less than 0.1 atm, and measurements at the highest temperatures had to be completed within two to three hrs.

Skeaff and Espelund (3) found in their work that for the pure $MgO-MgSO_4$ and $ZnO-ZnSO_4$ systems reproducible EMF values could not be obtained. This was attributed to poor kinetics, in particular to poor establishment of the SO_2-SO_3 equilibrium. In contrast, the systems with Mn, Fe, Co and Ni gave stable and reproducible values, possibly caused by these "colored" systems' ability to catalyze the gas reaction. On the addition of small amounts of iron or manganese oxide to the Mg and Zn systems they were able to obtain stable and reproducible values, however. The same observation was made in the present investigation for magnesium titanates, to which a few percent of hematite was added.

As it was feared, however, that iron oxide would dissolve in the oxide phases, thus altering the thermodynamics of the system, alternative catalysts were tried: platinum gauze and sodium sulfate. Whereas platinum gauze is not expected to affect the thermodynamics of the system that possibility exists for sodium sulfate. According to Ginsberg (6) the solid solubility of Na_2SO_4 in $MgSO_4$ is small. At temperatures below 814°C solid $MgSO_4$ coexists with the double sulfate $Mg_3Na_2(SO_4)_4$, whereas at temperatures above 814°C it coexists with a molten phase which at, e.g., 1000°C contains only about 10 mole% Na_2SO_4. In order to ascertain that solid $MgSO_4$ was always present as a separate phase during the experiments only about 4-5 wt% (3-4 mole%) Na_2SO_4 was added to the magnesium sulfate.

Figure 3 shows the observed equilibrium expression $\log p_{O_2} + 2 \log p_{SO_2}$ for samples 1-4 with the addition of either hematite, platinum gauze or sodium sulfate. It is apparent that whereas the samples with platinum gauze and sodium sulfate give closely similar results, the ones with hematite give slightly different values for the equilibrium expression, possibly due to solution of iron oxide in the magnesium titanates.

For the iron-magnesium titanates addition of hematite as a catalyst gave badly scattered and erratic data, whereas reproducible values were obtained with the addition of sodium sulfate. Also for the experiments with manganese and calcium small amounts of sodium sulfate were added.

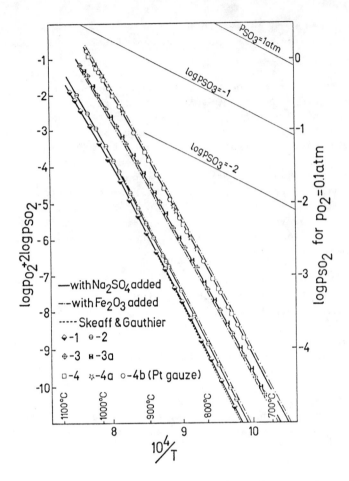

Fig. 3 Gas equilibria for sulfation of magnesia and magnesium titanates (samples 1-4) with different catalysts added. Skeaff and Espelund (3). Gauthier and Bale (7).

Results

The TiO_2-MgO-Fe_2O_3 system

Results for the TiO_2-MgO system are shown in Fig. 3, which includes the results of other authors (3,7) for the MgO-$MgSO_4$ equilibrium. All lines show a break at about 950°C which corresponds to a phase transformation in $MgSO_4$. Fig. 3 also includes lines for constant partial pressures of SO_3, and we see that for the magnesium titanates the SO_3 pressure becomes significant around 1000°C.

Assuming the different oxide phases to be stoichiometric the Gibbs energy change for the sulfation reactions may be calculated for temperatures below 950 C:

$$MgO + SO_2 + 1/2\ O_2 = MgSO_4 \quad (1)$$

$$\Delta G°_1 = -360577 + 250.81\ T\ (\pm 2100)\ J/mole$$

$$Mg_2TiO_4 + SO_2 + 1/2\ O_2 = MgSO_4 + MgTiO_3 \quad (2)$$

$$\Delta G°_2 = -361473 + 253.38\ T\ (\pm 2800)\ J/mole$$

$$2\ MgTiO_3 + SO_2 + 1/2\ O_2 = MgSO_4 + MgTi_2O_5 \quad (3)$$

$$\Delta G°_3 = -335842 + 243.70\ T\ (\pm 2600)\ J/mole$$

$$MgTi_2O_5 + SO_2 + 1/2\ O_2 = MgSO_4 + TiO_2 \quad (4)$$

$$\Delta G°_4 = -335020 + 249.82\ T\ (\pm 2400)\ J/mole$$

By further combination of these data the Gibbs energy for the formation of the magnesium titanates is calculated:

$2\ MgO + TiO_2 = Mg_2TiO_4 \quad \Delta G° = -24250 + 1.48\ T\ J/mole$
$(-25523 + 1.255T\ J/mole)$

$MgO + TiO_2 = MgTiO_3 \quad \Delta G° = -25146 + 4.05\ T\ J/mole$
$(-26360 + 3.138T\ J/mole)$

$MgO + 2\ TiO_2 = MgTi_2O_5 \quad \Delta G° = -25557 + 0.99\ T\ J/mole$
$(-27614 + 0.628T\ J/mole)$

The values in parenthesis are the ones listed by Kubaschewski (8).

For the sulfation of magnesium in titaniferous slags only the two-phase region TiO_2-M_3O_5, where M_3O_5 denotes the $MgTi_3O_5$-Fe_2TiO_5 solid solution, is of interest. Results for samples 5-9 (Table I) are shown in Fig. 4 together with the results for the iron-free sample 4. As samples 5 and 6, have closely the same M_3O_5 phase composition they give the same gas equilibrium. The same is true for samples 8 and 9. As the experiments were made at nearly one atm of SO_2 the left hand scale gives essentially the oxygen potential. This is seen to be as low as 10^{-10} atm. for sulfation of pure $MgTi_2O_5$ at 700°C

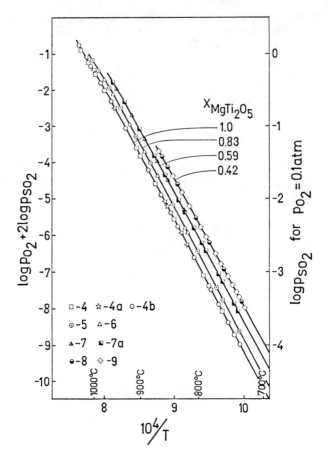

Fig. 4   Gas equilibria for sulfation of the M_3O_5 solid solution (samples 4-9).

and somewhat higher for the M_3O_5 solid solution. In practice, sulfation will be made under conditions where the oxygen potential is of the order of 0.1 atm. The question, therefore, arises whether the curves in Fig. 4 may also be directly applied to higher oxygen potentials, or whether they describe conditions where some of the trivalent iron has been reduced to divalent. In order to test this point sample 9 was, after the EMF measurement, analyzed for divalent iron, but no measureable amount was found. The curves are, therefore, taken to apply also for the M_3O_5 solution under fully oxidized conditions.

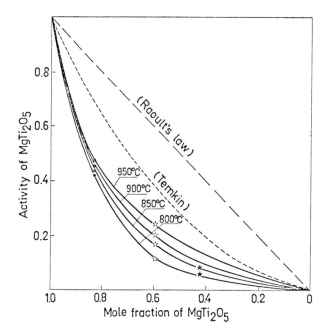

Fig. 5   Activity of $MgTi_2O_5$ in solid solution with Fe_2TiO_5. For Temkin relation see text.

Assuming that $MgSO_4$ and TiO_2 are present as essentially pure phases the chemical activity of $MgTi_2O_5$ in the M_3O_5 solid solution may be calculated from the difference between curves 4 and 5-9, and is shown in Fig. 5 for 800-950°C. It is apparent that the solid solution shows a pronounced negative deviation from Raoult's law. Raoult's law behaviour is, however, hardly to be expected for a solution where the components do not occur as distinct molecules but rather as a mixture of cations in an anion network. If it is assumed that the Ti^{4+}, Fe^{3+} and Mg^{2+} cations are statistically distributed on all cation sites, and if there is no specific cation interaction the activity of $MgTi_2O_5$ may be calculated from the Temkin relation:

$$a_{MgTi_2O_5} = (27/4) N_{Mg} (N_{Ti})^2$$

Here the cation fractions are:

$$N_{Mg} = \frac{n_{Mg}}{n_{Ti} + n_{Mg} + n_{Fe}}$$

$$N_{Ti} = \frac{n_{Mg}}{n_{Ti} + n_{Mg} + n_{Fe}}$$

The factor 27/4 is necessary to make the activity of pure $MgTi_2O_5$ come out as unity. In the solid solution $MgTi_2O_5$-Fe_2TiO_5 the number of cations, n_{Mg}, etc., are functions of the mole fraction $X_{MgTi_2O_5}$, viz:

$$n_{Mg} = X_{MgTi_2O_5}$$

$$n_{Ti} = 2 X_{MgTi_2O_5} + (1-X_{MgTi_2O_5})$$

$$n_{Fe} = 2(1-X_{MgTi_2O_5})$$

Inserting these numbers in the Temkin expression the Temkin activity is calculated and is included in Fig. 5. Even though the Temkin activity shows negative deviation from Raoult's law the experimental activities show an even larger deviation. This may be caused by specific interactions between the magnesium and iron cations.

The TiO_2-MnO-Mn_2O_3 system.

For this system only one oxide sample (No.10) with a molar ratio Mn/Ti $\approx$ 0.77 and with two different additions of $MnSO_4$ was studied, see Table II. The results are shown in Fig. 6, where the break at about 850°C corresponds to a phase transformation in $MnSO_4$.[*] Also shown is the oxygen potential for 1 atm of SO_2 for the Mn_3O_4-$MnSO_4$ equilibrium, as measured by Skeaff and Espelund (3).

The oxide sample had been prepared from a mixture of TiO_2 and Mn_3O_4 which was fired in air at 1200°C. During firing oxygen will be expelled to give $MnTiO_3$, which according to Grey et al. (9) does not oxidize in air at 1200°C. The possibility still exists, however, that some oxidation to trivalent manganese may occur during the EMF measurements at lower

---

[*] According to older literature and handbooks the melting point of $MnSO_4$ is 700°C. Recent work by Skeaff and Espelund and others have shown it to be at least 1000°C, however.

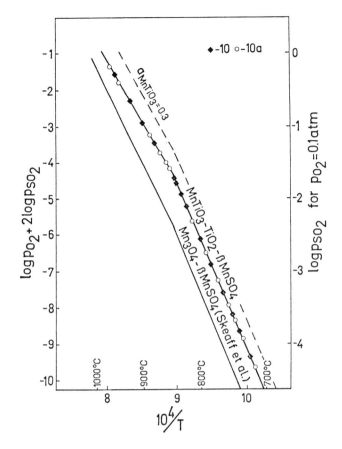

Fig. 6   Gas equilibria for sulfation of MnTiO$_3$ (sample 10) and calculated for a_{MnTiO} = 0.3. Skeaff and Espelund (3).

temperatures or during industrial sulfation. If so, this would be present as Mn$_2$O$_3$ in solid solution with MnTiO$_3$. To test this possibility the initial oxide preparation as well as the reaction mixture after EMF measurements were analyzed for their Mn^{3+} content. Whereas for the original oxide preparation no significant amount of Mn^{3+} could be detected, the analysis of the sample after EMF measurements showed that about 7 % of the manganese oxide was present as trivalent, corresponding to a solid solution of 4 mole% Mn$_2$O$_3$ and 96 mol% MnTiO$_3$. If the MnTiO$_3$ activity in the solid solution during the EMF measurements is estimated to be 0.96, combination with the results of Skeaff and Espelund gives for the stoichiometric reaction

$$Mn_3O_4 + 3\ TiO_2 = 3\ MnTiO_3 + 1/2\ O_2$$

$$\Delta G^o = 72027 - 50.16\ T\ (\pm 900)\ J/mole$$

As a further cheek the original oxide preparation was ignited in air at 750°C for 48 hrs, and it was found that about 78 % of the manganese had oxidized to trivalent, corresponding to a solid solution of about 76 mole% Mn_2O_3 and 24 mole% $MnTiO_3$. This was confirmed by X-ray diffraction analysis, which also showed that the amount of rutile phase had increased greatly due to the oxidation reaction:

$$2\ MnTiO_3 + 1/2\ O_2 = Mn_2O_3 + 2\ TiO_2$$

Thus it is clear that the measured gas equilibrium can not be directly applied to industrial sulfatization at high oxygen potentials. A possible way to handle this problem will be given in the discussion to this paper

The TiO_2-CaO-SiO_2 system.

In this system the two compositions shown in Table II with different additions of $CaSO_4$ were studied. The results are given in Fig. 7, from which the Gibbs energy change for the sulfatization reaction are calculated:

$$CaTiO_3 + SO_2 + 1/2\ O_2 = CaSO_4(\beta) + TiO_2$$

$$\Delta G^o = -383693 + 242.67T\ (\pm 1100)\ J/mole$$

$$CaTiSiO_5 + SO_2 + 1/2\ O_2 = CaSO_4(\beta) + TiO_2 + SiO_2$$

$$\Delta G^o = -308505 + 196.83T\ (\pm 1200)\ J/mole$$

The Gibbs energy change for sulfation of pure lime can not be measured by the present technique in the temperature range in question as this will lead to the formation of CaS. Instead it was calculated from data given by Turkdogan et al. (10)

$$CaO + SO_2 + 1/2\ O_2 = CaSO_4(\beta) \quad \Delta G^o = -461753 + 237.69T\ J/mole$$

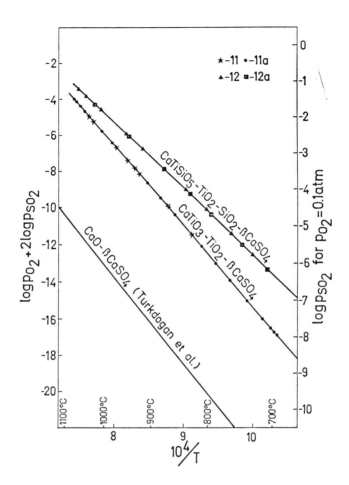

Fig. 7　Gas equilibria for sulfation of $CaTiO_3$ and $CaTiSiO_5$. Turkdogan et al.(10).

By combination of these data the Gibbs energy change was calculated for the reactions:

$CaO + TiO_2 = CaTiO_3$ $\Delta G^o = -78060 - 4.98\ T$ J/mole
                               $(-79916 - 3.35\ T$ J/mole$)$

$CaO + TiO_2 + SiO_2 = CaTiSiO_5$ $\Delta G^o = -153248 + 40.83\ T$ J/mole

The value in parenthesis is that listed by Kubaschewski (8).

Discussion

As previously mentioned sulfation of impurity elements may be carried out either with SO_3 or with a mixture of SO_2 and O_2. Of industrial interest would be a mixture of SO_2 and air. Assuming that air is present in sufficient surplus to make the oxygen potential after sulfation equal to 0.1 atm the corresponding equilibrium SO_2 pressures may be read on the right hand scales of Figs. 4 and 6-7. It is seen that for the sulfation of $CaTiO_3$ and $CaTiSiO_5$ (Fig 7) the residual SO_2 pressure is less than 0.01 atm below 1000°C. Sulfation of lime in titaniferous slags should, therefore, be thermodynamically easy, which agrees with the observation of Elger et al. (2).

For magnesium and manganese oxides sulfation is more difficult. For pure $MgTi_2O_5$ temperatures below about 850°C are needed to bring the residual SO_2 pressure below 0.01 atm. For solid solutions of $MgTi_2O_5$ in Fe_2TiO_5 even lower temperatures are needed. Starting, for example, with a slag which after oxidation contains 9 wt % MgO and 5 wt% Fe_2O_3, i.e., with a molar ratio $MgO/Fe_2O_3 = 7.13$, corresponding to $X_{MgTi_2O_5} = 0.88$, initial sulfation will be easy. As magnesia is removed from the M_3O_5 solid solution its mole fraction and activity decreases. For 90 % Mg removal, corresponding to a residual MgO/Fe_2O_3 molar ratio of 0.713 or $X_{MgTi_2O_5} = 0.42$, temperatures below about 810°C are needed to give SO_2 pressures below 0.01 atm. If, on the other hand, the oxidized slag had contained 10 % Fe_2O_3, then 90 % Mg removal would give an M_3O_5 phase with $X_{MgTi_2O_5} = 0.26$ and a $MgTi_2O_5$ activity of 0.02 (Fig. 5). In this case temperatures below 750°C would be needed to give SO_2 pressures below 0.01 atm.

For manganese oxide the application of the present results to industrial sulfation is more difficult. Small amounts of manganese oxide would most likely be dissolved in the M_3O_5 (pseudobrookite) phase. Its solubility in this phase is not very large, however. According to Grey et al.(11) the solubility in ferrous pseudobrookite ($FeTi_2O_5$) at 1200°C corresponds to only 10 mol% $MnTi_2O_5$. In ferric pseudobrookite (Fe_2TiO_5) and at lower temperatures it is likely to be even less. For higher manganese contents the stable phase in coexistence with rutile is likely to be a solid solution of $MnTiO_3$, Mn_2O_3 and Fe_2O_3. In this solution the activity of $MnTiO_3$ will be less than unity, and the gas equilibrium expression for the sulfation reaction:

$$MnTiO_3 + SO_2 + 1/2\ O_2 = MnSO_4 + TiO_2$$

will be higher than shown in Fig. 6. Thus if for p_{O_2} = 0.1 atm the activity of $MnTiO_3$ is taken to be for example 0.3 a temperature of less than 810°C is needed to give an equilibrium SO_2 pressure of less than 0.01 atm (Fig. 6). The fact that Elger et al.(2) found manganese to sulfate more easily than magnesium may then possibly be attributed to kinetic reasons, i.e., that manganese catalyzes its own sulfation.

If several impurities are present together their sulfates may react to give double sulfates or they may be mutually soluble in each other. Thus Elger et al.(1) report that magnesia and lime give the double sulfate $CaSO_4 \cdot 3MgSO_4$. Such reactions among the sulfates will lower their activity and make sulfation thermodynamically easier than for the simple systems. The same will be the case by the use of larger amounts of sodium sulfate.

In total it may be concluded that even though magnesium and manganese oxides are thermodynamically more difficult to sulfate than lime, and even though sulfation becomes more difficult with increasing iron content of the slag a reasonably good removal should be theoretically possible at temperatures around and below 800°C. The fact that the present authors so far have not succeeded in obtaining such results in practical tests would, therefore, be attributed to kinetic reasons.

Acknowledgements

The authors are grateful to the Royal Norwegian Council for Science and Technology (NTNF) for a fellowship to one of them (K.B.) and to Dr. D.G.C. Robertson of Imperial College, London, for constructive criticism of the manuscript.

References

1. G.W. Elger, R.A. Stadler, and P.E. Sanker, "Process for purifying a titanium-bearing material and upgrading ilmenite to synthetic rutile with sulfur trioxide," U.S.Patent (19)(11), 4,120,694 (45) Oct.17. 1978.

2. G.W. Elger, J.E. Tress, and R.R. Jordan, "Utilization of domestic low grade titaniferous materials for producing titanium tetrachloride," *Light Metals 1982,* AIME, Dallas, Texas, 1982, pp. 1135-1147.

3. J.S. Skeaff and A.W. Espelund, "An E.M.F. method for the determination of sulphate-oxide equlibria, results for the Mg, Mn, Fe, Ni, Cu and Zn systems," *Can.Met.Quart.* 12 (1973) pp. 445-454.

4. K. Borowiec and T. Rosenqvist, "Phase relations and oxygen potentials in the Fe-Ti-Mg-O system," *Scand.J.of Met.* 13 (1984)(in press).

5. T. Rosenqvist and A. Hofseth, "Phase relations and thermodynamics of the copper-iron-sulphur-oxygen system at 700-1000°C," *Scand.J.of Met.* 9(1980) pp. 129-138.

6. A.S. Ginsberg. "Ueber die Verbindungen von Magnesium-und Natriumsulfat," *Zeitschr.Anorg.Chem.* 61(1909) pp. 122-136.

7. M. Gauthier and C.W. Bale, "Oxide-sulfate equilibria in the Mg, Ni and Mn systems measured by a solid K_2SO_4 concentration cell," *Met.Trans.* 14B(1983) pp. 117-124.

8. O. Kubaschewski, "The thermodynamics of double oxides (a review)," *High Temperatures, High Pressures* 4(1972) pp. 1-12.

9. I.E. Grey and A.F. Reid, "Phase equilibria in the system $MnO-TiO_2-Ti_2O_3$ at 1473°K" *J.Sol.State Chem.* 17(1976) pp. 343-352.

10. E.T. Turkdogan, B.B. Rice, and J.V. Vinters, "Sulfide and sulfate solid solubility in lime, magnesia and calcined dolomite: Part I. CaS and $CaSO_4$ solubilily in CaO," *Met.Trans.* 5(1974) pp. 1527-1535.

11. I.E. Grey, A.F. Reid, and D.G. Jones. "Reaction sequence in the reduction of ilmenite:4- Interpretation in terms of the Fe-Ti-O and Fe-Mn-Ti-O phase diagrams," *Trans.Instn.Min.Met.* 83(1974) pp. C105-111.

VISCOSITY MEASUREMENTS OF INDUSTRIAL LEAD BLAST FURNACE SLAGS

R. Altman, G. Stavropoulos, K. Parameswaran, R. P. Goel

Central Research Department
ASARCO Incorporated
South Plainfield, New Jersey 07080, USA

Abstract

A Brookfield viscometer fitted with a stainless steel bob was used to measure the viscosity of industrial lead blast furnace slags between 1126 and 1298°C. Slag oxidation state was characterized by first annealing the slag samples at 750°C in sealed copper tubes in order to precipitate magnetite crystals. The magnetite concentration was then measured on an Outokumpu Saturated Magnetic Analyzer.

Using an Arrhenius-type equation, a viscosity model was developed by regression. The equation below accounted for 75% of the variability in the data used to generate the regression coefficients and, on average, could predict viscosity within $\pm$ 12% of the measured value.

$$\log \eta = 1.0600 \log(CR) + \frac{6801.2}{T(°K)} - 3.9881$$

where,

$$CR = \frac{\%\ SiO_2 + \%\ Al_2O_3 + \%\ MgO}{\%\ CaO + \%\ FeO + \%\ ZnO + \%\ S}$$

The greatest difficulty encountered in measuring viscosity was due to the occasional presence of solids in the slag which caused, at times, erroneously high viscosity measurements. Precautions to minimize this problem are outlined in the text.

Introduction

Viscosity of lead blast furnace slags can vary over a fairly wide range. While temperature is the principal determinant of slag viscosity, it is not the only influencing factor. Slag oxidation state and composition also affect viscosity. However, synthetic slags containing 3 or 4 elements cannot normally reproduce viscosities for industrial slags which typically have more than twice that number of major components. On the other hand, viscosity measurements made with industrial slags are complicated because:

1. Industrial slags are prone to react with the crucible because they are usually unsaturated in SiO_2, Al_2O_3 or MgO. Partial dissolution of the crucible constituents results in alterations in slag composition beyond the range of practical interest.

2. Crystallized phases such as zinc ferrite or magnetite can be present or even form during the time that viscosity measurements are made. Obviously, viscosity measurements made in the presence of entrained solids are extremely difficult if not impossible to interpret.

3. Upon reheating industrial slag, the zinc and lead concentrations usually decrease over time. This loss is a result of the tendency of zinc to vaporize and of lead prills to coalesce while the slag is molten.

4. Relatively high levels of sulfur in blast furnace slags interfere with the determination of slag oxidation state by wet-chemical methods.

5. Corrosion of the instrument placed in the slag to measure its viscosity (referred to as a "bob" in this paper) results in calibration loss over time and contributes markedly to measurement uncertainty.

This paper focuses on the method used and the problems encountered in measuring viscosities of slags produced at Asarco's Glover and East Helena lead smelters. From these viscosity measurements a model was formulated and tested with regard to its predictability using various statistical techniques.

Experimental

Viscometer

Viscosity measurements were made using a 4-speed, model LVF Brookfield viscometer. The shank and bob were fabricated in-house from 304 stainless steel, the shape and dimensions of which were modified somewhat from those given by Brookfield. Figure 1 shows the bob and crucible arrangement. Note that the bob is a solid piece with a conical top and bottom which facilitated slag drainage when the bob was removed from the melt for recalibration. The use of a solid bob instead of one having a hollow design had two advantages. First, surface area was minimized so that the bob corroded into the slag at an acceptably slow rate. Second, the solid bob was much easier to recalibrate after a campaign compared to a hollow bob which would always contain a large amount of hard-to-remove solidified slag.

As an aid to reduce shaft wobble the 3/16" shank leading from the bob was attached to a 10-inch stainless steel 0.040" wire extension by passing the hooked end of the extension wire through a 0.0465" double countersunk hole in the shank. Finally, this extension wire was hooked to a 0.040" x 3" extension wire which in turn was similarly attached to the viscometer shaft. This double-extension wire arrangement resulted in a high degree of flexibility and, thus, practically eliminated any wobble.

Furnace and Reactor Assembly

The custom designed furnace (Figure 2) measured 20" in diameter x 30" high and was insulated with a 12" core of castable bubble alumina surrounded by 4" of Fiberfrax Paper. Eight SiC heating elements were configured to generate 1 ohm resistance by forming two groups of elements connected in series with each group of four elements connected in parallel. Power was provided by a Halmar, Model PAl-2450AD 50 amp/208 volt single-phase SCR amplifier connected to a Barber Colman 560 digital set-point controller. A platinum, 10% rhodium thermocouple placed in the center of the furnace was used to control power input. The amount of power delivered varied in proportion to the difference between the control thermocouple and set-point voltages. At steady state this difference was quite small which allowed temperature to be controlled to within $\pm$ 1°C during a typical 6-8 hour campaign.

The reactor assembly (see Figure 2) consisted of a 3 1/4" OD x 2 7/8" ID mullite tube fitted with water-cooled brass collars at both ends which contained "O" rings to form a gas tight seal. The lower brass collar was machined to centrally position a 2 1/4" OD x 2" ID mullite tube which acted as support for a 304 stainless steel retainer crucible.

This crucible was held in position by notching its base so that it fit snugly into the mullite tube. A McDanel recrystallized alumina round bottom crucible, measuring 2 1/2" OD x 2 1/4" ID x 6" was fitted inside the retainer crucible, the bottom of which was tapered and partially filled with alumina powder to minimize point stress on the alumina crucible. The lower brass collar was also fitted with two access ports: one was used for an argon inlet, and the other held a back-up thermocouple which rested against the steel retainer crucible. The alumina thermocouple protection tube was held in position by an "O" ring squeezed between a threaded cap and coupling. The shaft of the coupling was welded to the face of the brass collar to completely seal the connection.

The brass collar on the top of the furnace was fitted with three access ports for slag sampling rod (left), viscometer shaft (center) and measuring thermocouple (right). The center port was connected to a "T" coupling to allow for a nitrogen purge which minimized air infiltration. The other two ports were fitted with couplings which held "O" rings to seal against the alumina thermocouple sheath or a cap when the port was not in use. The upper zone contained four mild steel radiation shields

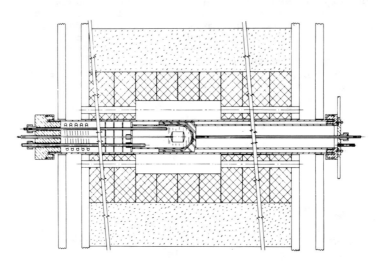

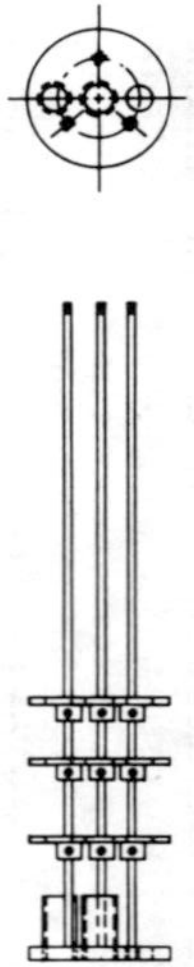

Figure 3
Upper zone radiation shields and support

Figure 2
Furnace and reactor assembly

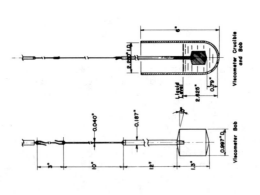

Figure 1
Bob and crucible arrangement

supported by three 3/16" stainless steel rods which screwed
into the base of the water-cooled brass collar (see Figure 3).
The bottom radiation shield rested directly on top of the
alumina crucible, thus acting as a cover. To minimize slag
build-up, the bottom radiation shield contained two 1" lips,
one of which screwed into the viscometer shaft access port,
and the other screwed into the slag sampling access port. A
water-cooled copper coil (see Figure 2), positioned between
the upper brass collar and the top radiation shield, was used
to condense zinc vapor in order to minimize its build-up on
the viscometer shaft and access ports.

Procedure

Prior to and immediately after slag viscosity measurements,
the viscometer was recalibrated in standard liquids provided
by Brookfield. Initial calibrations were made using the same
shank, bob and crucible as were actually used in later slag
viscosity measurements. Recalibrations were done using a new
crucible (as geometrically identical to the first crucible as
possible) with the original bob and shank after sand blasting
to remove slag. In some cases, a bob and shank were used in
as many as three consecutive campaigns. Figure 4 gives the
recalibration correction versus the number of hours the bob
and shank were used to make viscosity measurements. This
graph shows that calibration corrections are approximately proportional to the number of hours the bob had been in use.
Viscosity values appearing in this paper make use of this time-
proportional calibration change to correct viscometer deflection values taken during a 6-8 hour campaign.

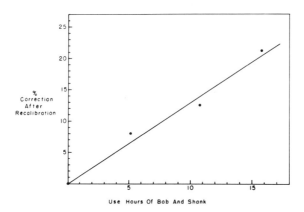

Figure 4. Calibration loss vs. time

A typical campaign was started by charging the alumina
crucible with 530 grams of slag. This amount of slag insured
that the bob was immersed 15-20 mm below the slag surface. The
reaction assembly was then aligned in the furnace so that the
shank and bob were free to rotate without obstruction. Argon
was then admitted through the bottom of the reaction tube at
about 80 cc/min as the furnace was heated slowly to 1250°C.
After about an hour at this temperature the bob was lowered to
the bottom of the crucible and then raised 20 mm to position it

approximately in the center of the melt. Viscometer deflections were noted at several RPM levels and at 10 and 20 mm from the bottom of the crucible. Figure 5 illustrates some typical viscometer readings. The slag temperature was monitored during this time by holding the thermocouple slightly above the slag surface. On several occasions the thermocouple was lowered to the bottom of the crucible to measure the temperature distribution within the slag. Since no significant variation was observed, all subsequent slag temperatures were estimated by holding the thermocouple 5-10 mm above the melt surface. Several grams of slag were then removed by dipping a sampling rod into the melt and quenching it in a water bath. The bob was then raised above the melt surface and the temperature was changed. After reaching the desired temperature and holding for 50-70 minutes, another slag sample was obtained and then a new set of viscosity and temperature measurements were made. From time to time small amounts of iron oxide and/or zinc oxide were added to the melt to either increase the oxygen potential of the slag or to replenish zinc which was lost through volatilization.

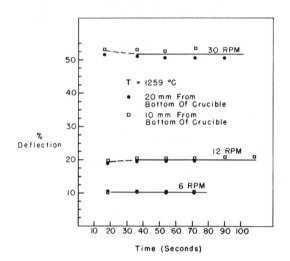

Figure 5. Viscometer deflection for various bob rotation speeds

Estimation of Slag Oxidation State

Determination of the oxidation state of iron silicate slag which contains a significant amount of sulfide sulfur is complicated because redox reactions occur between Fe^{+++} and $S^=$ during dissolution in acid. Consequently, an Outokumpu Saturated Magnetic Analyzer (SAT-MAG-AN) was used to give a <u>relative</u> estimate of slag oxidation state. The operating principle of this instrument relies on detecting an apparent weight increase of magnetic materials, such as magnetite, when exposed to a directional magnetic flux. However, this technique is hampered by several of its own complicating factors such as:

1. Bulk density variations due to changes in particle size distribution.
2. Magnetic flux induced magnetism and particle alignment which increases measured values if samples are re-analyzed.
3. Substantial variations in the composition of precipitated "magnetite" which can range from zinc or calcium ferrites all the way to aluminum spinels. Each of these can be present over a wide range of composition and exhibit quite different magnetic properties.
4. Incomplete magnetite precipitation from the slag matrix caused by a combination of the glassy nature of slag and rapid quenching which is done to avoid sample oxidation.

However, these problems can be diminished by:
1. Using a grinding procedure which consistently generates a similar particle size distribution and tapping the vial which contains the slag so that a given weight of sample occupies the same volume each time a measurement is made.
2. Tapping to inhibit particle re-orientation. Also, making only one measurement per sample vial reduces the chance that substantial particle realignment can occur.
3. Avoiding slags which exhibit wide variations in composition. Although there is little that can be done to control the compositions of the ferrite or spinel which precipitates from the slag, it is likely that a given class of slag would contain "magnetite" of similar composition. Microscopical studies of Asarco lead blast furnace slags reveal a fairly uniform distribution of iron, aluminum and particularly zinc ferrites. Therefore, the SAT-MAG-AN should **provide** a good _relative_ indication of the slag oxidation state since the _distribution_ of the various ferrites does not change appreciably.
4. Precipitating practically all of the ferric iron as ferrite. Granulated or quenched slag retains most of the ferric iron as part of the non-magnetic slag matrix; the remainder precipitates as magnetic secondary or dendritic crystals of zinc ferrite. For an accurate estimate of slag oxidation state practically all of the ferric iron must be separated from the slag as a ferrite or spinel. In order to make this separation, slag samples were annealed in evacuated copper tubes*

---

*Deoxidized copper tubes were used instead of quartz ampules because the zinc vapor released from the slag catalyzed devitrification which caused the ampules to rupture during cooling. Copper tubes were first deoxidized in graphite at 850°C for 72 hours in order to eliminate the possibility of oxygen transfer to the slag. Copper, however, is not capable of scavenging oxygen _from_ the slag, and after pretreatment with graphite to remove residual oxygen, would be inert with regard to oxygen transfer.

at temperatures below the liquidus for periods up to 88 hours and then analyzed on the SAT-MAG-AN. The purpose of this procedure was to provide enough mobility to the ions in the slag so that ferrite precipitation could proceed until the equilibrium level of residual Fe^{+++} in the slag matrix was established. In the temperature range 750°C-950°C, this residual Fe^{+++} concentration should be quite low. Figure 6 gives the results of the annealing experiments for three temperatures. The procedure followed in this paper and the results observed are similar to those published by Ligasacchi(1). As expected, the residual concentration of ferric iron in slag increased with temperature.

The annealing temperature selected was 750°C because at this temperature residual solubility of ferric iron in slag is low, but ion mobility is still sufficient to reach equilibrium. Based on these results all slags were annealed at 750°C for 25 hours. Statistical analysis of the data appearing in Figure 6 indicated a probable error of ± 3.3%.*

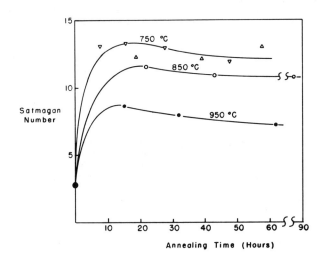

Figure 6. SAT-MAG-AN reading vs. annealing time

Results

Table I gives measured slag viscosities, compositions, temperatures and SAT-MAG-AN values. To a good approximation (usually within ± 3%) most slags exhibited Newtonian behavior. Those slags that deviated significantly gave a disproportionate viscometer deflection when bob rotation speeds exceeded 12 RPM. In addition, viscosity measurements obtained 10 mm from the bottom of the crucible often gave viscosity values 2-4 times

---

*All probable errors are calculated using "t" values obtained from "two-sided" tables at the 50% confidence level for the appropriate degrees of freedom.

Table I. Experimental Results

SiO$_2$	FeO	ZnO	CaO	MgO	Al$_2$O$_3$	Pb	S	Stmg*	Temp. °C	Visc. Poise	CR**
27.8	29.5	10.9	14.5	6.58	4.48	0.71	1.4	2.70	1250.0	1.63	0.6902
27.4	29.6	9.9	14.7	6.55	4.72	0.58	1.3	1.54	1223.1	2.21	0.6964
26.3	29.5	10.0	14.8	6.52	4.46	0.65	1.3	1.45	1200.7	2.61	0.6705
27.2	30.1	10.2	15.3	5.90	4.61	0.57	1.4	1.39	1177.7	3.25	0.6616
29.3	26.5	9.6	16.8	5.42	4.76	0.54	1.2	1.32	1249.6	1.90	0.7295
32.1	26.6	9.0	16.2	5.47	5.46	0.43	1.2	1.23	1227.7	2.32	0.8117
31.1	26.2	8.3	16.1	5.47	5.57	0.38	1.2	0.64	1201.1	2.99	0.8140
30.6	28.2	8.6	16.2	5.75	5.25	0.32	1.2	0.48	1175.3	4.17	0.7675
27.2	30.7	15.1	13.4	2.85	4.84	1.22	2.3	2.95	1250.9	1.40	0.5673
27.2	29.8	14.9	12.8	2.70	4.22	0.98	2.3	2.61	1275.5	1.74	0.5706
26.7	30.0	14.7	12.9	2.80	4.76	0.69	2.3	2.48	1199.5	2.16	0.5720
26.7	30.4	14.5	13.0	2.77	4.70	0.82	2.4	2.17	1176.1	2.57	0.5667
26.5	31.0	14.9	13.3	2.85	4.55	0.72	2.3	2.12	1152.2	3.05	0.5512
26.3	31.3	14.5	13.4	2.82	4.50	0.61	2.4	1.90	1125.9	3.76	0.5458
26.9	30.7	12.9	13.0	3.91	5.29	0.60	2.8	2.76	1249.2	1.50	0.6077
26.5	29.2	9.4	12.5	3.91	5.01	0.47	2.4	2.30	1225.1	2.00	0.6621
26.5	32.3	11.4	12.6	3.56	5.03	0.58	2.2	3.08	1200.3	2.39	0.5998
27.8	32.5	10.7	12.7	3.91	5.08	0.41	2.2	1.77	1175.3	2.70	0.6332
27.4	32.9	10.8	13.3	3.91	4.97	0.34	2.1	1.72	1150.2	3.45	0.6139
29.1	29.1	10.3	16.5	3.55	4.52	0.96	1.6	2.40	1251.8	1.99	0.6464
29.3	29.5	9.8	16.4	3.48	4.63	0.90	1.6	1.78	1224.0	2.56	0.6529
29.3	30.1	10.0	16.4	3.56	4.69	0.80	1.7	1.44	1199.8	3.09	0.6452
29.7	30.1	9.9	16.7	3.55	4.97	0.61	1.4	1.29	1176.1	3.86	0.6578
26.9	32.9	11.3	11.8	3.18	4.25	2.85	1.7	8.03	1252.0	1.47	0.5950
26.9	33.8	11.2	12.0	3.28	4.50	2.08	1.6	4.71	1227.6	1.68	0.5918
26.7	34.6	11.3	12.4	3.22	4.69	1.52	1.5	3.23	1199.6	2.03	0.5788
27.4	34.3	10.7	12.1	3.27	4.63	1.04	1.5	2.52	1181.1	2.50	0.6024
26.9	34.7	10.6	12.3	3.32	4.80	1.17	1.5	2.85	1252.3	1.72	0.5926
26.3	30.7	11.9	13.2	3.65	5.01	0.84	2.7	4.50	1250.9	1.42	0.5976
25.9	31.8	11.9	13.8	3.83	5.12	0.62	2.4	2.59	1227.9	1.89	0.5818
25.2	32.5	11.8	13.7	3.81	5.20	0.54	2.5	2.48	1207.8	2.25	0.5655
26.7	32.4	11.4	13.7	3.80	5.08	0.39	2.3	1.95	1177.7	2.85	0.5950
26.1	31.0	11.2	13.5	3.70	4.99	0.42	2.3	2.06	1203.2	2.64	0.5998
28.1	30.4	10.6	14.8	6.50	4.10	1.06	1.3	2.50	1250.0	1.64	0.6778
29.1	30.4	10.3	14.6	6.48	4.10	0.93	1.3	2.11	1223.1	2.12	0.7011
29.1	30.7	10.5	14.5	6.56	4.33	0.85	1.3	1.85	1200.1	2.56	0.7016
28.2	29.5	11.8	14.6	7.20	4.35	0.58	1.2	1.22	1249.6	1.91	0.6961
26.5	28.9	11.8	14.1	6.42	4.33	0.78	1.3	1.00	1227.7	2.33	0.6640
28.4	30.4	11.6	14.8	6.25	4.59	0.46	1.3	1.10	1201.1	3.01	0.6754
29.1	30.4	11.6	15.5	6.73	4.57	0.46	1.3	0.94	1175.3	4.23	0.6871
25.0	28.1	15.1	10.1	2.17	3.40	2.70	4.3	1.95	1250.9	1.48	0.5307
22.0	26.2	16.0	17.6	1.09	3.68	0.49	2.1	0.46	1229.3	1.33	0.4325
22.9	26.9	16.0	18.2	1.08	3.53	0.35	2.2	0.42	1177.0	1.92	0.4346
22.2	25.4	14.4	21.1	1.01	4.01	0.37	2.0	0.32	1244.9	1.62	0.4328
23.9	24.6	15.6	18.0	1.14	4.16	0.49	1.8	0.34	1218.1	2.12	0.4867
25.0	24.5	15.3	18.0	1.11	4.18	0.36	1.7	0.42	1195.7	2.16	0.5091
23.2	29.8	13.6	18.0	1.13	5.10	0.41	2.6	1.34	1199.5	1.34	0.4598
23.1	30.1	13.7	18.4	1.14	4.90	0.38	2.6	1.41	1187.3	1.45	0.4497
23.3	30.1	14.3	18.4	1.13	4.60	0.36	2.7	1.30	1176.5	1.58	0.4432
23.3	30.1	14.1	18.1	1.13	4.60	0.35	2.7	1.50	1222.3	1.27	0.4466
22.7	30.1	14.4	17.6	1.11	4.90	0.44	2.6	1.20	1298.0	0.96	0.4437
22.8	30.4	14.3	17.1	1.14	5.30	0.33	2.5	0.89	1252.1	1.13	0.4547
22.6	30.6	14.3	17.6	1.13	5.20	0.29	2.6	0.82	1196.5	1.76	0.4444
23.8	27.5	16.5	16.8	1.09	4.50	0.82	1.9	0.76	1225.2	1.16	0.4687

Table I. Experimental Results (Contd.)

SiO$_2$	FeO	ZnO	CaO	MgO	Al$_2$O$_3$	Pb	S	Stmg*	Temp. °C	Visc. Poise	CR**
23.6	27.8	16.7	16.8	1.09	4.40	0.69	1.9	0.50	1200.7	1.40	0.4603
23.5	27.8	16.6	16.4	1.08	4.50	0.58	1.9	0.62	1174.9	1.66	0.4638
23.5	28.0	16.7	16.9	1.06	4.50	0.72	1.9	0.60	1237.6	1.19	0.4576
23.9	28.3	15.6	17.3	1.06	5.10	0.41	1.9	0.32	1245.9	1.21	0.4764
21.8	26.1	17.2	17.5	1.04	4.25	0.38	2.2	0.29	1176.1	2.06	0.4300
23.4	26.8	15.5	18.6	1.18	4.80	0.89	2.2	1.93	1246.7	1.40	0.4656
22.6	27.5	15.0	18.9	1.14	4.80	0.66	2.1	0.62	1225.2	2.06	0.4494
23.1	27.8	14.9	18.6	1.14	4.70	0.53	2.1	0.46	1202.0	1.94	0.4565
22.8	27.8	14.8	18.6	1.09	4.80	0.49	2.1	0.50	1178.3	2.21	0.4532
22.8	27.8	14.6	18.4	1.11	4.80	0.53	2.2	0.35	1253.5	1.50	0.4557
25.2	27.1	11.1	18.3	1.08	7.50	0.37	1.8	1.16	1295.7	0.91	0.5794
23.9	27.8	13.1	17.8	1.01	6.80	0.45	1.7	0.55	1246.8	1.14	0.5250
24.2	28.2	13.7	17.5	0.99	6.20	0.33	1.8	0.64	1224.2	1.32	0.5129
23.6	28.3	14.0	17.4	0.99	5.70	0.31	1.9	0.55	1201.3	1.55	0.4917
23.8	28.8	13.5	17.7	0.98	5.50	0.24	1.8	0.66	1189.0	1.69	0.4900
23.2	28.8	13.9	17.2	0.96	5.30	0.30	1.9	0.81	1181.5	1.75	0.4767
23.6	28.8	14.2	17.4	0.96	5.60	0.22	1.9	0.77	1293.8	1.04	0.4841
23.6	30.3	13.2	16.8	0.99	5.70	0.21	1.9	1.02	1297.5	1.10	0.4870
23.3	30.0	13.4	17.0	0.98	5.90	0.21	1.9	0.40	1250.0	1.34	0.4844
22.9	25.6	14.9	19.7	1.19	6.40	0.37	1.7	0.98	1287.2	1.14	0.4926
22.1	26.7	15.4	19.2	1.11	5.20	0.65	1.9	0.28	1248.0	1.39	0.4495
22.5	26.7	15.2	19.4	1.13	5.10	0.34	1.9	0.28	1222.7	1.74	0.4546
22.7	27.0	15.0	19.3	1.13	5.10	0.27	2.0	0.32	1197.6	2.10	0.4570
22.8	27.3	15.3	19.7	1.11	5.00	0.24	1.9	0.31	1189.8	2.77	0.4503
22.8	27.5	14.4	19.6	1.09	5.00	0.25	2.0	0.29	1177.3	3.11	0.4550
22.9	27.2	14.2	19.9	1.08	5.00	0.31	1.9	0.37	1290.5	1.21	0.4585
24.8	28.8	9.3	21.0	1.16	7.30	0.18	1.7	1.55	1250.0	1.95	0.5474
25.0	29.2	7.9	20.7	1.11	7.60	0.12	1.6	1.84	1250.0	2.27	0.5675
24.2	29.1	12.9	17.2	2.66	4.62	1.78	1.9	4.99	1259.1	1.31	0.5152
24.1	29.5	13.0	17.6	2.59	4.78	1.04	1.5	1.85	1214.3	1.81	0.5109
24.0	29.9	12.9	17.9	2.60	4.76	0.76	1.6	1.25	1235.9	1.52	0.5034
23.1	30.7	12.5	18.4	2.70	4.88	0.64	1.6	0.98	1189.8	2.10	0.4854
24.4	31.0	12.1	18.2	2.63	4.94	0.50	1.6	0.78	1161.1	2.56	0.5083
25.8	28.2	13.1	18.4	2.66	4.82	1.06	2.0	3.78	1251.8	1.41	0.5394
25.5	29.0	13.1	18.5	2.46	4.89	0.88	2.0	2.68	1207.8	2.04	0.5248
25.0	29.1	13.3	18.7	2.48	4.97	0.82	2.0	2.25	1231.4	1.87	0.5143
25.4	29.3	12.8	18.7	2.39	4.94	0.66	2.0	1.69	1184.4	2.62	0.5212
25.5	29.9	12.5	19.0	2.43	4.83	0.49	2.0	1.38	1159.4	3.55	0.5167
25.2	28.4	14.2	16.8	2.09	4.42	2.18	1.9	5.65	1218.8	1.92	0.5173
25.4	29.1	14.2	17.1	2.11	4.44	1.58	1.8	3.62	1185.4	2.36	0.5137
25.7	29.5	14.1	17.4	2.13	4.47	0.92	1.7	1.65	1167.8	2.76	0.5152
25.4	29.7	13.6	17.8	2.14	4.58	0.99	1.8	1.82	1230.2	1.83	0.5107
25.0	30.3	13.6	17.7	2.13	4.70	1.70	1.8	1.69	1252.9	1.62	0.5021
25.5	30.3	13.5	17.5	2.11	5.00	0.81	1.7	1.64	1247.5	1.61	0.5176
23.7	31.1	13.3	17.1	2.44	5.07	0.68	2.4	4.75	1206.1	1.91	0.4884
24.1	31.2	13.5	17.3	2.42	5.10	0.68	2.4	3.38	1181.9	2.59	0.4910
24.5	31.6	13.0	17.5	2.43	4.82	0.61	2.4	2.64	1166.1	3.58	0.4922
24.7	31.6	13.3	16.9	2.38	4.81	0.62	2.4	2.48	1230.2	1.80	0.4967
24.3	31.8	12.8	17.0	2.35	4.96	0.54	2.3	2.16	1255.9	1.48	0.4947
24.5	32.2	12.4	17.2	2.34	5.14	0.42	2.2	2.02	1255.0	1.46	0.4997
23.6	31.6	13.9	17.4	2.45	4.83	1.09	2.6	3.45	1208.4	1.73	0.4715
23.6	31.7	13.8	17.1	2.44	4.76	0.99	2.7	2.52	1189.6	2.05	0.4717
24.0	31.2	13.6	17.0	2.44	4.52	0.70	2.6	1.61	1176.1	2.33	0.4807
23.9	31.7	13.5	17.5	2.44	4.68	0.69	2.5	1.28	1225.2	1.48	0.4758

Table I. Experimental Results (Contd.)

SiO$_2$	FeO	ZnO	CaO	MgO	Al$_2$O$_3$	Pb	S	Stmg*	Temp. °C	Visc. Poise	CR**
23.6	31.9	13.1	17.5	2.41	4.95	0.60	2.5	1.10	1247.9	1.37	0.4763
23.7	32.6	12.8	17.7	2.49	5.19	0.57	2.5	1.24	1250.0	1.33	0.4784
26.8	30.2	10.8	17.6	4.03	5.22	0.60	2.2	3.15	1248.0	1.53	0.5929
26.1	30.6	10.9	16.9	3.98	5.18	0.48	2.4	3.65	1207.0	2.79	0.5799
26.1	30.8	10.4	17.3	4.14	5.42	0.40	2.1	2.58	1223.1	3.11	0.5884
27.4	28.7	11.1	19.2	3.08	4.98	0.58	1.4	2.15	1250.9	1.71	0.5871
27.7	29.3	11.1	19.4	3.11	5.15	0.64	1.3	1.52	1196.9	2.23	0.5885
27.7	29.7	10.5	19.6	3.19	5.02	0.52	1.3	1.50	1226.0	1.97	0.5877
28.0	29.8	10.3	19.5	3.15	5.08	1.05	1.3	1.50	1181.1	2.85	0.5949
27.6	30.5	11.7	17.8	3.87	5.37	0.66	1.3	1.54	1152.1	3.84	0.6010
24.8	30.6	11.7	17.3	3.80	5.28	0.58	2.3	1.46	1259.1	1.33	0.5473
24.9	30.6	11.4	17.4	3.80	5.23	0.46	2.3	1.22	1208.5	2.51	0.5499
25.1	30.9	11.1	17.8	3.83	5.31	0.42	2.3	1.24	1229.3	2.83	0.5514
25.3	30.6	10.3	16.9	3.68	5.29	0.61	2.3	2.17	1252.5	2.10	0.5702
29.2	26.3	9.3	20.4	4.34	4.86	0.73	1.2	1.87	1253.3	2.27	0.6716
29.9	26.2	8.8	21.7	4.42	4.50	0.43	1.3	1.25	1255.0	2.39	0.6699
30.1	26.1	7.4	21.6	4.39	5.06	0.30	1.2	1.00	1233.4	2.71	0.7024
29.8	26.4	7.3	21.4	4.41	5.28	0.26	1.2	1.82	1208.2	3.59	0.7014
30.2	26.9	8.3	21.3	4.54	4.82	0.46	1.2	1.48	1212.8	3.81	0.6856
22.0	29.7	15.1	7.9	2.14	3.44	3.24	5.0	3.24	1199.4	2.10	0.4780
17.1	29.2	15.0	3.3	1.67	2.82	4.94	6.8	5.25	1176.1	2.55	0.3976

* Stmg: Outokumpo Saturated Magnetic Analyzer
**CR: Composition Ratio = (SiO$_2$+MgO+Al$_2$O$_3$)/(FeO+ZnO+CaO+S)

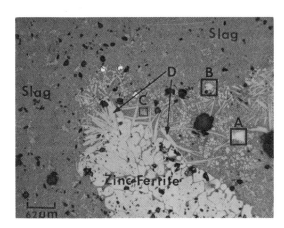

Figure 7. Photomicrograph and microprobe analysis of a slag sample taken about 10 mm from the bottom of the crucible. A: Fe, Zn chromate; B: Zn, Al, Mg ferrite; C: Zn, Fe aluminate; D: Zn ferrite + Zn oxide

those obtained from the center of the melt (20 mm from the crucible bottom). Microscopical examination of polished sections of samples taken from both depths indicated that the higher viscosity readings near the bottom of the crucible were caused primarily by the presence of zinc ferrite crystals and smaller amounts of iron, zinc, chromium, aluminum, and magnesium mixed oxides (see Figure 7). Samples taken from the middle zone showed practically no trace of oxide crystals. Chromium is released from the stainless steel bob that slowly corrodes into the melt and precipitates as chromium oxide which settles into primary crystals near the bottom of the crucible. Table II gives the slag analyses from the middle and bottom zones. As can be seen, the chromium concentration near the crucible bottom can be as high as ten times the concentration in the middle zone. From all of this evidence, it was concluded that some of the slags examined became saturated or were already saturated with zinc ferrite due to a combination of low temperature, high slag oxygen potential, and high ZnO concentration.

Aside from the experimental evidence, it seems unlikely that relatively small changes in composition would cause slags to exhibit occasional non-Newtonian behavior. It is far more probable that bob rotation speeds of 30 RPM and higher could have generated a disproportionate increase in viscometer deflection because agitation was sufficient to entrain zinc ferrite crystals which had previously collected near the bottom of the crucible. Dissolved lead oxide or metallic lead globules suspended in the slag during blast furnace operation also tended to settle near the crucible bottom as indicated by the very low lead percentages reported in Table I. These levels are only 24-40% of the initial lead concentrations. Chemical analyses of samples taken from the middle and bottom zones give further evidence that lead concentrates and may even coalesce near the bottom of the crucible (see Table III). Occasional precipitation and accumulation of metallic lead, zinc ferrite and chromium oxide on the bottom of the crucible would make viscosity measurements at the 10 mm depth unreliable. For these reasons only 6, 12, and 30 RPM deflection readings taken near the center of the melt (20 mm depth) are reported in this paper. Deflection values for 30 RPM were used only when they were in substantial conformity to the measurements at lower bob rotation speeds.

Figure 8 shows the relationship between the alumina content of remelted slags versus the time of sampling during a 6-8 hour campaign. As can be seen, the Al_2O_3 concentration in the slag remained almost constant despite prolonged contact with the alumina crucible. Apparently, the relatively modest temperatures coupled with the selection of a high density recrystallized alumina crucible were important factors which minimized Al_2O_3 dissolution.

Interpretation of Figures 8 and 9 leads to the conclusion that the stainless steel bob, which was in contact with the slag during much of the 6-8 hour campaign, probably had a reducing effect on Pb^{++}, Zn^{++} and Fe^{+++}. Note how lead and zinc content as well as SAT-MAG-AN readings decreased with time. To some extent this effect was minimized by making small periodic additions of zinc oxide and/or iron oxide during a campaign (see Figure 9). However, care was required to avoid adding

Table II. Chromium Dissolution and Segregation During Viscosity Experiments (East Helena Slags)

Bob position from crucible bottom (mm)	SiO$_2$	FeO	Al$_2$O$_3$	CaO	ZnO	Cr$_2$O$_3$
20	23.9	28.3	5.1	17.3	15.6	0.085
10	21.7	28.3	4.9	15.3	18.0	0.91
20	23.5	27.8	4.5	16.6	16.4	0.094
10	21.9	26.8	5.0	15.7	17.1	1.1
20	23.5	28.0	4.5	16.9	16.7	0.11
10	22.1	29.8	4.5	14.9	17.6	0.53

Table III. Segregation of Glover Slag Constituents During Viscosity Measurements

Bob position from crucible bottom (mm)	SiO$_2$	FeO	ZnO	CaO	MgO	Al$_2$O$_3$	Pb	S	Temp (°C)	Viscosity (Poise)
20	27.2	30.7	15.1	13.4	2.85	4.84	1.22	2.2	1251	1.4
10	25.0	28.1	15.1	10.1	2.17	3.40	2.70	4.3	1251	1.5
20	31.1	26.2	8.3	16.1	5.47	5.57	0.38	1.2	1201	3.0
10	22.0	29.7	15.1	7.9	2.14	3.44	3.24	5.0	1199	2.1*
20	30.6	28.2	8.6	16.2	5.75	5.25	0.32	1.2	1175	4.2
10	17.1	29.2	15.0	3.3	1.67	2.82	4.94	6.8	1176	2.6*

*High lead and sulfur levels near the bottom of the crucible indicate the possibility of entrained matte in the slag.

so much zinc or iron oxide that they could not be digested substantially during the relatively short time periods (30 minutes) between sets of viscosity measurements.

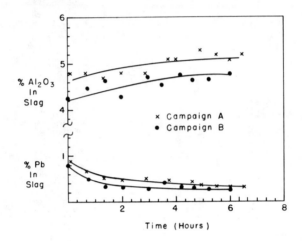

Figure 8. Alumina and lead content in slags vs. time

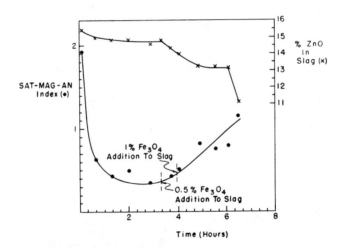

Figure 9. SAT-MAG-AN reading and zinc content in slag vs. time

Slag viscosities were re-measured on three separate occasions to determine reproducibility. Between 1170 and 1210°C the probable error was estimated to be ± 10.7% and between 1210 and 1250°C the probable error was about ± 6.2%. The higher uncertainty at temperatures close to, or within, the liquidus is another indication of the occasional presence of entrained solids. However, excellent reproducibility was obtained for at least two sets of viscosity measurements at 1178°C. So it cannot be said that solids were present in all cases when viscosity measurements were made below 1210°C.

Data Correlation and Model Predictability

Unlike synthetic slags which usually have no more than 3 or 4 constituents, industrial slags may have 5 to 8 major components that typically account for no more than 80-90% of the slag composition. As a consequence, it was not practical to analyze the relationship between viscosity and composition by changing the level of each slag constituent one at a time. Instead, statistical analysis (regression) was used to formulate an analytical expression which would quantify the influence temperature, slag composition and oxygen potential have on viscosity. The form of expression used to make this regression was based on the well known Arrhenius equation which has been used successfully to relate viscosity of a pure liquid, such as molten silica, to temperature(2),

$$\eta = A \exp\left(\frac{E}{RT}\right) \quad (1)$$

where η is absolute viscosity
R is the universal gas constant
T is absolute temperature
E is the activation energy of flow
A is a constant

Rewriting expression (1) in log form which would make it suitable for a multi-linear regression gives,

$$\log \eta = \log A_1 + A_2 \left(\frac{1}{T}\right) \quad (2)$$

In general, A_1 and A_2 should vary with composition and/or temperature and, in fact, a large number of functional forms were used in an effort to obtain the best fit to the raw data as well as explain the relative influence of individual slag constituents on viscosity. Unfortunately, there was a significant degree of co-dependence among the composition variables probably brought about, in part, by the direct relation between the quantity of gangue present in the blast furnace charge and the flux required to form a fluid slag. This high degree of collinearity would make it difficult to apportion the influence of each chemical component on viscosity because the variation of each component was, in turn, functionally related to one or more of the other slag constituents. Since it was not possible to show the individual effect of each slag constituent, all composition values were lumped together to define a single variable. The best results were obtained when slag composition was described as a weight ratio of slag constituents,

$$CR = \frac{\% SiO_2 + \% Al_2O_3 + \% MgO}{\% CaO + \% FeO + \% ZnO + \% S} \quad (3)$$

In this expression slag species which tend to increase viscosity are in the numerator and those which reduce viscosity are in the denominator. SAT-MAG-AN values appeared to have no influence on viscosity over the range measured in this study. No improvement was found in the correlation when A_2 was included as a second composition variable or when A_1 was expressed as a complex function of CR such as a polynomial. The equation selected which gave the best overall fit to the raw data was,

$$\log \eta = 1.0600 \log(CR) + \frac{6801.2}{T(^\circ K)} - 3.9881 \qquad (4)$$

Judging from the fit between the regression expression and the raw data, equation (4) has a probable error of about $\pm$ 12%.

Figure 10 demonstrates the quality of the fit between the raw data and the regression expression. The raw data points were selected to conform to a narrow temperature range so that a two dimensional plot would be adequate to demonstrate how well the model relates viscosity to composition and temperature. Although a few points appear to be out-of-line with the rest of the data, the general trend shows that lowering temperature and/or raising the composition ratio result in higher viscosity.

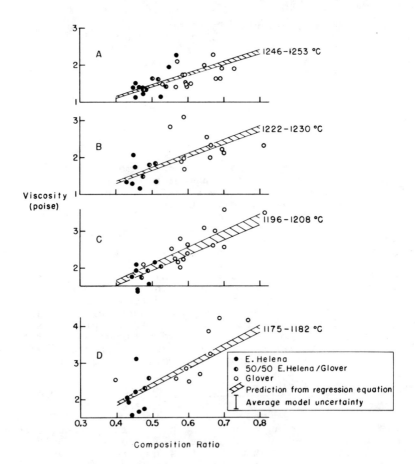

Figure 10. Viscosity vs. composition ratio over various temperature ranges

Figure 11 gives a comparison between the viscosity estimates obtained from equation (4) and Battle's correlation(3). At 1200°C there appears to be about a 3/4 poise difference which decreases to about a 1/2 poise disparity at 1250°C. Battle estimated a ± 1 poise error in his expression and that of the present study (equation (4)) is about ± 0.3 poise on average. Thus, agreement between the two studies is well within the estimated error of each model.

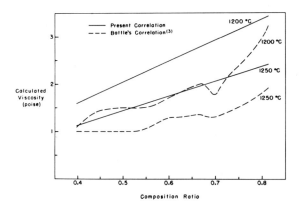

Figure 11. Comparison of present correlation with Battle's model(3)

The relative importance of temperature and composition ratio on viscosity can be determined by differentiating equation (4) as follows:

$$d(\ln \eta) = \frac{\partial(\ln \eta)}{\partial(CR)} d(CR) + \frac{\partial(\ln \eta)}{\partial T} dT \quad (5)$$

that is, $\quad d(\ln \eta) = \frac{d\eta}{\eta} = \frac{1.0600}{(CR)} d(CR) - \frac{15660.3}{T^2} dT \quad (6)$

and equating each term of the differential expression to an assumed change in viscosity. If, for example, the corresponding percent change in CR is sought at constant temperature for a 10% increase in η, equation (6) gives,

$$\frac{\Delta(CR)}{(CR)} \times 100 = \frac{100}{1.6} \times \frac{\Delta \eta}{\eta} = \frac{100 \times 0.1}{1.6} = 9.43\% \quad (7)$$

Similarly, the corresponding percent change in temperature at constant CR can be found for a 10% increase in η by using equation (8),

$$\frac{\Delta T}{T} \times 100 = -\frac{(100 \times T)}{15660.3} \times \frac{\Delta \eta}{\eta} = -6.386 \times 10^{-4} \, T \, \% \quad (8)$$

However, in this case the percent change in temperature is a linear function of temperature, ranging from -0.926% at 1177°C to -0.988% at 1275°C. These calculations show that, on average, temperature influences viscosity 9.9† times more than does the composition ratio.

Discussion

Probably the greatest difficulty in measuring lead blast furnace slag viscosities is due to the occasional presence of solids such as spinels, magnetite or other ferrites. It is difficult if not impossible to eliminate all of these solids without raising the temperature well beyond the range of interest. About the only thing that can be done is to allow a reasonable time for settling (~1 hr) and make repeated measurements in the center of the melt at low bob speeds so as not to entrain those solids which have already settled near the bottom of the crucible. Repeat measurements should also make it relatively easy to spot spurious data caused by entrained solids in the slag.

It may seem surprising that the regression analysis gave a better fit to the raw data when MgO was included with acidic species in the composition ratio term (see equation (3)). However, the tendency of MgO to raise viscosity at low concentrations in slag has been noted by Williams et al(4) and Ouchi and Kato(5). In addition, Toguri et al(6) found, as in this study, that slag oxygen potential did not have an important influence on viscosity.

There is a noticeable difference between the viscosity of lead blast furnace slags generated at Asarco's Glover Plant and those from East Helena. The higher viscosities produced by Glover slags are probably a result of a substantially higher SiO_2 and MgO content compared to East Helena slags. Mixtures of 50% Glover and 50% East Helena slags produced intermediate viscosities.

From a production viewpoint, low viscosity slags are preferable because they should contain less entrained lead and should be easy to skim from the furnace. However, other slag properties are also important. First, the slag should melt over a narrow temperature range to avoid stickiness in the furnace shaft. Second, the slag "melting point" should not be excessively high to avoid maintaining furnace temperatures much above 1200°C. Plans are underway to determine the liquidus range of low viscosity slags. Those slags which exhibit the best combination of viscosity and liquidus temperature range should prove to be the best candidates for improved blast furnace operation.

---

†$\dfrac{\text{Temperature effect on viscosity}}{\text{Effect of Composition Ratio on Viscosity}} = \dfrac{9.43}{0.957} = 9.85$

Summary and Conclusion

A Brookfield viscometer fitted with a stainless steel bob was used to measure the viscosity of industrial lead blast furnace slags between 1126 and 1298°C. Approximately 1 lb. of slag was held in an alumina crucible under a protective argon atmosphere. Slag samples were taken before and occasionally after each viscosity measurement. It was found that contact with the crucible caused the Al_2O_3 content to increase by about 1 percentage point. More importantly, partial dissolution of the stainless steel bob resulted in an 8-10% calibration error in the measured viscosity toward the end of a 6-8 hour campaign. This error was eliminated substantially by assuming a proportional correction based on the time the measurements were made.

Slag oxidation state was characterized by first annealing the slag sample at 750°C in sealed copper tubes in order to precipitate magnetite crystals. The magnetite concentration was then measured on an Outokumpu Saturated Magnetic Analyzer.

Using an Arrhenius-type equation a viscosity model was developed by regression. The equation below accounts for 75% of the variability in the data used to generate the regression coefficients and, on average, could predict viscosity within ± 12% of the measured value.

$$\log \eta = 1.0600 \log(CR) + \frac{6801.2}{T(°K)} - 3.9881$$

where,

$$CR = \frac{\%\ SiO_2 + \%\ Al_2O_3 + \%\ MgO}{\%\ CaO + \%\ FeO + \%\ ZnO + \%\ S}$$

The greatest difficulty encountered in measuring viscosity was due to the occasional presence of solids in the slag which caused erroneously high viscosity measurements. Precautions to minimize this difficulty are outlined in the text.

Glover slags exhibited a significantly higher viscosity compared to East Helena slags. The cause is probably related to the substantially higher SiO_2 and MgO content in Glover slags. There is substantial evidence from this study as well as from the literature that increasing either SiO_2 or MgO tends to increase slag viscosity.

Besides viscosity, the temperature range of the slag liquidus is another property that should be studied before specific recommendations can be made concerning optimum slag composition for efficient blast furnace operation. Plans are underway to measure the liquidus temperature range of selected slags which exhibited low viscosities.

Acknowledgement

The authors would like to thank ASARCO Inc. for permission to publish this paper.

References

(1) A. Ligasacchi, "A Study of Magnetite and Magnetic Compounds in Copper Reverberatory Smelting," AIME Transactions, vol. 233(10), (1965), pp. 1848-1856.

(2) J. Elliot, M. Gleiser, and V. Ramakrishna, Thermochemistry for Steelmaking, vol. II, p. 661; Addison-Wesley Publishing Co., Inc., 1963.

(3) T. P. Battle, Activities and Viscosities of Lead-Smelting Slags, M. S. Thesis (1983), Dept. of Metallurgical Engineering, Colorado School of Mines, Golden, Colorado.

(4) P. Williams, M. Sunderland, and G. Briggs, "Viscosities of Synthetic Slags in the System $CaO-FeO-SiO_2-MgO$," Trans. Inst. Min. Metall. Sect. C., 92, June 1983, pp. C105-C109.

(5) Y. Ouchi, and E. Kato, "Viscosities of Ternary Lead-Silicate Melts and Activities of PbO in These Melts," Canadian Metallurgical Quarterly, Vol. 22, No. 1, 1983, pp. 45-51.

(6) J. M. Toguri, G. H. Kaiura, and G. Marchant, "The Viscosity of the Molten $FeO-Fe_2O_3-SiO_2$ System," Extractive Metallurgy of Copper, Vol. I, pp. 259-273. Pyrometallurgy and Electrolytic Refining International Symposium, 1976, J. C. Yannopoulos, and J. C. Agarwal, Editors.

THE EFFECT OF IRON ACTIVITY AND OXYGEN PRESSURE ON THE COPPER AND IRON TRANSFER IN FAYALITE SLAG AT 1200°C

T. Nakamura, B. Chan and J.M. Toguri

Department of Metallurgy and Materials Science
University of Toronto
Toronto, Ontario, Canada. M5S 1A4

Abstract

The cleaning of copper containing fayalite slags using a Cu-Fe alloy was studied under an oxygen pressure ranging from 10^{-7} to 5×10^{-11} atm at 1200°C.

The reduction of copper from fayalite slags by Cu-Fe alloys was found to be very effective. For example, at a P_{O_2} of 10^{-7} atm, the copper content in the slag was reduced from 4.1 to 0.4 wt. percent after 90 min. of reaction time using a Cu- 5.27 wt. percent Fe alloy.

The copper reduction rates were found to be independent of P_{O_2} in the gas phase when using Cu-Fe alloys which exhibit an iron activity of greater than 0.4. A mechanism for slag cleaning is proposed based on the concept that the Cu-Fe alloy lowers the oxygen potential of the slag according to Eq.1, followed by the reduction of cuprous oxide by wustite as shown in Eq.2.

$$\underline{Fe} + (Fe_3O_4) = 4(FeO) \qquad (1)$$

$$3(FeO) + 2(CuO_{0.5}) = 2\underline{Cu} + (Fe_3O_4) \qquad (2)$$

Here $\underline{X}$ represents element X in the alloy phase while (X) represents the element in the slag phase.

The copper and iron transfer behaviour obtained in this study are explained by a stepwise equilibrium model.

Introduction

To maximize energy usage and to minimize the pollution to the environment, recent copper smelting operations have tended to produce a higher grade copper matte. However, such practice results in a much higher copper content in the smelting slag than previously experienced. Converter slags have always been treated for copper recovery and in the traditional process the converter slag was recirculated to the smelting furnace for this purpose. With the recent trend in smelting, leading to higher copper loss in slag, it would appear that a separate slag cleaning process will become more important.

It is now well established that there are two forms of copper loss to slags. These are the mechanically entrained copper and the chemically dissolved copper. While numerous studies exist on the dissolved copper in slags and there is general agreement amongst the investigators on the amount of this copper, the amount and nature of the mechanically entrained copper is uncertain. Nevertheless, two methods are currently employed for cleaning copper containing slags (1,2). One method involves the techniques of mineral processing whereby grinding and flotation steps are applied to separate the copper-rich matte and/or metallic copper phase from slowly cooled copper-containing slags. The other method is based on a high temperature process. The oxygen potential of the molten slag is reduced using reductants such as carbon, pyrite and SO_2 in an electric furnace. By reducing the oxygen potential, both dissolved and entrained copper in the slag are reduced. The reduction of the dissolved copper by the lowering of the oxygen potential is well established thermodynamically (3,4,5). The decrease in the entrained copper is due to a decrease in the magnetite phase which is known to entrap copper and matte particles in fayalite slags.

Although the thermodynamics of slag cleaning are well established for fayalite based slags, the knetics of the process are not well known. Consequently, the present study was undertaken to investigate the reduction rates of copper in fayalite slag by using Fe-Cu alloys at 1200^0C and to study the mechanism of copper reduction by iron, especially the effect of iron activity and oxygen pressure.

Experimental

Sample Preparation

Three Cu-Fe alloys with compositions of about 95, 98 and 99.5% Cu were prepared by mixing the required amounts of high purity copper (99.99%) and electrolytic iron (99.9%) in an alumina crucible. This crucible was placed inside a larger graphite crucible and the alloy was melted under an argon atmosphere using an induction furnace at 1300^0C. The prepared alloys were then machined to the required size for the experiments.

The fayalite slag was designed to contain about 4 wt % Cu. The master slag was made by melting a mixture of Fe, Fe_2O_3 and SiO_2 using a gas furnace without atmospheric control. The master slag was ground in a ball mill to produce a homogeneous well-mixed sample source. The 4 wt % copper containing slag was prepared by melting the powdered master slag and copper metal in a silica crucible under a P_{O2} controlled atmosphere at 1250^0C.

Apparatus

The apparatus used for the present study is shown schematically in Fig.1.

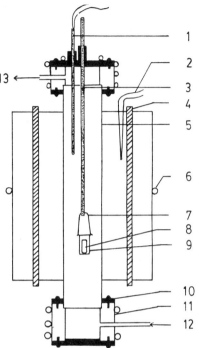

1. thermocouple
2. thermocouple to controller
3. suspension rod
4. Glo-bar element
5. reaction tube
6. watercooled copper tube
7. Molybdenum basket
8. silica crucible
9. Molybdenum basket
10. top o-shaped disc
11. middle quench chamber
12. CO/CO_2 gas
13. off gas to burner

Fig.1: Glo-bar Resistance Furnace

An alumina reaction tube, 50 mm in diameter and 1000 mm long, was placed in a silicon carbide resistance furnace. A water-cooled brass cap sealed the top of the reaction tube. Through this cap a thermocouple protection tube and a sample support rod were inserted into the reaction chamber. A water-cooled brass quench chamber was located at the bottom of the reaction tube. A gas inlet was placed in this chamber. The temperature profile of the furnace indicated a constant zone ($\pm 2.5^0 C$) of 50 mm length at $1200^0 C$.

The oxygen pressure for this study was obtained by use of a mixture of CO and CO_2. Since the oxygen pressures required ranged from 5×10^{-11} to 10^{-7} atm at $1200^0 C$, the ratio of the CO/CO_2 had to be varied widely. Therefore, two stages of gas mixing were necessary in order to obtain the desired ratio as shown in Fig.2

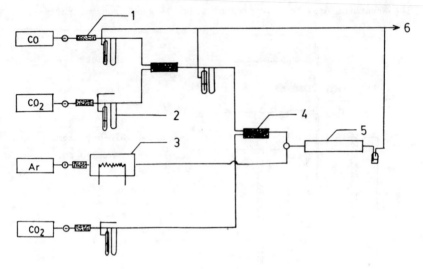

1. drierite
2. flowmeter
3. getter furnace
4. mixer
5. reaction tube
6. to burner

Fig.2: Schematic Representation of Gas Train

Each gas was used after drying with $CaSO_4$. Argon gas was purified by a copper getter furnace.

Experimental Procedure

Fayalite slag containing 4 wt % copper and the desired Cu-Fe alloy were placed in a silica crucible which in turn was placed in a Mo crucible. The Mo crucible, containing the sample, was suspended at the top part of the reaction tube under an argon atmosphere during the heating period. The argon gas was changed to a CO/CO_2 gas mixture when the furnace temperature reached about 1000^0C. When the required temperature of 1200^0C was attained, the sample was transferred to the constant temperature zone of the furnace. At the completion of the experiment, the CO/CO_2 mixture was replaced with Ar gas and the sample was plunged quickly into the quench chamber by pushing the suspension rod down. The crucible broke during cooling and the slag and alloy were readily recovered in separated form.

The iron content of the alloy and the copper content of the slag were analyzed by atomic absorption. Total Fe and silica content of the sample were analyzed by titration with $K_2Cr_2O_7$ and gravimetrically respectively.

Results

The copper content of the original slag was found to be 4.1 wt % Cu and the iron contents in the original alloys were analyzed to be 5.27, 1.86 and 0.5 wt %. The analytical value of the original slag and alloy are shown in Table I.

Table I. Analytical Values of the Original Alloys and Slag

		Cu wt %	Fe wt %	SiO_2 wt %
Cu-Fe alloy	I	99.50	0.50	-
	II	98.18	1.82	-
	III	94.73	5.27	-
Slag		4.1	42.4	37.5

The experimental results are summarized in Table II.

Table II. Summary of the Experimental Conditions and Results

#	Initial Conditions					Results			
	wt % Fe in alloy	P_{O_2}/atm	reaction time/min	alloy mass/g	slag mass/g	wt%Fe in alloy	wt%Cu in slag	wt%Fe in slag	wt%SiO_2 in slag
1	0.5%	5×10^{-11}	20	34.4	3.4	0.33	2.9	-	-
2			40	43.2	4.3	0.30	1.4	-	-
3			90	45.1	4.5	0.30	1.3	-	-
4			240	36.1	3.6	0.29	1.5	-	-
5		10^{-7}	20	30.2	6.0	0.09	2.7	-	-
6			40	30.4	6.1	0.09	2.2	-	-
7			90	31.0	6.2	0.09	2.4	-	-
8			240	30.1	6.1	0.05	1.9	-	-
9		10^{-9}	90	30.1	6.2	0.30	1.8	42.8	38.1
10	1.86%	5×10^{-11}	20	44.1	4.5	1.54	0.6	-	-
11			40	45.1	4.5	1.61	0.9	-	-
12			90	43.1	4.3	1.61	0.5	-	-
13			240	48.1	4.8	1.43	0.3	-	-
14		10^{-7}	20	29.7	5.9	1.27	1.5	-	-
15			40	30.2	6.1	1.09	0.5	-	-
16			90	32.4	6.5	1.05	0.4	-	-
17			240	31.0	6.2	0.76	0.3	-	-
18		10^{-9}	90	30.9	6.2	1.32	0.4	-	-
19	5.27%	5×10^{-11}	20	42.5	4.3	4.26	0.6	-	-
20			40	41.1	4.1	3.74	0.2	-	-
21			90	39.5	4.0	3.67	0.4	44.0	37.6
22			240	43.5	4.4	4.21	0.3	-	-
23		10^{-7}	20	30.4	6.1	4.25	0.7	-	-
24			40	30.7	6.1	4.10	0.4	-	-
25			90	30.4	6.1	3.50	0.3	44.1	38.2
26			240	30.2	6.0	3.94	0.2	-	-
27		10^{-9}	90	31.0	6.0	4.22	0.4	44.2	37.4

In general, the results indicate that the reduction rate of copper in the slag and the transfer rate of iron from the alloy are predominantly controlled by the iron content in the original alloy.

Discussion

Recent trends indicate that electric furnaces are being employed for slag cleaning instead of mineral processing techniques (6,7). The chemistry of slag cleaning of fayalite-based slags in electric furnaces is similar to that of the process of copper loss. As indicated earlier, many thermodynamic studies have been carried out on copper losses to slag (8,9,10). From these studies, it is known that slag cleaning is accomplished by reducing the oxygen potential in the slag phase. From a thermodynamic point of view, various reductants such as coke, pyrite, pyrrhotite, iron, SO_2 and natural gas have been proposed for this purpose. From a kinetic and mechanistic point of view, very little information exists on the reduction of copper from slags.

Imris (11) has reported the rate of copper reduction from fayalite slags when treated with pyrite. Both alumina and iron crucibles were used and the results showed that the copper reduction rates using iron crucibles were much faster than those using alumina crucibles. This suggests that the copper reduction rate with iron is faster with pyrite and that Fe is an effective reductant.

The Mechanism of Copper Reduction With Cu-Fe Alloy

It is possible to consider two mechanisms for reducing copper in fayalite slags with Cu-Fe alloy. One is the direct cementation reaction occurring at the slag-alloy interface between Cu^+ in slag and Fe in alloy, as shown by equation (1)

$$\underline{Fe} + 2(Cu^+) = (Fe^{2+}) + 2\underline{Cu} \tag{1}$$

$\underline{X}$ represents the element X in the alloy phase and (X) represents the element in the slag phase.

The other is that initially iron at the interface reduces the magnetite in the slag, according to equation (2), followed by the reduction of Cu^+ with Fe^{2+} near the interface by the reaction given in equation (3).

$$\underline{Fe} + (Fe^{3+}) = 2(Fe^{2+}) \tag{2}$$

$$(Fe^{2+}) + (Cu^+) = \underline{Cu} + (Fe^{3+}) \tag{3}$$

In general, the results of the present study showed that the mole numbers of iron transferred were larger than the moles of iron calculated from the moles of Cu transferred according to equation (1). One exception was observed when the experiment was performed using a 0.5 wt % Fe alloy at a P_{O_2} of 5×10^{-11} atm when the reverse results were obtained. Thus the mass balance suggests that the copper reduction by a Cu-Fe alloy proceeds according to equation (2) and (3).

A hot thermocouple technique developed by Yanagase (12) was employed to confirm the mechanism. A small amount of fayalite slag containing about 3 wt % Cu was melted on a Pt/Pt-Rh(13%) thermocouple at 1250^0C in Ar. A very thin rod (< 0.5 mmϕ) was pushed into the slag phase. After a certain reaction time, the current to the thermocouple was cut off to effect a rapid quenching of the melt. The solidified sample was analyzed using SEM with EDX to observe the copper distribution around the iron rod. Although the copper content of the slag near the iron rod increased slightly, a copper layer could not be observed at the interface. Thus, it does not appear that the reduction of copper with iron occurs by a direct cementation mechanism as given by equation (1).

Rate Model of Cu and Fe Transfer Behaviour in Slag Cleaning

The reactions of Cu and Fe transfer in slag cleaning are considered as follows;
at the slag-gas interface (interface I)

$$6(FeO) + O_2 (g) = 2(Fe_3O_4) \qquad (4)$$

at the metal-slag interface (interface II)

$$\underline{Fe} + (Fe_3O_4) = 4(FeO) \qquad (5)$$

$$3(FeO) + 2(CuO_{0.5}) = 2\underline{Cu} + (Fe_3O_4) \qquad (6)$$

Because of the inherent difficultires in defining the respective concentration gradients, a stepwise equilibrium model was applied. This type of model was developed by Kellogg for calculating the zinc fuming process (13). Mackey and Nagamori applied a similar model to the magnetite reduction in slag using a gas reductant (14).

Initially, the equivalent oxygen potential gradient across the slag-gas boundary was used to define the magnitude of the driving force for oxygen to cross the boundary.

$$J_{O2} = R_{O2} \cdot (P_{O2}(slag) - P_{O2}(g)) \qquad (7)$$

J_{O2}: the oxygen flux across the boundary
R_{O2}: the rate constant for oxygen transfer
$P_{O2}(slag)$: P_{O2} in the slag phase
$P_{O2}(gas)$: P_{O2} in the gas phase

where $P_{O2}(slag)$ is related to slag composition through equation (8):

$$K_4 = \frac{a^2(Fe_3O_4)}{a^6(FeO) \cdot P_{O2}(slag)} \qquad (8)$$

K_4 : equilibrium constant of equation (4)
$a(Fe_3O_4)$: activity of magnetite in the slag phase
$a(FeO)$: activity of wustite in the slag phase

Similarly the equivalent iron activity and cuprous oxide activity gradients can be used to describe the magnitude of the driving force for iron and copper across the slag-metal boundary.

$$J_{Fe} = R_{Fe} \cdot (a_{\underline{Fe}} - a(Fe)) \qquad (9)$$

$$J_{Cu} = R_{Cu} \cdot (a^*_{CuO_{0.5}} - a_{CuO_{0.5}}(slag)) \qquad (10)$$

R_{Fe} : rate constant of iron transfer
R_{Cu} : rate constant of copper transfer
$a_{\underline{Fe}}$: iron activity in the Cu-Fe alloy

$a(Fe)$: iron activity in the slag phase
$a^{*}CuO_{0.5}$: activity of $CuO_{0.5}$ at the interface of slag-alloy
$aCuO_{0.5}$: activity of $CuO_{0.5}$ in bulk slag

where a_{Fe} is represented by equation (11), assuming application of the regular solution model for the Cu-Fe alloy and using data obtained at 1600°C(16) since the iron activity in the alloy at 1200°C was not available.

$$a_{Fe} = 0.21 \text{ wt \% Fe} \quad (11)$$

$a(Fe)$ is related to $a(Fe_3O_4)$ and $a(FeO)$ through equation (12):

$$a(Fe) = \frac{a^4(FeO)}{K_5' \cdot a(Fe_3O_4)} \quad (12)$$

K_5' is the equilibrium constant of equation (5'):

$$(Fe) + (Fe_3O_4) = 4(FeO) \quad (5')$$

$a^{*}CuO_{0.5}$ is obtained as follows using equation (13):

$$Cu + 1/4\, O_2 = CuO_{0.5} \quad (13)$$

$$a^{*}CuO_{0.5} = K_{13} \cdot a_{Cu} \cdot P_{O_2}^{1/4} \quad (13')$$

K_{13}: equilibrium constant of equation (13).

For the present study, it can be assumed that $a_{Cu} = 1$ and P_{O_2} in equation (7) is a known experimental variable.

$a_{CuO_{0.5}}$ (slag) is represented by equation (14):

$$a_{CuO_{0.5}} = A(wt\%Cu) \quad (14)$$

A is a function of slag composition and temperature. Reported values of A vary between 1/25 and 1/37; 1/30 was selected for this work.

The relationship between both the $a(FeO)$ and $a(Fe_3O_4)$ with wt%FeO, wt%Fe_2O_3 and wt%SiO_2 must be known in order to perform the required calculations. Equation (15) and (16) were obtained using thermodynamic data at 1200°C reported by Diaz(16).

$$a(FeO) = -0.01644 \cdot X_{Fe2O3} - 0.0254\, X_{SiO_2} + 1.3620 \quad (15)$$

$$a(Fe_3O_4) = 0.00659 \cdot (X_{Fe2O3})^2 - 0.0146 \quad (16)$$

Copper and iron transfer during the slag cleaning reaction can be calculated by a stepwise method on a computer using equation (7),(9) and (10) and the mass balance in the system. The rate constants, R_{O_2}, R_{Fe} and R_{Cu}, were selected so as to obtain a good fit between the calculated and the experimental results. In the case of R_{O_2}, when P_{O_2} in the gas phase was higher than that of the original slag, the value of R_{O_2} was 0.1 while when the P_{O_2} in the gas phase was lower than that of the original slag, a value of 0.01 was obtained for R_{O_2}. This suggests that the oxidation rate of the slag by the gas phase is faster than the reduction rate of the slag, which was also pointed out by Jimbo et al (17).

7×10^{-3} was selected for both R_{Fe} and R_{Cu}. The change in composition of the slag and alloy was calculated every second and the corresponding new set of activities was determined.

The experimental results obtained for a P_{O_2} of 10^{-7} atm are shown in Fig.3 along with the calculated lines.

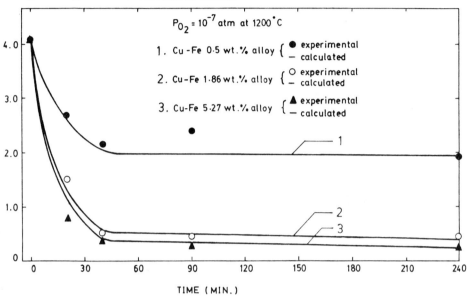

Fig.3(a). Copper Reduction Rate Using Cu-Fe Alloy Under a P_{O_2} of 10^{-7} atm in Fayalite Slag

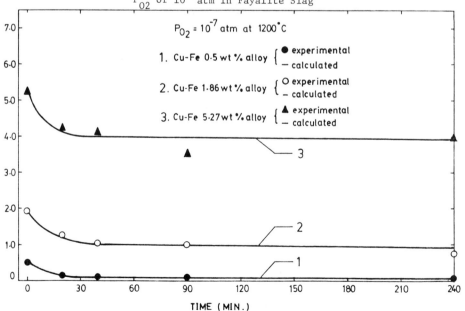

Fig.3(b). Iron Transfer in Cu-Fe Alloy Under a P_{O_2} of 10^{-7} atm

Even though some assumptions were necessary, the calculated lines are in good agreement with the experimental results. The results show that the copper reduction rate depends on the iron content in the alloy. A very low copper content can be expected even under a high P_{O_2} such as 10^{-7} atm when using a Cu- 5.0 wt%Fe alloy. Similarly the results obtained under a P_{O_2} of 5×10^{-11} atm are shown along with the calculated lines in Fig.4(a) and (b).

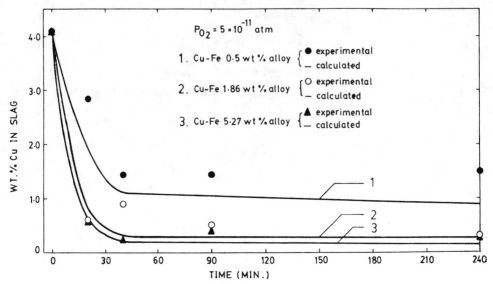

Fig.4(a). Copper Reduction Rate in Fayalite Slag Under a P_{O_2} of 5×10^{-11} atm

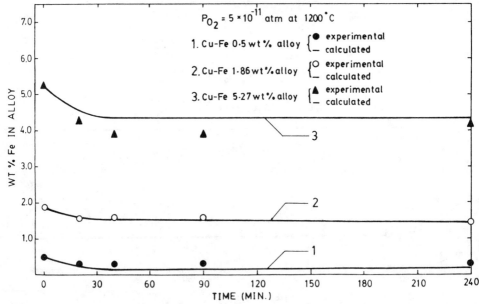

Fig.4(b). Iron Transfer Rate in Cu-Fe Alloy Under a P_{O_2} of 5×10^{-11} atm.

The trends in the copper and iron transfer behaviour are almost the same as in Fig.3 except for the data obtained when using a Cu- 0.5 wt%Fe alloy. The reduction in the copper content by using a Cu- 0.5 wt%Fe alloy under a P_{O_2} of 5×10^{-11} atm for 4 hours shows a decrease from 4.1 wt%Cu to 1.5 wt%Cu which is slightly lower than that obtained under a P_{O_2} of 10^{-7} atm. The iron content of the alloy under a P_{O_2} of 5×10^{-11} atm is higher than that under a P_{O_2} of 10^{-7} atm. This suggests that P_{O_2} in the gas phase has an influence on the copper reduction rate when Cu- 0.5 wt%Fe alloy is used, probably due to the weak reducibility of this alloy.

The copper content in the slag after 90 min. of reduction with a Cu-Fe alloy is plotted against the P_{O_2} in Fig.5.

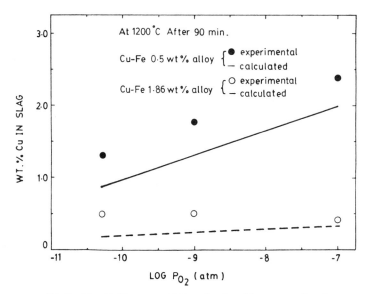

Fig5. The Influence of P_{O_2} in Gas Phase on the Copper Content After 90 min. of Reaction Time

The calculated lines are also shown. This figure indicates the influence of P_{O_2} on the reduction rate. The slag copper content decreases with decreasing P_{O_2} in the gas phase when using a Cu - 0.5 wt%Fe alloy as the reductant. On the other hand, when a Cu- 1.82 wt%Fe alloy is used for a reductant, the P_{O_2} in the gas phase does not affect the slag copper content. These results are well described by the calculations developed in this work.

The effect of the iron activity in the Cu-Fe alloy on the copper content in the slag after 240 min. is shown in Fig.6.

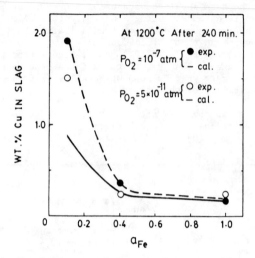

Fig.6. The Influence of Iron Activity in Alloy on the Copper Content After 240 min.

The solid and dotted lines represent the calculated values at $P_{O_2} = 5 \times 10^{-11}$ and 10^{-7} atm respectively. The iron activity shows a strong effect on slag copper reduction below 0.4. Excellent agreement between the calculated and experimental results is observed.

Conclusions

Slag cleaning reactions in fayalite slag with Cu-Fe alloys were studied at $P_{O_2} = 5 \times 10^{-11}$ to 10^{-7} atm at 1200°C and the Cu and Fe transfer behaviour were discussed using the stepwise equilibrium model developed by Kellogg.

Cu-Fe alloys were found to be an effective reductant for the slag cleaning process, especially when the activity of Fe in the alloy was above 0.4. In this case, the P_{O_2} in the gas phase showed almost no effect on the copper reduction rate.

The mechanism of slag cleaning with Cu-Fe alloys was considered as follows. Initially magnetite in fayalite slags were reduced by the Cu-Fe alloy at the slag-alloy interface. Cuprous oxide was then reduced by wustite which was produced by the magnetite reduction.

Based on the proposed mechanism, the Cu and Fe transfer behaviour during slag cleaning have been described by using a stepwise equilibrium model. Excellent agreement between the calculated and experimental results has been obtained.

Acknowledgements

The authors are grateful to the Natural Science and Engineering Research Council of Canada for financial support of this project.

References

(1) J.C. Agarmal, P.R. Ammann, F.C. Brown, J.J. Kim and S.N. Sharma, "Process Analysis for Recovery of Metal Values From Copper Smelter Slags", Extractive Metallurgy of Copper, Vol.1, edited by J.C. Yannopolous and J.C. Agarmal, 1976, pp. 351-368.

(2) S.C.C. Barnett, "The Methods and Economics of Slag Cleaning", Min.Mag. (5), 1979, pp. 408-417.

(3) M. Ruddle, B. Taylor and A. Bates, "The Solubility of Copper in Iron Silicate Slags", Trans.Inst.Min.Met. 75, 1966, pp. 1-22.

(4) J.M. Toguri and N.H. Santander, "The Solubility of Copper in Fayalite Slags at 1300°C", Can Met.Quart., 8(2), 1969, pp. 167-171.

(5) A. Yazama, "Thermodynamic Consideration of Copper Smelting", Can.Met.Quart., 13(3), pp. 443-453.

(6) P.R. Ammann, J.J. Kim, P.B. Crimes and F.C. Brown, "The Kennecott Slag Cleaning Process", in Extractive Metallurgy of Copper, Vol. 1, edited by J.C. Yannopoulos and J.C. Agarmal, 1976, pp. 331-350.

(7) H.P. Rajcevic and W.R. Opie, "Development of Electric Furnace Slag Cleaning at a Secondary Copper Smelter", J. Metals. 34(3), 1982, pp. 54-56.

(8) J.C. Yannopoulos, "Control of Copper Losses in Reverberatory Slags - A Literature Review", Can.Met.Quart., 10, 1971, pp 291-307.

(9) F. Schnalek and J. Imris, "Slags From Continuous Copper Production" in Advances in Extractive Metallurgy and Refining, edited by M.J. Jones, The Institution of Mining and Metallurgy, London, 1972, pp. 39-62.

(10) M. Nagamori, "Metal Loss to Slag: Part I- Sulfidic and Oxidic Dissolution of Copper in Fayalite Slag From Low Grade Matte", Metal.Trans. 5(3), 1974, pp. 531-538.

(11) I. Imris, "Thermodynamics of Reducing Copper Losses in Slags", Copper Metal.: Pract.Pap.Meet. edited by M.J. Jones, The Inst. Min. Metall. London, pp. 18-22, 1975.

(12) K. Morinaga, T. Kanagase, Y. Ohata and Y. Ueda, "Principle and Apparatus of Hot Thermocouple", TMS Paper Selection A-79-18, 1978.

(13) H.H. Kellog, "A Computer Model of the Slag-Fuming Process", Trans.Am.Inst.Min.Engrs. 239, 1967, pp. 1439-1442.

(14) P.J. Mackey, "The Physical Chemistry of Copper Smelting Slags- A Review", Can.Met.Quart. 21(3), 1982, pp. 221-260.

(15) L. Timberg, J.M. Toguri and T. Azakami, "A Thermodynamic Study of Copper-Iron and Copper-Cobalt Liquid Alloys by Mass Spectrometry", Met.Trans.B. 12(2), 1981, pp. 275-279.

(16) C. Diaz, The Thermodynamic Properties of Copper-Slag Systems, INCRA Monography, On Metallurgy of Copper, Vol IV, International Copper Research Association, New York, 1974, pp. 84.

(17) I. Jinbo, S. Goto and O. Ogama, "Equilibria Between Silica-Saturated Iron Silicate Slags and Molten Cu-As, Cu-Sb and Cu-Bi Alloys", <u>Metal. Trans</u>. 15 B (3), 1984, pp. 535-541.

Process Development

INFLUENCE OF ORGANIC ADDITIVES ON ELECTROLYTIC TIN DEPOSITION AND DISSOLUTION IN STANNOUS SULFATE ELECTROLYTE

Professor Dr.-Ing. Dr. h.c. Roland Kammel
Institut für Metallurgie - Metallhüttenkunde -
Technische Universität Berlin

Dr.-Ing. Uwe Landau
Fa. Schempp & Decker GmbH & Co. Industriegalvanik KG, D-1000 Berlin 37

Dipl.-Ing. Bernhard Szesny
Fa. Hermann C. Starck Berlin, Werk Goslar, D-3380 Goslar 1

Stannous sulfate electrolytes are widely used nowadays in electrolytic tin production. Characteristic features of this electrolysis process are that dense cathode deposits can only be obtained by additions of organic additives and that so called black mud is formed on the anode surface which leads to an increase in cell-voltage. In order to prevent excessive cell-voltages or oxygen evolution the black mud has to be removed periodically.

Studies in laboratory scale have been performed about the effect of different organic additives on electrolytic tin deposition and the formation of surface layers on pure and industrial tin anodes. Results reveal that organic addition agents are effective in reducing cathode roughening as well as extending the covering time of the anode by black muds.

Introduction

To receive high grade tin qualities from crude tin, an electrolytic refining process is employed, especially to remove contents of lead, bismuth, antimony and noble metals. Many different types of electrolytes were tried, but at present, electrolytes based on stannous sulfate and sulfuric acid are prefered. The problem with electrolytic tin deposition from acidic electrolytes is that even at low current densities needle-shaped nonadherent tin deposits are formed which are causing short circuits between the electrodes. By the addition of organic additives, particularly animal glue, cresylic and phenolsulfonic acid dense cathode deposits free from dendrites can be achieved, but the electrolytes are getting quite expensive because of these additives (1). It has been shown, however, that the favourable effect of the additives on tin electrocrystallization can not be related to the technical cresylic and phenolsulfonic acid itself but is caused by the contents of the by-product dioxydiphenyl sulfone (2). In electroplating industry this inhibiting additive has been used to ensure smooth and dense tin deposits until lately even more effective additives were developed.

Another tin electrolysis problem in acidic electrolytes arises by the formation of black muds which are accumulated on the anodes. These layers have to be brushed off at least every two days to prevent increase in cell-voltage. At too high cell-voltages, the tin quality deteriorates by the electrolytic codeposition of impurities and even oxygen may be evolved at the anodes. It has been shown (3) that the increase in cell-voltage is caused by precipitated stannous sulfate continuously covering the anode surface. This passivating layers are formed unavoidably on tin anodes in electrolytes and at current densities generally nowadays used in electrolytic tin refining plants. Therefore, it is of technical importance to develop methods which enable to extend the time of characteristic exponential increase in cell-voltage. In the present paper, the influence of different organic additives on tin electrocrystallization and the covering time of the anode by stannous sulfate have been studied by electrochemical methods.

Experimental

For cathodic and anodic experiments, electrolytes with different tin contents have been employed. Electrolyte A for cathodic experiments contained 26.7 g/l Sn whereas electrolyte B, used to study the covering time of the anode with stannous sulfate, was saturated with stannous sulfate and contained 78.3 g/l Sn. Both electrolytes were containing sulfuric acid with a concentration of 135.9 g/l. The saturated electrolyte B was chosen to prevent redissolution of stannous sulfate crystallites precipitated on the anode which would complicate the interpretation of covering time experiments. Conductivities of both solutions at electrolyte temperature of 298 K were $0.42 \Omega^{-1}$ cm^{-1} (A) and $0.40 \Omega^{-1}$ cm^{-1} (B). Organic additives employed in cathodic and anodic experiments respectively are summarized in Table I.

Table I. Organic Additives Used in Cathodic and Anodic Experiments

cathodic experiments	4,4'-dihydroxydiphenyl sulfone	$C_{12}H_{10}O_4S$	Merck
	polyethylene glycol 10000	$H(C_2H_4O)_nOH$	Merck
	Triton X-100	$C_{34}H_{62}O_{11}$	Hoechst AG
anodic experiments	methylene-blue	$C_{16}H_{18}ClN_3S \cdot 2H_2O$	Merck
	quinoline-yellow	$NaC_{18}H_{10}O_2N$	Ferak
	dodecylsulfate Na-salt	$CH_3(CH_2)_{11}OSO_3Na$	Ferak

Potentiodynamic and galvanostatic conditions were performed with a Wenking-Potentiostat ST72. The electrode potential was measured by a Luggin capillary. A pure tin rod (99.9999 % Sn) served as a reference electrode so that the overpotential could be evaluated directly. Tin cylinders (5 mm dia.) were imbeded in resin thus the cross-section served as electrode. During electrolysis the electrode surface was in vertical position. For cathodic and anodic experiments, pure tin (99.9999 % Sn, Zinsser) was used as electrode material whereas industrial anodic tin (98.123 % Sn, 0.017 % Fe, 0.78 % Pb, 0.224 % Cu, 0.174 % Sb, 0.073 % Ag and 0.6 % Bi) from ENAF-Tin-Refinery (Bolivia) was employed only in anodic experiments. Preparation of tin electrodes included polishing by diamondpaste (10 μm) and ultrasonic cleaning in destilled water. Pure tin additionally was electropolished in a methanolic solution containing 2 mol of sulfuric acid whereas industrial tin anodes could not be prepared correspondingly. Anode and cathode compartments were separated by a liquid junction to prevent intermixing of reaction products. Stirring of the electrolyte only was provided by natural convection flow.

Cathodic Results

Tin Deposition from Additive-free Electrolyte

The cathodic current potential curve of tin deposition without addition agents from electrolyte A is given in Figure 1. This polarization curve as well as the following ones were measured in potentiodynamic mode. The potential was changed at a constant rate to a preselected potential and then reversed to the starting potential. As shown by the steep slope of the current-potential curve, tin is electrodeposited with little activation polarization. This gives rise to coarse and dendritic type tin deposits which are manifested by the broad hysteresis in the current-potential curve. Even as the potential sweep in R is reversed current-density increase because of the growth of tin dendrites. Electrocrystallization of tin only starts from few nuclei as shown in Figure 2a for a low current-density of -50 A/m^2. Cathodic current densities are marked by a negative sign. The tetragonal crystals are growing isolated from each other.

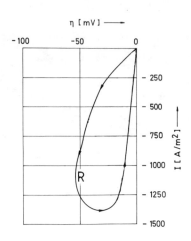

Figure 1 - Cathodic current-potential curve of tin deposition from additive-free electrolyte A (potentiodynamic mode, $d\eta/dt = \pm 30$ mV/min)

Even at a higher current density of -100 A/m^2 and a calculated mean deposit thickness of 50 μm, the base material still is to be seen, Figure 2b.

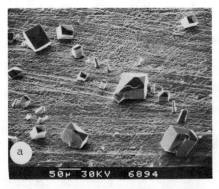 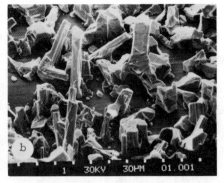

Figure 2 - Scanning electron micrograph of tin deposits from additive-free electrolyte A: a) I = -50 A/m^2, t = 20 min; b) I = -100 A/m^2, t = 100 min.

Tin Deposition from Additive-containing Electrolytes

To receive dense cathodic tin deposits, the number of nuclei has to be increased as well as the growth of single crystallites to be inhibited. In the literature, different addition agents are mentioned (4, 5) which are inhibiting the tin electrodeposition. In this paper, the effect of dihydroxydyphenyl sulfone (DS), polyethylene glycol 10000 (PG) and Triton X100 (TX) has been studied. Dihydroxydiphenyl sulfone had first to be solved in dimethylformamide (DMF) before adding to the electrolyte. DMF itself did not show any influence on the current-potential curve. Addition of all three additives exhibit an inhibiting effect as shown in Figure 3 by the shift of the respective current-potential curve to less cathodic current densities and more negative overpotentials compared to additive-free solution. The current-potential curves are given at a concentration where the inhibiting effect of the addition agents nearly has reached its maximum value.

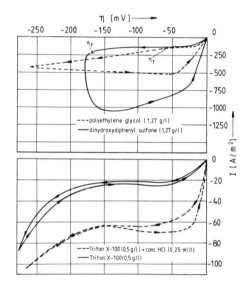

Figure 3 - Cathodic current-potential curve of tin deposition from additive containing electrolyte A (potentiodynamic mode, $d\eta/dt = \pm$ 30 mV/min)

Scanning micrographs in Figure 4a, b and c reveal the positive effect of the inhibiting agents on electrocrystallization of tin at -100 A/m^2 (deposit thickness 50 μm). Dense deposits are obtained that are free of any dendritic growth. Whereas by additives of DS and PG the electrodeposits are more or less disordered, tin crystallites by TX-addition show preferential orientations.

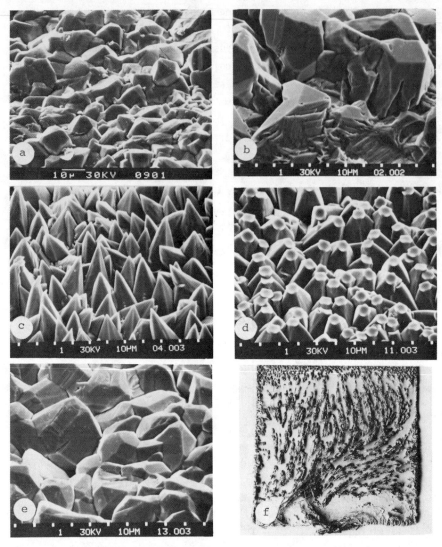

Figure 4 - Scanning electron micrograph of tin deposits (50 μm) at -100 A/m^2 from additive-containing electrolyte A:
a) 1.27 g/l   dihydroxydiphenyl sulfone
b) 1.27 g/l   polyethylen glycol 10000
c) 0.5 g/l   Triton X-100
d) 0.5 g/l   Triton X-100 + 0.5 ml/l conc. HCl
e) 0.5 g/l   Triton C-100 + 6.0 ml/l conc. HCl
f) 0.06 g/l   Triton X-100.

It has been reported that only small amounts of hydrochloric acid exhibit a positive effect on cathodic tin deposits (6). In Figure 3 resulting current-potential curve is shown if hydrochloric acid is added e.g. to TX. Even as the current-potential curve is shifted to higher cathodic current densities, only a very small hysteresis is obtained indicating little roughening of the cathode surface. In combination with the inhibitors studied in this paper, hydrochloric acid becomes a catalyzer. The catalyzing action is increased with higher hydrochloric acid concentrations. Scanning micrographs of tin deposits from TX-containing bath in Figure 4d and e exhibit a flattening of the top of tin crystallites at low hydrochloric acid concentrations and a dense and rather smooth deposit at higher ones. Crystallites at higher hydrochloric acid concentrations show nearly any preferential orientation.

It has been estimated as a characteristic feature of potentiodynamic polarization measurements in additive-containing solutions that a broad hysteresis in the current-potential curve suddenly occurs when the cathode potential exceeds a certain value n_t to negative potentials as shown in Figure 3 for PG and DS. This characteristic potential has not been reached for TX-addition (Figure 3), thus only a very small hysteresis is obtained. At a constant potential sweep rate, this characteristic potential is dependent on concentration as well as on the kind of the addition agent. By higher additive concentrations, n_t is shifted to more negative potentials. At the same additive concentration, n_t is obtained e.g. for DS at about -170 mV but for PG only at about -60 mV. For less than half of this concentration, even at -300 mV, n_t still is not yet obtained with TX-additive. The broad hysteresis in the current-potential curve is due to a sudden increase of the cathode surface. In Figure 4f a dendritically tin deposit is shown on a cathode that has been coated just within the range of drastic surface roughening in a rectangular electrolytic cell at -100 A/m^2 and TX-concentration of 0.06 g/l. Slight stirring of the electrolyte by a magnetic stirrer is pictured by the orientation of tin dendrites.

Anodic Results

Anode Covering Times in Additive-free Electrolyte

The covering time of pure tin and industrial tin anodes with stannous sulfate was studied in electrolyte B by galvanostatic mode. During electrolysis, stannous sulfate crystallites are formed on the anode and grow until the whole anode surface is covered by a non-conducting salt film. The

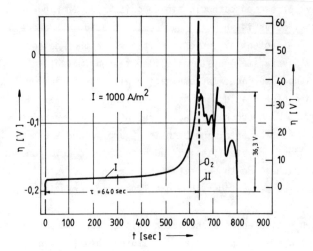

Figure 5 - Typical potential time characteristic during covering of a pure tin anode with non-conducting salt film at 1000 A/m^2 in electrolyte B.

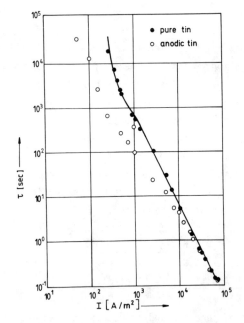

Figure 6 - Covering time - current density relationship for pure and anodic tin anodes in electrolyte B.

closure of the last pores in the salt film under galvanostatic conditions is characterized by a steep increase of the anode potential as shown in Figure 5 for a typical potential time diagram. The time between switching on the electric current and steep increase in anode potential is denoted by covering time τ. Covering time dependence on geometrically current density is presented in Figure 6 for both anode qualities. Each value in the diagram represents the mean of ten single measurements. A linear relationship between current density I and covering time τ in form of $\tau.I^m = K$ holds for pure tin anodes down to about 1000 A/m^2. By the method of least squares values for constants m and K were evaluated as 2.0 and $7.7 \cdot 10^6$ [s.(A/m^2)^m]. At about 230 A/m^2 at the given experimental conditions no complete covering of pure tin anodes was obtained. Surface layers formed on pure tin anodes exhibit a light-grey colour.

Covering times of industrial tin anodes, especially at lower current densities, are very much reduced compared to pure tin, Figure 6. A linear relationship as indicated above was only obtained at very high current densities. Covering times contrary to pure tin could be evaluated down to very low current densities. Surface layers formed on industrial tin anodes show a dark-grey to black colour thus indicating different deposit properties.

Anode Covering Times in Additive-containing Electrolytes

Covering time τ_m in presence of organic additives was measured with pure tin anodes at 1000 A/m^2 for different additive concentrations. The dependence of covering time on current density was studied for two selected additives with industrial tin anodes. By the quotient of covering times in additive-containing and additive-free electrolyte $\tau_m/\tau > 1$ elongation of covering time is indicated. Concentration dependence of τ_m/τ is shown in Figure 7a to c for quinoline-yellow, methylene-blue and dodecylsulfate Na-salt. Already at low concentrations of quinoline-yellow ($\geq$ 0.08 g/l) and methylene-blue ($\geq$ 0.018 g/l), the covering time of pure tin anodes is elongated by the factor of about 1.4 compared to the pure electrolyte. Addition of dodecylsulfate Na-salt at the given current density for nearly all concentrations does not show any effect on covering time. Distinct elongation of the covering time for industrial tin anodes was obtained at low current densities by addition of 0.4 g/l quinoline-yellow or dodecylsulfate Na-salt respectively, Figure 8. At current densities less than 250 A/m^2, the covering time is increased by a factor of about 3.6 for quinoline-yellow and 2.6 for dodecylsulfate Na-salt. With increasing current density, the elongating effect of these additives on covering time decreases.

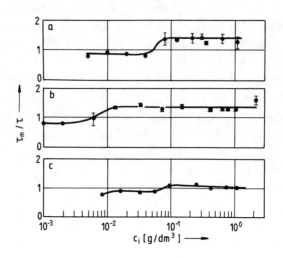

Figure 7 - Concentration dependence of τ_m/τ at 1000 A/m^2
a) quinoline-yellow; b) methylene-blue; c) dodecylsulfate Na-salt.

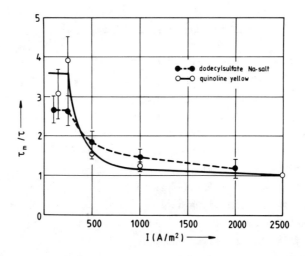

Figure 8 - Current density dependence of τ_m/τ at an additive concentration of 0.4 g/l.

Conclusion

The experimental results with both additive-free and additive-containing electrolytes indicate that cathodic tin deposition as well as anodic tin dissolution can be altered very favourably by organic addition agents. Additives employed in cathodic experiments exhibit different maximum inhibition ability, increasing in the following order, PG, DS and TX. The inhibiting effect can be reduced by hydrochloric acid as it was already mentioned by Meibuhr et al (4). With increasing inhibiting effect of the addition agent more oriented tin crystals are formed. Dense and non-dendritic tin deposits were obtained by the inhibitors employed in this study unless the inhibiting power became too low or a certain potential in the current potential curve was exceeded. Beyond this potential, dramatic roughening of the cathode occurs. Ksouri and Wiart (7) dedicate this roughening of the tin deposits to a stationary current potential curve showing no polarization at a certain electrode potential. Stationary current potential curves, apparently exhibiting no polarization at a certain electrode potential were found by previous investigations about the copper deposition in presence of organic additives (8, 9). According to these results the stationary current potential curve is unsteady, and two states, corresponding to two current densities limiting the unsteady region, coexist at the same electrode potential as outlined in Figure 9. Thus by electrolyzing within current density range I_a and I_b of the unsteady region, at the same electrode potential η_t rough metal deposits will always be produced, and the degree of roughening is dependent on the width of the unsteady region. In branch a and b on the other hand macroscopically even deposits are formed. By potentiodynamic mode of measuring employed in this study, an unsteady region in the current potential curve only is indicated by the broad hysteresis due to surface roughening that suddenly occurs at a certain potential. In electrolytic tin refining, current densities will have to be adjusted well above

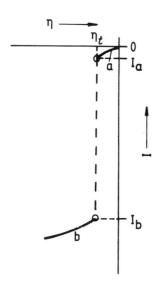

Figure 9 - Outline of a stationary cathodic current potential curve exhibiting an unsteady region.

this critical region otherwise despite organic addition agents rough tin deposits will be obtained.

Experimental results of anodic tin dissolution in electrolytes, saturated by stannous sulfate principally indicate that the time until both pure and industrial tin anodes are covered completely by a non-conducting stannous sulfate layer can be elongated by organic additives, especially at low current densities. According to Osterwald (10), covering time may be subdivided into three stages that characterize (a) increase of metal ion concentration at the electrode surface until the solubility product of the surface layer is achieved, (b) supersaturation necessary for nuclei formation and (c) formation and growth of nuclei. In saturated solutions, employed in the present study only the last two steps have to be considered. Distinct elongation effect at low current densities indicate that growth of stannous sulfate nuclei is probably inhibited by organic additives. Similar effects by quinoline-yellow and methylene-blue were observed by Marc (11) and Reich et al (12) with growth of sodium sulfate and thallium bromide respectively. Inhibition effect according to these authors is produced by adsorption of organic molecules at growth sites of the crystal lattice thus sterically hindering growing lattice planes.

Shorter covering times with industrial tin dissolution compared to pure-tin dissolution in additive-free electrolyte is caused by a porous layer growing between stannous sulfate crystallites. This porous layer, responsible for the characteristic dark colour of the deposits, is initiated by different impurities of anodic tin (3). Electrolyte exchange by natural or forced convection flow at the electrode surface is hindered by these pores thus the stannous sulfate layer is closed earlier at industrial tin anodes.

Further experiments will have to prove whether organic additives of the present study or even other ones also work at less concentrated stannous sulfate electrolytes. If the elongating effect of organic additives is confirmed again, then it should be possible to reduce brushing procedures in electrolytic refineries. As organic additives tested in the present study may be effective at the corresponding counterelectrode as well their respective influence will have to be investigated also. Perhaps, it is possible to find out an organic additive that is simultaneously favourable at the cathode as well as at the anode.

Acknowledgements

The authors are expressing their gratitude to the Deutsche Forschungsgemeinschaft, Bonn - Bad Godesberg, for financial support. They also acknowledge most thankful the experimental assistance of Dipl.-Ing. Halife Tepe and Dipl.-Ing. A. Juan Leon.

References

1) Peter Paschen, "Aktuelle Probleme der Zinnmetallurgie", Erzmetall 29 (1976) pp. 14-18.

2) H. J. Steeg, "Die elektrolytische Abscheidung von Zinn- und Blei-Zinn-Legierungen", Metall 30 (1976) pp. 834-839.

3) Roland Kammel, Uwe Landau, and Bernhard Szesny, "Anodic Layer Formation on Tin in Stannous Sulfate Electrolytes", to be published.

4) S. Meibuhr, E. Yeager, A. Kozawa, and F. Hovorka, "The Electrochemistry of Tin - I. Effects of Nonionic Addition Agents on Electrodeposition from Stannous Sulfate Solutions", J. Electrochem. Soc. 110 (1963) pp. 190-202.

5) H. D. Hedrich, and Ernst Raub, "Die inhibierte Elektrokristallisation des Zinns aus sauren Elektrolyten", Metalloberfläche 31 (1977) pp. 97-103.

6) T.S. Mackey, "The Electrolytic Tin Refing Plant at Texas City, Texas", Journal of Metals 22 (1969) pp. 32-43.

7) M. Ksouri, and R. Wiart, "Ursache der Bildung bestimmter schwammartiger oder dendritischer galvanischer Abscheidungen", Metalloberfläche 32 (1978) pp. 63-68.

8) Roland Kammel, Uwe Landau, and Manfred Mayer, "Entstehung von unebenen Kupferniederschlägen durch organische Badzusätze", Erzmetall 36 (1983) pp. 465-471.

9) Roland Kammel, Uwe Landau, and Manfred Mayer, "Einfluß von Badzusätzen auf das Verhalten von Kupferelektroden in sauren kupfersulfathaltigen Elektrolyten", paper presented at the International Symposium "Refining Processes in Metallurgy" ("Raffinationsverfahren in der Metallurgie"), 20./22.10.1983, Hamburg, Proceedings A 3, pp. 21-34.

10) Jörg Osterwald+, "Keimbildung und Kristallwachstum bei der anodischen Bildung von Silbersulfatdeckschichten auf Silber in Schwefelsäure", Z. Elektrochemie 66 (1962) pp. 492-496.

11) R. Marc, "Über die Kristallisation aus wässrigen Lösungen", Z. Phys. Chemie 67 (1909) p. 470.

12) R. Reich, and M. Kahlweit, "Zur Kinetik des Kristallwachstums in wäßrigen Lösungen I und II", Berichte der Bunsengesellschaft 72 (1968) pp. 66-74.

ELECTROLYTIC EXTRACTION OF MAGNESIUM FROM COMMERCIAL ALUMINUM ALLOY SCRAP

Basant L. Tiwari, Ram A. Sharma, Blake J. Howie

Electrochemistry Department
General Motors Research Laboratories
Warren, Michigan 48090-9058

An electrolytic process, based on a three-layer concentration cell, to extract magnesium from commercial aluminum alloy scrap has been examined. The process was repeatedly tested at 1025 K with two cells of different sizes, 0.6 kg and 5 kg feed aluminum, to have a better understanding of the operating parameters and evaluate the scale-up potential. In both cells, the magnesium content of the scrap was selectively reduced from 1.1 to less than 0.1 wt.% at anodic current density as high as 1.1 A/cm^2 with current efficiency exceeding 85%. Magnesium was recovered in the form of salt-coated globules. Comparison of performances for the two cells suggests that the process can be scaled up. With further development, this process could become a viable alternative for demagging secondary aluminum.

Introduction

The production of secondary aluminum casting alloys such as 380 and 319 usually requires the removal of excess magnesium (termed "demagging") in order to meet specifications. These alloys are widely used in automobiles, and their magnesium content is specified to be below 0.1 wt.% (1). Fortunately, magnesium is more reactive than aluminum and other alloying elements present in the aluminum, and, therefore, it can be selectively removed from the molten aluminum by controlled oxidation. Chlorination (2-5) processes are most widely used by the secondary smelters for demagging the casting alloys. These processes are based on the favored reaction of magnesium with chlorine, producing a magnesium chloride dross which being lighter separates from molten aluminum. While the process is reasonably efficient at high magnesium content, it has certain drawbacks. Unless the process is well controlled, an unacceptable environmental condition may result in the plant (5). In addition magnesium is lost in the dross which sometimes also poses disposal problems.

The electrolytic process described in this paper, on the contrary, recovers magnesium in the form of salt-coated globules and apparently causes no environmental problems. The process is, in general, based on two specific properties of magnesium: higher oxidation potential (more reactive) and lower density than the corresponding values for aluminum. And the process consists of covering the molten aluminum scrap with an electrolyte (a mixture of alkali and alkaline earth metal halides) and passing a current between molten aluminum acting as an anode and an inert cathode dipped into the electrolyte. On applying a potential, magnesium being more reactive dissolves first in the electrolyte from the aluminum melt, and concurrently deposits on the cathode. Because of lower density, magnesium floats on the electrolyte and, thus, it is separated from the aluminum. In this case the electrochemical unit works as a concentration cell, and the potential can be determined using the overall cell reaction,

$$Mg \text{ (in Al)} = Mg \text{ (cathode)} \tag{1}$$

$$E = \frac{RT}{nF} \ln a_{Mg} \text{ (in Al)} \tag{2}$$

where E is the cell potential; F, the Faraday constant; R, the gas constant; T, the melt temperature in Kelvin; n, 2; a_{Mg}, the activity of magnesium in the alloy with respect to pure liquid magnesium. The potential at 0.1 wt.% magnesium in the alloy (the final magnesium content permissible in castings) is calculated to be approximately -0.30 V at 1000 K, using an activity coefficient for magnesium of 0.90 (6). This small potential value is characteristic of a concentration cell and suggests that most of the electrical energy required to carry out the process would be consumed in overcoming the ohmic resistance of the cell.

Most of the electrolytic processes reported in the literature work on the concentration cell principle and have been tested so far only on laboratory scale with synthetic aluminum-magnesium alloy. Some of these processes were not developed originally for demagging purposes and differ in their choice of electrolyte compositions and cell design. A brief description of these processes follows.

Belyaev and co-workers (7) used a three-layer electrochemical cell for removing magnesium from high-magnesium (30 wt.%) aluminum alloy, and established the following refining conditions: electrolyte composition (wt.%),

10-18 $MgCl_2$, 35-50 KCl, 35-40 NaCl, 10-20 $BaCl_2$, 1-2 CaF_2; melt temperature, 975-995 K; current density, 1 A/cm^2. These investigators (8) also used a bipolar electrolytic cell for extracting magnesium. The main advantage of using a bipolar design is the possibility of producing a cathodic magnesium that is less contaminated with aluminum than that obtained with the conventional three layer cell. However, the applied potential of a bipolar cell would be higher than the applied potential of the conventional three-layer cell for the same electrolyte thickness and current density. The higher potential is due to a long path for the ionic migration in the bipolar cell.

Cleland and Fray (9) reported the application of packed-bed electrodes in an electrolysis cell to refine lead-zinc alloy and indicated the possibility of using this cell to also remove magnesium from aluminum alloy. The main advantage of using packed-bed electrodes is that they provide larger surface area for electrode reactions to occur and thus facilitate the flow of a larger current through the cell. But the process appears expensive because it would require high maintenance cost to repair and replace the metal pumps and reactor which deteriorate from wear and tear caused by flowing metal. In addition, more experimental work is needed to develop the process for commercial application.

More recently, Tsumura described (10) an electrochemical process to remove magnesium from aluminum alloy. As described before, the process involves covering the molten aluminum with a layer of molten electrolyte mixture (50 wt.% NaCl + 50 wt.% KCl, or 59 wt.% $MgCl_2$ + 20 wt.% KCl + 20 wt.% NaCl + 1 wt.% NaF) and passing an electric current between the molten aluminum and the electrolyte to remove magnesium. According to Tsumura, enough potential was applied to evolve Cl_2 bubbles at the aluminum-electrolyte interface. The chlorine reacted with magnesium producing $MgCl_2$ in the electrolyte. This mechanism may be valid with the first electrolyte mixture (50 wt.% NaCl + 50 wt.% KCl), but is doubtful with the second electrolyte mixture which contains 59 wt.% $MgCl_2$. With the second electrolyte, the cell would probably work as a concentration cell.

Burkhardt described (11) a process for purifying aluminum and other light metals using an electrochemical cell. The cell is made of a molten metal cathode (Cu, Fe) as bottom layer, $BaCl_2$-based (80 wt.% $BaCl_2$, 20 wt.% $CaCl_2$) electrolyte as middle layer, and aluminum alloy to be purified as anode top layer. Extremely fine bubbles of $AlCl_3$ were generated at the aluminum-electrolyte interface by electrolytic decomposition of halide salts. These bubbles slowly rise through the melt and on their way react with magnesium, collect inclusions of oxides and nitrides, and remove hydrogen, thus purifying the molten aluminum.

We first successfully tested the electrolytic process in a small cell by reducing the magnesium content of 0.6 kg synthetic aluminum-magnesium alloy from 1.5 wt.% to less than 0.1 wt.% at current densities as high as 1.66 A/cm^2 with current efficiency near 100% (12). Magnesium in the form of salt-coated globules was recovered as a by-product. The results suggested that, with further development, this process could become a viable alternative for demagging scrap aluminum. Accordingly, we decided to further test the process for selective removal of magnesium from commercial scrap and evaluate the scale-up potential using a larger cell. The results of our investigation are described in this paper.

Experiment

Materials

Aluminum Alloys. The chemical compositions of the aluminum alloys used in this study are given in Table I. Alloy A was supplied in the form of scrap chips which were premelted and cast into a slab of 3.8 cm thickness. Alloy B was in the form of 5.5 kg slabs from which pieces were cut to perform the tests. Alloy C was similar to Alloy B in composition, but it was prepared in our laboratory.

Table I. Composition of Major Alloying Components in Aluminum Alloys

Alloy	Chemical Analysis, wt.%				
	Si	Cu	Mg	Fe	Zn
A	9.55	2.12	0.55	0.75	0.57
B	8.87	3.50	0.95	0.43	0.38
C	9.00	3.30	1.10	0.50	0.60

Electrolytes. The following criteria were considered in selecting the electrolytes used in this investigation; each candidate must be stable in contact with aluminum alloy and magnesium, have density lower than aluminum and higher than magnesium, have melting temperature lower than aluminum and magnesium, have high ionic conductivity, and be readily available at a low cost. Two electrolytes were selected based on the above requirements. Electrolyte I, whose composition was obtained from the literature (13), contains 45 wt.% $CaCl_2$, 30 wt.% NaCl, 15 wt.% KCl, and 10 wt.% $MgCl_2$, and it was used successfully in the previous study (12). Electrolyte II was formulated by considering the phase diagrams of the halides, Fig. 1, and may also be considered an improvement over the first electrolyte. Figure 1a shows that the addition of about 10 wt.% $MgCl_2$ in $CaCl_2$ does not significantly change the melting point of $CaCl_2$, and therefore, for our purpose the mixture behaves like $CaCl_2$. Figure 1b, however, shows that a small addition of KCl (15 wt.% to a melt containing about 68 wt.% $CaCl_2$ and 32 wt.% NaCl, marked by point A in the diagram) also does not significantly lower the melting point, and, therefore, KCl may be eliminated from electrolyte I. Hence, the composition of the alternative electrolyte II was chosen as 58 wt.% $CaCl_2$, 32 wt.% NaCl, and 10 wt.% $MgCl_2$, closely represented by point A with 10 wt.% $MgCl_2$ substituted for $CaCl_2$.

The electrolytes were prepared by mixing weighed amounts of anhydrous magnesium chloride (99% purity), anhydrous calcium chloride (98% purity), sodium chloride (99% purity), and potassium chloride (99.9% purity) in a helium-atmosphere glove box. Electrolyte II is simpler, as it excludes KCl, and denser (1.88 g/cm^3 compared to 1.76 g/cm^3, calculated using the densities of the electrolytes' constituents (16)) than electrolyte I. The electrolytes I and II melt at approximately 775 K and 780 K, respectively. The standard free energies of formation for the constituent chlorides indicate that the electrolytes are stable in contact with both magnesium and the aluminum alloys containing copper, silicon, zinc and iron (17).

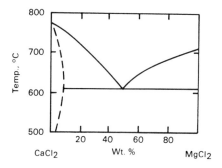

(a) System CaCl$_2$-MgCl$_2$ (18)

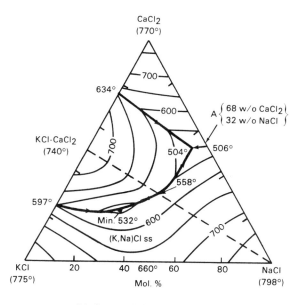

(b) System KCl-NaCl-CaCl$_2$ (19)

Figure 1. Phase Diagrams of Halides.

Electrochemical Cells

Two laboratory cells capable of refining 0.6 kg and 5 kg aluminum alloy were used. The cell design had molten aluminum alloy as the bottom layer, salt electrolyte as the middle layer, and transferred magnesium as the top layer. The small cell had an aluminum anode depth of 10 cm and an anode area of 25.5 cm^2. Corresponding values for the large cell were 20 cm and 111.2 cm^2, respectively. In each cell, electrolyte depth was 4.5 cm.

Figure 2 shows the assembled cells and the graphite crucibles fitted with alumina sleeves which were used to electronically insulate the cathode from the anode. Internal diameters of the large and the small crucibles were 11.9 cm and 5.7 cm, respectively. These crucibles were placed inside the respective stainless steel vessels to hold the aluminum alloy, electrolyte, and the transferred magnesium, as shown schematically in Fig. 3. The stainless steel vessels were heated in a tubular resistance furnace, equipped with a temperature controller, to melt the aluminum and electrolyte.

Procedure

A weighed amount of alloy was added to the graphite crucible which was then placed inside the stainless steel vessel. The vessel was heated to about 775 K, with its lid open and an inert gas (helium or argon) flowing inside to remove any water absorbed in the crucibles from the previous test. Then the vessel was closed, with its lid holding the cathode current collector and the thermocouple, and the inert gas flow was maintained. Next the vessel was heated to about 1025 K to melt the aluminum and then cooled to about 575 K to solidify the molten aluminum. This was done to prevent any mixing of the electrolyte in the molten aluminum, should they have been melted together. A weighed amount of electrolyte was placed on top of the solidified aluminum, and the vessel was again heated to 1025 K to melt both phases. The cathode was lowered into the electrolyte and the thermocouple was positioned.

Magnesium was transferred to the cathode by passing current for a predetermined time. The cell was cooled, and the refined aluminum and transferred magnesium were usually separated by dissolving the entrained electrolyte in water. However, the salt-coated globules were also easily separated from the electrolyte by crushing the electrolyte. Representative samples of the refined aluminum were analyzed for magnesium and other alloying elements such as silicon, copper, zinc, and iron. The recovered magnesium was dissolved in HCl and the solution was analyzed for magnesium and aluminum.

The effective number of coulombs utilized in transferring magnesium was calculated from the difference between magnesium contents of the unrefined and the refined aluminum, using Faraday's law. The current efficiency of the process was calculated by dividing this quantity by the total number of coulombs passed through the cell.

Experimental Problems and Their Solutions

Graphite-Disk Cathode Degradation. The graphite-disk cathode broke many times during some electrorefining experiments. This breakage was found to be associated (Table II) with high calculated current efficiency ($\geq$98%), high aluminum transfer to the cathode and the electrolyte, higher open circuit cell potential (1.25 V), and a floating potential between the steel vessel and ground. We later observed that the furnace wiring was touching the steel vessel. Because of this short-circuiting, more current had

(b) Alumina Sleeve-Fitted Graphite Crucibles

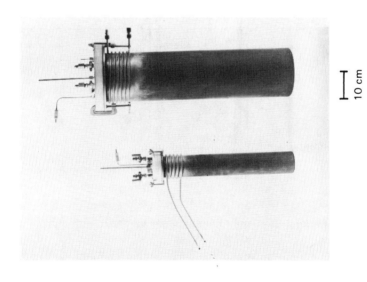

(a) Assembled Cells

Figure 2.

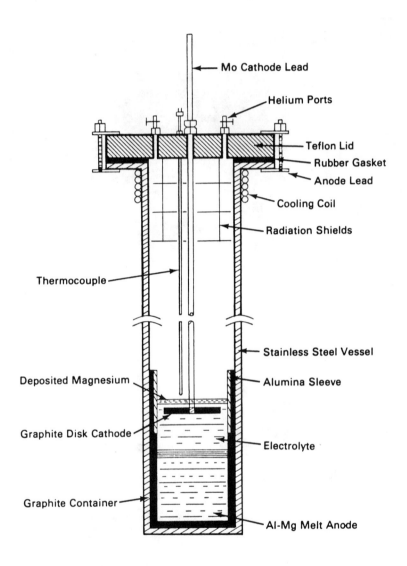

Figure 3. Schematic Diagram of the Electrolytic Cell.

undoubtedly been passing through the electrodes than was measured by the meter, increasing the cell potential to a value where an excessive amount of $AlCl_3$ and $SiCl_4$ may have been formed at the anode according to the following reactions:

$$2/3\ Al + 2Cl^- \rightarrow 2/3\ AlCl_3 + 2\bar{e} \qquad (3)$$

$$Si + 4Cl^- \rightarrow SiCl_4 + 4\bar{e} \qquad (4)$$

Table II. Some Pertinent Details for Evaluating Cathode Disk Breakage Problem

Expt. No.	Anodic Current Density (A/cm^2)	Current Efficiency (%)	Al in Magnesium Residue (wt.%)	Open Circuit Potential (V)	Cathode Disk Broke
		Small Cell			
14	0.92	100	97.5	1.25	yes
15	0.94	117	82.2	1.25	yes
20	1.00	97	25.2	0.75	no
		Large Cell			
16	1.10	98	48.1	1.25	yes
17	1.10	125	33.3	1.25	yes
12	1.10	86	23.5	0.87	no

These compounds could react with graphite forming a product like Al_4SiC_4, which was detected by x-ray diffraction. The formation of such a solid compound inside the disk pores might have introduced stress in the disk, causing it to crack. Table II shows that the disk broke during the experiments having open circuit potentials (measured just after opening the electrical circuit) higher than 0.9 V, a potential high enough to evolve $AlCl_3$ (12).

The short-circuiting problem was eliminated by using a new furnace with an alumina tube to avoid the contact of furnace wiring with the steel vessel. Both the steel vessel and the furnace were also grounded to prevent any leakage of current through the electrodes. And, the graphite disk was replaced with a molybdenum disk for studying the influence of current density and electrolyte composition.

Water. Both $CaCl_2$ and $MgCl_2$ contained a significant amount of water which caused problems during the experiments. Water reacted with $MgCl_2$ producing hydrochloric acid which, in turn, vigorously reacted with the steel vessel and the radiation shields. The water also reacted with the graphite cathode and the crucible as both were found severely corroded. To eliminate the water problem, both the $CaCl_2$ and $MgCl_2$ were dehydrated before using them to make the electrolyte mixture. Calcium chloride was slowly heated under vacuum to 425 K and was held at this temperature for one hour. It was further heated to 875 K to remove all the water. Since magnesium chloride reacts with water, it was slowly heated under flowing Cl_2 to 475 K and was held at this temperature for overnight. It was further heated to 875 K to remove the rest of water. The above treatment removed all water from the chlorides and no water problem was found in the subsequent experiments.

Results and Discussion

Several experiments were conducted to confirm the feasibility of the electrolytic demagging process. Using both the small and the large cells, magnesium content of scrap aluminum was selectively reduced to the maximum level, 0.1 wt.%, permitted in the casting alloys. By comparing the performance of the cells, the scale-up potential of the process was evaluated. In addition, the influences of current density, amount of electrical charge passed, and electrolyte density on the performance of the process were examined. The experimental conditions and the results are summarized in Table III.

Table IV presents the typical analyses of aluminum alloys refined using both cells under similar conditions of temperature and current density. By comparing the chemical analyses of the feed aluminum and the refined aluminum one may conclude that the magnesium was selectively removed, while the other alloying elements remained in the refined aluminum.

Table IV. Typical Composition of Refined Aluminum[1]

	Chemical Analysis, wt.%				
	Mg	Si	Cu	Fe	Zn
Feed Aluminum	0.95	8.87	3.50	0.40	0.40
Refined Aluminum Small Cell	0.10	9.10	3.30	0.41	0.40
Large Cell	0.04	8.72	3.52	0.47	0.40

(1) Experimental Conditions: Current density, 0.40 A/cm^2; Electrolyte depth, 4.5 cm; Temperature, 1025 K

Experimental Findings: Applied Cell Potential, 1.6 V; Current Efficiency, 86.0%

An important result obtained from Table IV is that under the similar conditions of temperature (1025 K), electrolyte depth (4.5 cm) and anodic current density (0.40 A/cm^2), magnesium was selectively removed from both the cells, and the applied cell potentials (1.6 V) and the current efficiencies (86%) of both the cells were the same. The result, therefore, suggests that the increased depth of anode, from 10 cm in the small cell to 20 cm in the large cell, did not interfere with the removal process. Also, the refined aluminum in the large cell showed a surprisingly uniform distribution of magnesium throughout the 20 cm depth (0.03 wt.% at the top and 0.04 wt.% at the bottom), suggesting that the mixing during electrorefining was fast enough to facilitate the rapid transfer of magnesium from bulk to aluminum-electrolyte interface. These findings, therefore, suggest that the process can be easily scaled up.

Table III. Summary of the Experimental Conditions and Results

Expt. No.	Alloy[3]	Electro.[1]	Anodic Current Density (A/cm^2)	Amount of[4] Coulomb (%)	Applied Cell Potential (V)	Open Circuit Potential (V)	Mg in Refined Aluminum (wt.%)	Current Efficiency (%)
				Small Cell				
6	A	I	0.40	110	1.60	0.69	0.04	85.5
7	B	I	0.40	110	1.50	–	0.10	85.5
14	C	I	0.92	97	4.86	1.25	0.06	99.7
20	C	I	1.00	90	2.86	0.75	0.13	97.0
25	C	II	0.98	90	3.60	0.85	0.11	102.0
26	C	II	0.98	90	3.60	0.83	0.12	97.2
27	C	II	0.94	90	3.46	0.75	0.13	103.0
31	C	II	0.98	73	3.15	0.75	0.38	99.1
33	C	II	0.98	90	3.36	0.76	0.23	87.2
35	C	I	0.98	90	3.57	0.86	0.11	99.3
				Large Cell				
10	B	I	0.40	110	1.6	0.75	0.04	85.1
11	B	I	0.73	110	2.8	0.75	0.04	85.6
12	B	I	1.10	110	3.6	0.87	0.06	86.0
16	C	II	1.10	100	4.2	1.25	0.18	98.0
17	C	II	1.10	100	4.2	1.25	0.12	125.0
23	C	I	0.88	86	3.56	0.89	0.23	94.1

Temperature = 1025 ± 10 K

(1) Depth: 4.5 cm, Weight: Electrolyte I: 202 g (small cell), 880 g (large cell)
              Electrolyte II: 215 g (small cell), 940 g (large cell)
(2) Anodic Area: 25.5 cm^2 (small cell), 111.2 cm^2 (large cell)
(3) Weight: Small Cell: 585 g, 10 cm depth; Large Cell: 5050 g, 20 cm depth
(4) Percent of faradaic equivalent (of Mg) passed through cell.

Most of the magnesium removed from aluminum was found floating as globules on the electrolyte (Fig. 4), although some were usually seen condensed on the cooler part of the stainless steel vessel. The salt-coated magnesium globules were easily separated from the electrolyte by crushing the electrolyte. The magnesium globules, thus separated, often contained 98.6 wt.% magnesium with the balance aluminum.

Current Density

The effect of current density and the amount of total electrical charge (expressed as percent of the faradaic electrical charge required to remove all magnesium) on current efficiency were studied using both cells. The tests were performed with anodic current density ranging from 0.40 to 1.10 A/cm^2. As can be seen from Table V, the overall current efficiency varies between 85 and 100%. In the majority of the experiments when the amount of charge passed through the cell, small or large, is 90% of the faradaic amount, current efficiency averages 97.0% and magnesium content of the refined aluminum averages 0.15 wt.%. However, the final current efficiency drops to 85% when an excessive amount of charge ($\geq$100% theoretical) is passed through the cells (Expt. Nos. 6, 7, 10, 11 and 12) to reduce the magnesium content of the refined aluminum below 0.15 wt.%, indicating that the current efficiency averages only 40% when magnesium content is being reduced from 0.15 to 0.04 wt.%.

Table V. Effect of Current Density and Amount of the Refining Current on Current Efficiency

Expt. No.	Anodic Current[1] Density (A/cm^2)	Amount of[2] Charge Passed (%)	[Mg] in Refined Al (wt.%)	Current Efficiency (%)
		Small Cell		
6	0.40	110	0.04	85.0
7	0.40	110	0.04	85.0
20	1.00	90	0.13	97.0
25	0.98	90	0.11	102.0
26	0.98	90	0.12	97.0
27	0.94	90	0.14	103.0
31	1.00	73	0.38	99.0
33	0.91	90	0.23	87.0
35	0.98	90	0.11	99.0
		Large Cell		
10	0.41	110	0.04	85.0
11	0.73	110	0.04	86.0
12	1.10	110	0.06	86.0
23	0.88	86	0.23	94.0

(1) Anode Areas: 25.5 cm^2 (small cell), 111.2 cm^2 (large cell)

(2) Percent of the faradaic equivalent (of Mg) passed through the cell.

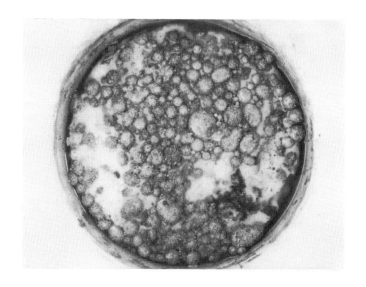

Figure 4. Mg Globules Floating on the Electrolyte.

As the magnesium content of the molten aluminum anode approaches 0.15 wt.% (this value, however, depends on current density and hydrodynamic condition of the melt), more and more current is utilized in transferring aluminum to the cathode according to the following possible reactions.

$$2/3 \text{ Al (anode)} + 2\text{Cl}^-(\text{electrolyte}) \rightarrow 2/3 \text{ AlCl}_3(\text{electrolyte}) + 2\bar{e} \qquad (5)$$

$$2/3 \text{ AlCl}_3(\text{electrolyte}) + \text{Mg(cathode)} \rightarrow \text{MgCl}_2(\text{electrolyte}) + 2/3 \text{ Al(cathode)} \qquad (6)$$

$$\text{MgCl}_2(\text{electrolyte}) + 2\bar{e} \rightarrow \text{Mg(cathode)} + 2\text{Cl}^- \qquad (7)$$

The $AlCl_3$ formed by reaction (5) passes into the electrolyte and reacts with magnesium at the cathode, producing $MgCl_2$ in the electrolyte and aluminum at the cathode. Magnesium chloride, however, is simultaneously reduced at the cathode releasing chloride ion (Cl^-) for reaction (5). During the refining process when the magnesium content of the bulk aluminum drops to 0.04 wt.% (concentration of magnesium at the aluminum-electrolyte interface should be practically negligible), most of the current is utilized in transferring aluminum. Therefore, for efficient performance of the process, magnesium content of the aluminum should be aimed not much below 0.15 wt.%

Electrolyte Density

The effect of electrolyte density was studied in the small cell using electrolytes I and II; a better separation of magnesium globules was seen with electrolyte II. Because of the lower density, some magnesium globules were found submerged in the electrolyte I (Fig. 5) whereas, because of higher density, all the globules were seen floating in electrolyte II.

Electrolyte Resistivity

In a concentration-type cell, most of the applied cell potential is used in overcoming the ohmic resistance of the electrolyte, and, therefore, the specific resistance of the electrolyte must be evaluated accurately. The total cell potential may be split in three parts: i) reversible cell potential, which will be independent of the cell current as long as the cell reactions remain unchanged; ii) concentration overpotential, which initially increases rapidly with increase in cell current and finally becomes constant; and iii) ohmic potential, which varies linearly with the cell current. Therefore, the sum of the constituent potentials would initially increase rapidly with increase in the cell current and eventually become linear as evident in Fig. 6. The slope of the potential-current curve above 5A was, therefore, used to estimate the specific resistivity (0.48 ohm-cm) of the electrolytes; the experimental value compares well with calculated specific resistivities of the first and the second electrolytes -- 0.45 and 0.44 ohm-cm, respectively -- calculated using the specific resistivities of the electrolytes' constituents (16). As expected, the figure also shows that, at a given current, only 0.5 V of the applied potential is used in removing magnesium, the rest is used in overcoming the ohmic resistance of the cell.

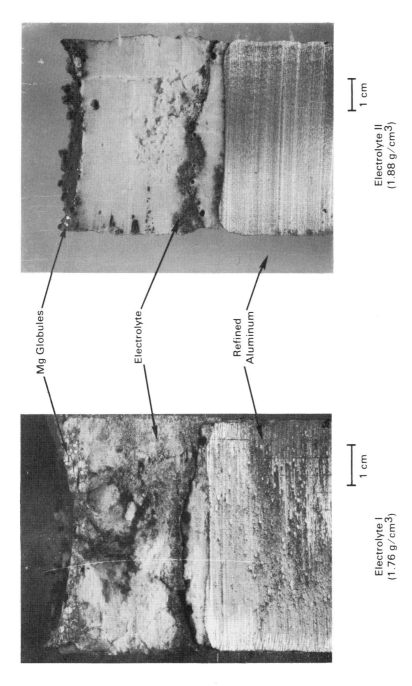

Figure 5. Effect of Electrolyte Density on Separation of Magnesium Globules.

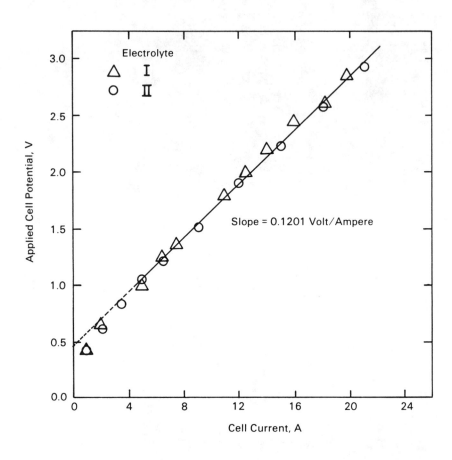

Figure 6. Effect of Cell Current on Applied Cell Potential of Small Cell; Electrolyte Depth 4.5 cm, Cathode Disk Area 18.1 cm^2, 1025 K.

Summary and Conclusions

Two electrolytic cells capable of demagging 0.6 kg and 5 kg aluminum were used to selectively reduce the magnesium content of scrap aluminum to the maximum level, 0.1 wt.%, allowed in the casting alloys, at anodic current density as high as 1.1 A/cm^2 with current efficiency exceeding 85%.

Both selective removal of magnesium from aluminum scrap and scale-up potential of the electrolytic demagging process were established. Experimental results indicated that the open circuit potential of the cell should be kept below 0.9 V to prevent excessive transfer of aluminum to the cathode and the electrolyte. Also, cell efficiency declines drastically when the magnesium content of the refined aluminum is reduced to less than 0.15 wt.%, and a better separation of magnesium globules was achieved with a denser electrolyte.

The electrolytic process appears to cause no environmental problems and recovers magnesium in the form of salt-coated globules. Perhaps these globules could be used to desulfurize iron and steel melts (18) and alloy aluminum (19). With further development, the electrolytic process could become a viable alternative for the removal of magnesium from molten aluminum.

Acknowledgments

The authors wish to acknowledge Ross Johnson of GM Central Foundry Division, Bedford Plant, for taking interest in this work and supplying scrap aluminum for the tests.

References

1. "Aluminum Standards and Data," 194, Aluminum Association (1978).

2. M. C. Mangalick, Light Metals, The Metallurgical Society of AIME, 2, 613 (1974).

3. Alcoa Technology Marketing Division, Report No. 1, June 1972, Pittsburgh, PA.

4. B. L. Tiwari, J. Metals, 34 (9), (1982), pp. 54-58. Also see Light Metals, The Metallurgical Society of AIME, (1982), pp. 889-902.

5. M. R. Smith, Conservation and Recycling, 6 (1-2), (1983), pp. 33-40.

6. M. M. Tsyplakoya and K. L. Strelets, Zhurnal Prikladnoi Khimii, 42 (11), (1969), pp. 2498-2503.

7. A. I. Belyaev, A. Ya. Fisher and A. G. Nikitin, Tsvetn. Metal., 9 (4), (1966), pp. 84-86.

8. A. G. Nikitin, A. I. Belyaev and A. Ya. Fisher. Tsvetn. Metal., 11(4), (1968), pp. 62-64.

9. J. H. Cleland and D. J. Fray, Advances in Extractive Metallurgy, The Institute of Mining and Metallurgy, London, (1977), pp. 141-146.

10. Y. Tsumura, "Demagging Process for Aluminum Alloy without Air Pollution," U.S. Patent No. 4183745, Jan. 1980.

11. H. Burkhardt, "Process for Purifying and Transporting Light Metal," U.S. Patent No. 3335076, Aug. 1967.

12. B. L. Tiwari and R. A. Sharma, <u>J. Metals</u>, <u>36</u>(7), (1984), pp. 41-43. Also see <u>Light Metals</u>, The Metallurgical Society of AIME, (1984), pp. 1205-1216.

13. N. Sevryukov, B. Kuzmin, and Y. Chelischev, "General Metallurgy" translated from the Russian by B. Kuznetsov, Peace Publishers, Moscow.

14. "Phase Diagrams for Ceramists," M. K. Reser, Editor, The American Ceramic Society, 1969.

15. "Phase Diagrams for Ceramists - 1969 Supplement," M. K. Reser, Editor, The American Ceramic Society, 1969.

16. "Molten Salt Handbook," G. J. Janz, Academic Press, New York, 1967.

17. JANAF Thermochemical Tables, Second Edition, NSRDS-NBS-37, June 1971.

18. A. M. Smillie and A. Huber, "Desulfurization of Iron and Steel and Sulfide Shape Control," Edited by W. G. Wilson and A. McLean, Iron and Steel Society of AIME, (1980), p. 77-89.

19. E. Myrbostad and K. Venas, <u>Light Metals</u>, The Metallurgical Society of AIME, (1983), pp. 921-930.

ELECTROCHEMICAL EVALUATION OF ZINC SULPHATE ELECTROLYTE

CONTAINING COBALT, ANTIMONY AND ORGANIC ADDITIVES

Thomas J. O'Keefe* and Mark W. Mateer**

*Professor of Metallurgical Engineering
Materials Research Center
University of Missouri-Rolla
Rolla, Missouri 65401
USA

**Materials Engineer
ARCO
Dallas, Texas 75221
USA

Summary

The effects of cobalt, β-naphthol, glue and antimony on the polarization behavior of zinc sulfate electrolyte were investigated. It was found that β-naphthol was beneficial in reducing the detrimental effects of cobalt while antimony and excess glue were each harmful. Since the nucleation overpotential was insensitive to cobalt, other regions of the polarization curve were used to indicate electrolyte balance. The low current density area occurring in the reverse or anodic sweep direction was often an indicator when an unstable condition existed. The two parameters evaluated were the degree of inflection in the polarization curve in the low current density region during scanning and the rate of current rise observed when the sample was potentiostated at -0.781 Volts (SHE). Using this technique, a two-additive system was shown to be desirable to control cobalt and antimony containing electrolytes. The structure of the deposits was examined using scanning electron microscope techniques.

Introduction

Major advances have been made recently in evaluating the quality of zinc electrolyte by electrochemical methods. Galvanostatic and cyclic voltammetry techniques can indicate the level of active organics, such as glue, in solution based on cathodic overpotential measurements (1,2). The major objective of such measurements is to insure a proper balance between certain harmful impurities, such as antimony, arsenic or germanium and the glue. Certain inorganic impurities cause a depolarizing effect, whereas the organic proteins polarize the initial zinc deposition. In both cases, the degree of overpotential change noted is proportional the relative quantitites of the species present. A good degree of electrolyte control, and a resulting improvement in current efficiency, can be maintained for simpler systems by monitoring the overpotential and adjusting the amounts of additives accordingly.

The measurement of a single overpotential value is adequate where a single additive, such as glue, is used to counteract a member of the group of impurities which cause a depolarizing effect on the initial activation overpotential. Unfortunately, not all of the harmful impurities affect the polarization behavior in this way. The class of impurities with low hydrogen overvoltages (Ni, Co, Cu, Ag, etc.) do not influence, to any degree, the initial overpotential value. Their presence can have very undesirable effects on current efficiency in certain instances and additional reagents are sometimes used to reduce the active concentration of these impurities in solution. An example would be the use of β-naphthol in electrolyte with high levels of cobalt (3). Thus, as the number of additives that are used in the system increases, the ability to estimate ultimate behavior during electrolysis becomes increasingly complex. The objective of this research was to address the problems associated with the electrochemical characterization of a multiple additive system. The specific goals were to determine the polarization behavior of zinc electrolyte containing various combinations of glue, antimony, cobalt and β-naphthol and the possibility of finding a means of indicating a properly balanced electrolyte to supplement the initial or nucleation overpotential value now used.

Experimental

Solution Preparation

A stock solution of neutral (pH = 4.0 - 5.0) purified zinc sulphate was obtained from Cominco Ltd. of Trail, B.C. for use in this study, see analysis in Table I.

Table I. Analysis of Neutral Purified Solution ($mg\ell^{-1}$)

Zn	146000.
Mn	2060.
Cd	0.3
Sb	0.03
Pb	1.8
Fe	5.5
Ge	<0.01
Co	0.4
Ni	0.35
Cl^-	75.
F^-	1.3

Cobalt stock solution was prepared by dissolving cobaltous sulphate in distilled water to a concentration of 1000 mgℓ^{-1}. β-naphthol was dissolved in dilute NaOH to obtain a 2000 mgℓ^{-1} stock solution. Solid pearl glue was dissolved in distilled water to a concentration of 1000 mgℓ^{-1}. To obtain 10 mgℓ^{-1} antimony, antimony potassium tartrate was dissolved in distilled water. The organic additives were refrigerated to prevent decomposition and new stock solutions were made at regular intervals.

The test solutions of 50 gℓ^{-1} zinc and 150 gℓ^{-1} H_2SO_4 plus additives were prepared by taking the required amount of neutral purified solution and reagent grade sulphuric acid, mixing them with distilled water, then including the additive solution and adjusting to 200 mℓ volume.

Cyclic Voltammetry

The cyclic voltammetry experiments were conducted in a pyrex "H" cell employing an Aℓ working electrode, a Pt counter electrode and a mercurous sulphate reference electrode. The electrolyte for the reference electrode consisted of 200 gℓ^{-1} H_2SO_4, giving a potential of +0.669 V versus SHE. A constant temperature bath was used to maintain the desired temperature of 35°C. A Petrolite Potentiodyne Analyzer (Model M-4100) was used to generate the cyclic voltammograms, which were recorded as a log current density versus potential plot. The experimental apparatus has been described previously (4).

The working electrodes were prepared by polishing on 600 grit paper, water washing in an ultrasonic cleaner after polishing and then rinsing in distilled water. The electrodes were then placed in the "H" cell and the potential scan started immediately.

The voltammograms were obtained by varying the potential from -0.631 V versus SHE_2 to a more cathodic potential where the total current density was 50 mA/cm^{-2}. At this point, the direction was reversed, sometimes after holding at the high current density, and driven to the original starting potential. A scan rate of 0.5 mV sec^{-1} was used as it gave the best results for the tests being conducted.

In addition to the transient studies, tests were performed similar to those previously described except the cycle was stopped and potentiostated at -0.781 V versus SHE for various time intervals (up to 10 minutes) during the reverse or anodic sweep. The change in current with time was monitored at this potential and morphologies of the deposits were checked using SEM techniques.

Results and Discussion

Cobalt Effects

The nature of the polarization curves of relatively pure zinc sulphate electrolytes has been discussed previously (4,5). For that reason, only solutions containing a significant amount of cobalt and other additives are considered.

The polarization curve for an electrolyte containing 50 gℓ^{-1} zinc, 150 gℓ^{-1} H_2SO_4 and 5 mgℓ^{-1} cobalt is shown in Figure 1. The cathodic sweep begins at a potential of -0.631 V versus SHE and is labeled point A. The current increases up to point B where it is thought the adsorption of zinc ions onto the surface of the cathode restricts further reaction of hydrogen

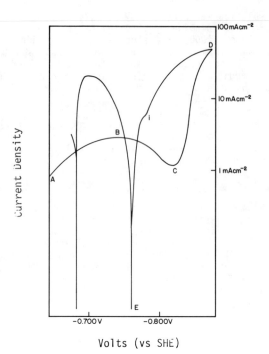

Figure 1. Cyclic voltammogram for zinc sulphate electrolyte containing 50 gℓ^{-1} Zn, 150 gℓ^{-1} H_2SO_4 and 5 mgℓ^{-1} cobalt.

ions or other reducible species in solution. The current decreases until point C where the reduction to metallic zinc begins. When the cathodic current density reaches 50 mA/cm^{-2}, called point D, the sweep direction is reversed and the current remains cathodic until -0.761 V versus SHE where the total current becomes anodic. The latter potential approximates a pseudo-equilibrium potential for the Zn/Zn^{+2} couple, and indicates the point at which anodic zinc dissolution predominates over any cathodic reaction, such as hydrogen evolution. Continuing the sweep produces an anodic current which persists until all of the previously deposited zinc has been oxidized into the solution.

The initial polarization behavior is observed to be independent of cobalt concentration in the range studied, which was zero to ten ppm. The deposition overpotential for solutions containing 0.5 or 10 ppm Co is approximately -0.840 V.

One noticeable effect of cobalt in the electrolyte is the appearance of an inflection in the curve, labeled point i, to indicate a region of possible instability for zinc deposition. No such inflections are present in curves generated using the purer reference solutions. When the inflection is present, there appears to be an increase in hydrogen evolution from the deposit and the bubbles appear to have a much smaller diameter.

The region of instability was investigated more extensively using a continuous zinc film deposited at 50 mA/cm^{-2} for two minutes, then allowing the potential to go to point i and holding for a fixed time interval.

Figure 2 shows the structure of a deposit made from an electrolyte, to which no cobalt was added, after holding two minutes at point i. As can be seen, there is little evidence of corrosion or deterioration of the zinc, as the crystals are well formed, with sharp edges, and have the appearance of a normal deposition structure. When 5 mgℓ^{-1} of cobalt is added and a similar testing procedure is followed, a structure such as shown in Figure 3 is obtained. Definite signs of dissolution or zinc corrosion are

Figure 2. Zinc deposit obtained for electrolyte containing 50 gℓ^{-1} Zn and 150 gℓ^{-1} H$_2$SO$_4$. Potential held at point i for 2 min. (3000X).

Figure 3. Zinc deposit obtained for electrolyte containing 50 gℓ^{-1} Zn, 150 gℓ^{-1} H$_2$SO$_4$ and 5 mgℓ^{-1} cobalt. Potential held at point i for 2 min. (3000X).

revealed with large irregular pits present between the crystals, which now show less distinct morphological features. Deposits held for similar short times at more reducing potentials, outside the region of instability, show no signs of undergoing such resolution in the presence of cobalt. The importance of modifications in the polarization curve in the lower current density regions is thus clearly shown. When a curve inflection is obtained, the zinc deposit can be expected to begin actively dissolving at potentials that are usually sufficiently cathodic to maintain zinc deposition.

Antimony and Glue Effects

When antimony is present in an active form, an initial depolarization during the cathodic sweep and lowered current efficiency are noted, as reported previously (4-7). The addition of low levels of antimony, 10-80 parts per billion, to the reference electrolyte also caused a slight inflection in the polarization curve at a similar potential to that noted for cobalt.

For solutions containing both cobalt and antimony, a distinct peak occurred in the instability region and was accompanied by an increase in hydrogen evolution. The peak is shown in curve A of Figure 4. From all indications, this peak is evidence of an undesirable synergism that results from cobalt-antimony interactions. It seems noteworthy that the condition is more pronounced at low current densities, or regions on the cathode surface that would be at lower negative potentials than the bulk electrode. In Figure 5 the extensive resolution resulting from the combination of antimony and cobalt is illustrated. The height of the current peak and the amount of hydrogen evolved are directly dependent on the quantity of cobalt in solution if the antimony content is held constant. Thus, as the cobalt content of the solution increases, there is a steady increase in the height of the peak.

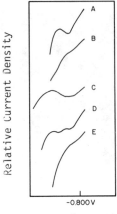

Figure 4. Low current density region of a zinc sulphate cyclic voltammogram containing 50 gℓ^{-1} zinc, 150 gℓ^{-1} H_2SO_4 and A) 5 mgℓ^{-1} cobalt and 0.08 mgℓ^{-1} antimony; B) 5 mgℓ^{-1} cobalt, 10 mgℓ^{-1} glue and 0.08 mgℓ^{-1} antimony; C) 5 mgℓ^{-1} cobalt and 20 mgℓ^{-1} glue; D) 5 mgℓ^{-1} cobalt, 0.08 mgℓ^{-1} antimony and 5 mgℓ^{-1} β-naphthol; E) 5 mgℓ^{-1} cobalt and 40 mgℓ^{-1} β-naphthol.

Figure 5. Zinc deposit obtained for electrolyte containing 50 gℓ^{-1} Zn, 150 gℓ^{-1} H_2SO_4, 0.08 mgℓ^{-1} Sb and 10 mgℓ^{-1} cobalt. Potential held at point i for 2 min. (3000X).

Glue is commonly used to reduce the damaging effect of antimony on current efficiency. The apparent detrimental interaction between antimony and cobalt can also be decreased by adding glue to the solution. The peak is diminished until it is again only an inflection, while the initial overpotential is polarized to a value indicative of a balanced glue-antimony combination (8). Curve B of Figure 4 shows the decrease in this peak when glue is present. The zinc is finer grained and although there is some pitting, it is much less pronounced than when no glue is added to the cobalt-antimony mixture.

Accompanying the improvement in the structure of the deposit and the decrease in the degree of instability seen on the polarization curve is a corresponding increase in the initial overpotential of the zinc. For a given amount of cobalt, the glue was effective in preventing excessive zinc dissolution and hydrogen evolution in the region of instability. The glue is thought to counteract any residual antimony present, thus minimizing the synergism between cobalt and antimony. A proper balance between antimony and glue is therefore still necessary and is determined by measuring the initial overpotential.

The existence of the inflection point in curve B of Figure 4 indicates that cobalt remains active even when glue is present. Continued additions of glue do not eliminate the inflection, indicating that glue is not sufficient to counter the effects of the higher cobalt levels used in this study. The resulting deposits are characterized by a structure typical of glue additions with many small pits evident over the entire deposit (see Figure 6). The polarization curves showed the inflection expanding into a broad peak, extending well into potential regions normally characterized by anodic zinc currents. There was also a large increase visible in the amount of hydrogen evolved in this region. With 5 mgℓ^{-1} cobalt and 20 mgℓ^{-1} glue, the peak is so broad that no anodic zinc current was detected, a condition resulting from the partial cathodic current exceeding that for the anodic zinc dissolution. Curve C in Figure 4 shows the back sweep portion of the polarization curve made from electrolyte to which 5 mgℓ^{-1} cobalt and 20 mgℓ^{-1} glue are added.

Figure 6. Zinc deposit obtained for electrolyte containing 50 gℓ^{-1} Zn, 150 gℓ^{-1} H$_2$SO$_4$, 5 mgℓ^{-1} glue and 10 mgℓ^{-1} cobalt. Potential held at point 1 for 2 min. (3000X).

Improper glue contents are thus seen to create a broad potential range of apparent deposit instability, a condition capable of maintaining moderately high cathodic current densities in a region where zinc will dissolve. Hydrogen evolution is thought to be the primary reduction reaction occurring but the actual mechanism responsible, or the reason for enhanced hydrogen ion reduction when excessive glue is present with cobalt, is not known. Additional research is necessary in order to gain more insight into the interaction of glue and cobalt.

At first appearance, these results seem to conflict with previous work on cobalt which indicated reduced activity of cobalt and improved current efficiency with the addition of glue (6,8). The nature of the zinc electrolytes used and the sensitivity of the polarization behavior to minute differences in concentrations of impurities such as antimony, germanium and arsenic must be considered in each case. The strong dependence of the cobalt effect on antimony level has been exhibited and it becomes evident that the composition of the electrolyte becomes the deciding factor in determining the effects of additives. An electrolyte containing cobalt and antimony could show an improved deposit after a glue addition, whereas if the antimony were initially absent the glue appears to be harmful. A knowledge of the initial polarization potential is necessary to maintain proper glue-antimony balance but not sufficient to guarantee electrolyte quality or good current efficiency under all conditions. However, one other portion of the curve which offers some insight into the system behavior appears to be the low current density region encountered in the reverse sweep direction. In all instances where apparent anomalous or differences in results are encountered, the structure and morphology influences should be considered. Thus, variations in deposit characteristics should always be included in any evaluation of electrolyte impurities, as the results are often very structure sensitive.

Effects of β-naphthol

An attempt was made to make use of the region of instability in providing an additional indicator of the existence and severity of impurities capable of catalytic hydrogen evolution, an example of which is cobalt. If the above hypothesis is correct, then the presence of an additive capable of counteracting cobalt should show a decrease in observed current in the area of instability.

The use of β-naphthol and α-nitroso β-naphthol in the removal of cobalt ions from solution is well known (3,9). At the lower pH range of the electrolytic cells, however, α-nitroso β-naphthol is reported to lose its effectiveness while β-naphthol does not.

The addition of β-naphthol to the starting electrolyte causes a slight amount of polarization and grain refinement, similar to but to a much lesser degree than glue. Nucleation is increased with a corresponding decrease in crystal size and the appearance of a more uniform deposit. There is also no change seen in the polarization curve at point i and the zinc electrode potential at point E (cathodic to anodic current) is unchanged. If cobalt and β-naphthol are added in equal concentrations, the β-naphthol again acts in the manner of a small glue addition. The deposit is refined and the pits are much smaller but more numerous, making an estimate of differences in pitting with and without β-naphthol very difficult. The hydrogen peak on the polarization curve is also extended to slightly more positive potentials.

An improvement is seen when the β-naphthol is increased relative to the cobalt. Figure 7 shows a deposit with 5 mgℓ^{-1} cobalt and 20 mgℓ^{-1} β-naphthol. Though some improvement over Figure 3 is seen, the pitting is still very evident. If 40 mgℓ^{-1} β-naphthol is added to 5 mgℓ^{-1} cobalt, the deposit improves greatly. The pits are almost eliminated and the crystals are well formed, as seen in Figure 8. Curves D and E of Figure 4 represent the effects of β-naphthol additions on the polarization curves in the presence and absence of antimony respectively. In each case, the effect of the organic is to reduce the hydrogen peak. Curve D shows the flattening of the peak seen in Curve A into a plateau, reducing the maximum current but still indicating an unstable condition of some form. Curve E indicates how the inflection shown in Figure 1 can be reduced with β-naphthol. Thus β-naphthol seems effective for reducing the influence of cobalt, but does not completely eliminate the undesirable characteristics noted in the low current density region.

Figure 7. Zinc deposit obtained for electrolyte containing 50 gℓ^{-1} Zn, 150 gℓ^{-1} H$_2$SO$_4$, 5 mgℓ^{-1} cobalt and 20 mgℓ^{-1} β-naphthol. Potential held at point i for 2 min. (3000X).

Figure 8. Zinc deposit obtained for electrolyte containing 50 gℓ^{-1} Zn, 150 gℓ^{-1} H$_2$SO$_4$, 5 mgℓ^{-1} cobalt and 40 mgℓ^{-1} β-naphthol. Potential held at point i for 2 min. (3000X).

Potentiostatic Studies

Because the cobalt can best be detected from the polarization curve in the region of instability, a method was sought to measure the effect of the impurity in this area. When the deposits are held at point i, the current is observed to rise with time. The amount of increase seems to be directly related to the magnitude of the hydrogen peak on the polarization curve and the extent of the dissolution seen on the deposit.

The dissolution mechanism is thought to be one involving local cell corrosion with cobalt acting as a cathode (2). As the potential of the deposit becomes more positive, the zinc deposition current decreases and is less able to cover and suppress the codeposited cobalt as a site for hydrogen discharge. The potential of point i, -0.781 V versus SHE, is assumed to be the potential at which the impurity activated hydrogen discharge becomes favorable under the conditions used in this study. The enhanced hydrogen evolution provides sufficient shifting in the partial polarization curves to cause the zinc reaction to change from deposition to dissolution.

The deposit morphology is also important in determining the onset of this condition as the low points will become electron deficient and tend to become anodic, while cathodes will appear on the upper surface, particularly those rich in cobalt. A reaction sequence of this type could be expected to generate a pitting type of surface corrosion defect. As zinc dissolves, more cobalt will be exposed, providing more cathodic area to increase even further the rate of pit formation and growth. The effect is somewhat autocatalytic in nature. The localized influence of increased hydrogen gas evolution on mass transport may also accelerate or modify the nature of the reactions occurring.

The rise in current upon holding at point i is thought to be due to increased hydrogen evolution caused by the low overvoltage sites provided by the cobalt using all of the applied direct current to discharge hydrogen. The current rise should then be directly related to the amount of active cobalt present on the surface, giving an indication of the extent of possible local cell corrosion on the deposit and the overall quality of the electrolyte.

By using the current rise technique as a measure of stability, the effect of cobalt and other impurities on the behavior of the zinc electrolysis can be more clearly observed. Table II lists selected current rise data obtained experimentally for a number of different combinations of additives.

Table II. Current Rise Resulting from Potentiostating Selected Deposits at -0.781 V (SHE) for 1 Minute

	Solution Components			Current Rise mA/cm^2
Co ppm	Sb^{+3} ppb	Glue ppm	β-naphthol ppm	
0	0	0	0	0.2
10	0	0	0	10.0
0	80	0	0	1.0
0	0	0	5-40	0.1
0	0	30	0	0.6
10	0	0	20	7.0
10	0	10	0	15.0
10	80	0	0	19.0
10	60	10	0	14.0
10	60	10	40	11.0
10	60	10	80	4.5

Antimony additions cause a slight increase in current, but the magnitude is not nearly indicative of the expected effect that 0.08 mg ℓ^{-1} antimony would have on the deposit. However, it must be remembered that the presence of this concentration of antimony is readily detected by monitoring the initial polarization. The overall results seem to indicate that the mechanism involved for the cobalt and the antimony may be substantially different, with cobalt being the type of impurity that catalyzes H_2 evolution because of overvoltage considerations. Since no large increase in cathodic current occurs when antimony is present, a large increase in hydrogen evolution due to a catalyzed reaction on a low overpotential surface would not be expected. The effect of antimony is not to be assumed to be insignificant just because of the lower current rise. More likely, the response may be due to a different mechanism responsible for the dissolution of the zinc cathode.

In contrast to antimony, the addition of cobalt to the electrolyte always produces a significant current rise. Table II shows that of the single additions made to the cobalt containing solutions, only β-naphthol was capable of reducing the current rise. For both 5 and 10 mg ℓ^{-1} cobalt, additions of β-naphthol to these solutions give a gradual decrease in current. The last three entries of Table II show the ability of β-naphthol to reduce the action of the cobalt. The current rise is reduced considerably but not eliminated by the additions of β-naphthol, even at 80 mg ℓ^{-1} concentration. Although a proper balance of glue and antimony can also show an improvement individually, both cause increases in current.

The indication is that when these two are balanced, the synergism that each has with cobalt when present individually, is minimized.

In a few cases, a balanced glue-antimony solution with lower cobalt contents gives the smallest current rise, even better than when β-naphthol is added to the electrolyte. The β-naphthol does help considerably when a higher concentration, such as 10 mgℓ^{-1} cobalt, is present. The test appears to detect the differences in active cobalt caused by the additive more readily when the initial concentration of cobalt is higher. The research also indicated the need to attain a threshold amount of an additive such as β-naphthol before an improvement is noted. Beyond this threshold, increased additions of β-naphthol cause continuing decreases in current rise.

The current rise data show the strong, undesirable effects of antimony and glue, individually, in cobalt containing solutions. The results might be expected for antimony, but the glue influence was not expected. In solutions without cobalt, a slight current rise might be predicted because excess glue, in the 20 mgℓ^{-1} range, is known to cause decreased current efficiency. However, the current rise increases by a factor of three when 20 mgℓ^{-1} glue is added to a solution containing 5 mgℓ^{-1} cobalt, showing an undesirable synergism between these two, at least in the low current density range. Overall, an evaluation of the current rise obtained upon potentiostating in the instability region seems to provide an additional valuable piece of data on the electrolyte quality, particularly where multiple additives such as antimony-glue-β-naphthol are employed.

The beneficial effects of β-naphthol can also be shown by means of Figure 9, which shows a plot of three different cobalt containing solutions and their response to the addition of β-naphthol. It appears that the organic reduces the activity of the cobalt up to a ratio of approximately 8 mgℓ^{-1} β-naphthol to 1 mgℓ^{-1} cobalt. As could be expected, the best combination of deposit structure and current rise from a cobalt containing solution results from a large addition of β-naphthol, while the initial overpotential had been balanced into the optimum range for the starting electrolyte. This deposit is shown in Figure 10. The current rise for this deposit is 1.5 mA/cm^{-2}, as compared to 4 mA/cm^{-2} for cobalt alone.

Additional interpretation is needed to determine an exact quantitative meaning of the current rise data. The manner in which the current rise tests were conducted makes the technique predict large changes in deposit behavior but is inconclusive for very small changes. The implication is then that the final judgement on the acceptability of an electrolyte must not be made from the current rise data only. However, it does seem to be possible using this method to judge the effectiveness of corrective measures taken to improve electrolytes containing multiple additives.

Nucleation Overpotential Measurements

Attempts were made to correlate the initial overpotential obtained at 10 mA/cm^2 with the current rise data. One problem encountered was the greater difficulty in obtaining consistently reproducible values with the mixed system than experienced previously with less complex electrolytes. There did seem to be some trends, but the lack of sensitivity of this particular group of impurities and additives to overpotential was the reason for searching in other regions of the polarization curve for an indication of electrochemical activity.

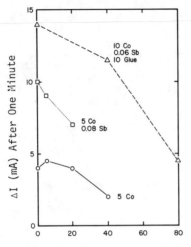

Figure 9. Current rise upon holding for one minute versus concentration of β-naphthol.

Figure 10. Zinc deposit obtained for electrolyte containing 50 gℓ^{-1} Zn, 150 gℓ^{-1} H_2SO_4, 5 mgℓ^{-1} cobalt, 10 mgℓ^{-1} glue, 0.08 mgℓ^{-1} Sb and 40 mgℓ^{-1} β-naphthol. Potential held at point i for 2 min. (3000X).

The β-naphthol causes only a slight amount of polarization for the concentration range studied in this investigation. As noted previously, Co gave a similar small response, thus making it more difficult to rely solely on a measurement of overpotential to indicate impurity activity. The antimony and glue additions gave their usual responses of depolarization and polarization respectively. Very roughly, there did appear to be an optimum overpotential for the various combinations of species. However, what seemed to evolve was not a single relationship between overpotential and current rise, but rather a series of curves with each curve representing a basic type of electrolyte. For example, the response of a 5 ppm Co and a 10 ppm Co electrolyte to the various additives and impurities was similar, but at the same overpotential the higher Co containing electrolyte gave a higher current rise. Thus, it appears that a characterization of a mixed electrolyte using overpotential might well be possible, but additional work is needed to more clearly establish the rather complicated interactions occurring in solution and their effects on the zinc and the efficiency of the deposition.

Conclusions

None of the measures taken in this study were able to return a cobalt containing electrolyte to the polarization behavior exhibited by the reference solution. Altering certain process conditions, such as decreasing the acid content, lowering the temperature or changing the current density used to form the deposit could possibly help to diminish the differences observed. In any event, β-naphthol and cobalt do not appear to be similar to glue and antimony where a proper organic addition can completely eliminate the effect of the impurity.

In addition, it must be recognized that this is a short duration test and the evaluation is being made on a very thin deposit. Other overriding factors may have to be considered when attempts are made to extrapolate the results to standard, long term depositions. For the most part, the experiments were not only conducted to find a means of monitoring the electrolyte quality, but also to gain a better understanding of the fundamentals underlying the system. The results show that different parts of the cyclic voltammogram might be used in gaining a better understanding of the system, while simultaneously leading to the eventual development of a more comprehensive method of electrolyte control. The possibility of incorporating additional steps, such as the potentiostated current rise technique, into the standard polarization measurements now employed appear promising and probably warrant additional, in-depth studies. Even when other regions of the curve are used, it would still appear to be desirable to monitor and maintain an optimum initial nucleation overpotential value.

References

1. R. C. Kerby, H. E. Jackson, T. J. O'Keefe, and Y. M. Wang, Met. Trans., 8B (1977) p. 661.
2. R. C. Kerby and C. J. Krauss, Lead-Zinc-Tin '80, Ed., J. M. Cigan, T. S. Mackey and T. J. O'Keefe, Las Vegas, NV, AIME 187.
3. M. G. Kershaw and C. J. Haigh, Aus. I.M.M. Conference, Tasmania, p. 308, May 1977.
4. B. A. Lamping and T. J. O'Keefe, Met. Trans., Vol. 7B, Dec. 1976, p. 551.
5. D. R. Fosnacht and T. J. O'Keefe, J. Appl. Electrochem., 10 (1980) 495.
6. G. C. Bratt, Aus. I.M.M. Conference, Tasmania, p. 277, May 1977.
7. D. J. Mackinnon and J. M. Brannen, J. Appl. Electrochem., 7 (1977) 451.
8. D. R. Fosnacht, Ph.D. Dissertation, University of Missouri-Rolla, Rolla, MO (1978).
9. I. M. Kolthoff and E. Jacobsen, J. Amer. Chem. Soc., Vol. 79, (1957) 3677.

PYROHYDROLYSIS OF NICKEL CHLORIDE SOLUTION IN A 30-INCH DIAMETER FLUIDIZED-BED REACTOR

Mahesh C. Jha, Bruce J. Sabacky, and Gustavo A. Meyer*

AMAX Extractive Research & Development, Inc.
Golden, Colorado 80403

Abstract

AMAX has developed a chloride refining process to produce pure nickel and cobalt metals from mixed sulfide precipitates. Metal chloride solutions, generated in a hydrochloric acid leaching step and purified in a solvent extraction circuit, are converted to metal oxides and HCl gas by pyrohydrolysis. A pilot plant consisting of a 30-inch diameter fluidized bed reactor and other ancillary units was designed, constructed, and operated for over one year. Process design was aimed at minimizing energy consumption, maximizing conversion efficiency, and achieving simplicity of operation and control. The operation provided valuable information on process scale-up parameters, process control strategies, and materials of construction. Several process variables such as bed temperature, solution composition, solution feeding system, bed particle size distribution, and fines recycle system were investigated to determine their effects on process efficiency and the physical and chemical characteristics of the product.

*Now with Newmont Exploration Ltd., 44 Briar Ridge Road, P.O. Box 310, Danbury, Connecticut 06810

Introduction

During the last decade, AMAX developed and demonstrated a novel hydrometallurgical process for the production of high purity nickel and cobalt metals from low grade oxide ores such as laterites. The process consists of leaching, concentration, and refining steps. The first two steps constitute the AMAX Acid Leach Process which is described in numerous U.S. Patents and publications (1,2,3). The refining step, referred to as the AMAX Chloride Refining Process, is disclosed in U.S. Patent 4,214,901 (4) and described in a paper contained in the proceedings of the Third International Hydrometallurgy Symposium (5).

The AMAX Chloride Refining Process consists of three major steps: leaching of sulfide concentrate with hydrochloric acid, solution purification by solvent extraction, and conversion of the purified chloride solution to metal oxide and HCl gas by pyrohydrolysis. Details of the first two steps have been described in previous publications (5,6,7). The objective of this paper is to describe the pilot plant demonstration of the pyrohydrolysis step of the refining process. A 76-cm (30-inch) diameter fluid bed pyrohydrolysis unit (also called hydrolyzer in this and other publications) was designed and constructed as an integral part of the AMAX Chloride Refining Pilot Plant. The operation of this plant for over a year proved the fluid bed pyrohydrolysis step to be a viable method for the conversion of nickel chloride solutions to low chlorine, low sulfur nickel oxides in a controllable range of particle sizes from one millimeter granules to a relatively fine powder.

The paper begins with a brief discussion of the principles underlying nickel chloride pyrohydrolysis and the reasons for selecting a fluid bed reactor for conversion of nickel chloride solutions. The objectives for design and operation of the pyrohydrolysis pilot plant are presented next followed by a description of the plant. Operating experiences are then described, and finally, performance of the fluid bed pyrohydrolysis unit is discussed in terms of process variables such as fluid bed temperature, feed solution composition, and fluid bed particle size distribution and their effects on fuel consumption and conversion efficiency.

Chloride Pyrohydrolysis in Nickel Refining

There were two major reasons for selecting a chloride route to refine sulfide concentrates produced by the AMAX Acid Leach process. First of all, sulfur in the concentrates is recovered as SO_2 or H_2S, both of which are required reagents in the ore leaching and concentration steps. No sulfate waste streams are produced. And secondly, the chloride system offers a conveniently accomplished means of purification by solvent extraction as disclosed in U.S. Patent 4,221,765 (8). The AMAX Chloride Refining Process was thus developed to capitalize on these two advantages and possibly introduce further improvements on existing chloride refining technology utilized by nickel producers such as Falconbridge, Nippon, and Sumitomo (9,10,11).

The pyrohydrolysis of nickel chloride solution is an integral part of the AMAX Chloride Refining Process. Incorporation of this unit operation in the chloride refining flowsheet results in several important advantages over other chloride refining schemes.

1. Pyrohydrolysis of purified chloride solutions directly produces HCl instead of Cl_2 produced in the electrolysis step of other processes(11,12). This eliminates the need for conversion of Cl_2 to HCl prior to its recycle to the dissolution step.

2. The pyrohydrolysis reactor receives concentrated nickel chloride solution instead of hydrated crystals as in the Falconbridge process(9). This results in a mechanically simple and easily controlled unit operation. Off-gases from the reactor are quenched with purified solution from the solvent extraction section, thus utilizing the sensible heat leaving the reactor to concentrate the feed solution.

3. The pyrohydrolysis reactor can convert solutions varying in impurity content and thus produce oxide ranging from utility grade to Class I specification. This flexibility allows response to market demands not attainable with the electrolytic or carbonyl refining methods which are confined to the production of higher grades of metal.

4. A fluid bed pyrohydrolysis reactor can be controlled to produce an oxide product of desired particle size distribution according to market demands or the requirements of subsequent processing such as reduction to metal. This flexibility is achieved through the use of an open or closed circuit dust and fines recycle system and will be discussed in a later section of this paper.

Figure 1 shows a schematic diagram for the AMAX Chloride Refining Process. The diagram depicts how off-gases from the pyrohydrolysis step are used to concentrate the purified feed solution from the solvent extraction step. It also shows how HCl regenerated in the pyrohydrolysis section is recycled to the dissolution step, resulting in virtual elimination of any waste stream from the refinery.

Thermodynamics of Nickel Chloride Pyrohydrolysis

The reaction for pyrohydrolysis of nickel chloride is shown below. With increasing temperature, the reaction proceeds favorably in the direction shown and at $800°C$, the standard free energy change for the reaction is about -4500 cal/g-mole $NiCl_2$

$$NiCl_2 + H_2O \longrightarrow NiO + 2HCl$$

The effect of temperature on the free energy change of reaction as well as on the free energy of formation of reaction components is shown in Figure 2. The data for nickel compounds are taken from a Bureau of Mines compilation devoted entirely to nickel(13). The data for other compounds are taken from a standard text(14). The pyrohydrolysis reaction is endothermic. If nickel chloride solutions are processed, energy for the evaporation of water and residual HCl as well as for the pyrohydrolysis reaction has to be supplied to the reactor.

If combustion of fuel oil (86% C and 14% H) and pyrohydrolysis of a 45 weight percent $NiCl_2$ solution proceed in a single reactor, at least 9 moles of water vapor and 19 moles of inert gases (CO_2, N_2, and excess O_2) will be generated for every mole of $NiCl_2$ processed. If vaporized, the nickel chloride would exert a partial pressure of 29 mm Hg at a reactor

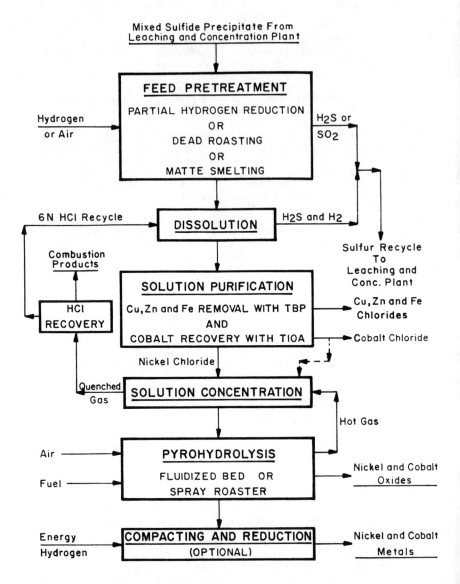

Figure 1. Schematic Diagram of the AMAX Chloride Refining Process

pressure of 700 mm Hg (atmospheric pressure in Golden, Colorado). As shown in Figure 2, this pressure is below the vapor pressure of nickel chloride at temperatures above 800°C. This analysis indicates that the pyrohydrolysis of nickel chloride probably proceeds via a vapor phase reaction with the formation of nickel oxide on existing surfaces by a heterogeneous mechanism or with homogeneous nucleation of a very fine NiO powder.

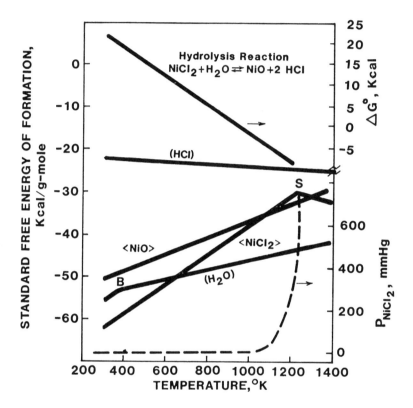

Figure 2. Thermodynamic Data for Pyrohydrolysis of Nickel Chloride

When given the standard free energy of reaction for pyrohydrolysis and the standard free energy of sublimation for $NiCl_2$, an equilibrium constant can be determined for the relationship shown below:

$$K = \frac{P^2_{HCl}}{P_{NiCl_2} \cdot P_{H_2O}}$$

A conversion fraction X can then be determined for the process conditions stated above

$$K = \frac{4X^2}{(1-X)(9-X)}$$

Thus, for a temperature of about 850°C (K=15), a 97 percent conversion at equilibrium can be expected with a small increase in the conversion fraction for increasing temperature. As a result, changes in reactor temperature between 800 and 900°C should have only a minor effect on reactor performance with respect to thermodynamic considerations. An increase in temperature may, however, enhance the formation of fine NiO via a homogeneous reaction mechanism.

Material and Energy Balances for Nickel Chloride Pyrohydrolysis

Material and energy balance diagrams for the chloride pyrohydrolysis process are shown in Figures 3 and 4, respectively. The primary component of the process is the pyrohydrolysis reactor which operates at conditions necessary to drive the conversion reaction to the right as explained in the previous section. Fuel and combustion air are reacted to provide sufficient energy at the required reaction temperature (800-900°C) to evaporate water and HCl in the feed solution, hydrolyze the metal chloride, and compensate for heat losses. Solid oxide product and gaseous HCl, H_2O, and combustion products exit the reactor at its operating temperature.

Although the pyrohydrolysis reactor accomplishes the required conversion of chloride solution to solid oxide product, additional process components are needed to utilize the sensible heat in the gas stream leaving the reactor, concentrate the feed solution, recover hydrochloric acid, and scrub off-gases prior to venting them to the atmosphere. Concentrating incoming solution from the solvent extraction section is accomplished by contacting it with reactor off-gases in a quenching system. In doing so, water and HCl are evaporated to produce a reactor feed of maximum metal chloride concentration and reactor off-gases are cooled to about 90-95°C. Gases leaving the quenching system contain all the products of fuel combustion plus essentially all HCl and H_2O present in the metal chloride solution coming from the solvent extraction section. (See Figure 1)

Gases from the quenching system are passed through an adiabatic absorber where they are contacted with a countercurrent flow of process water from the scrubbing system. Hydrochloric acid of azeotropic concentration (20 weight percent HCl) is produced. Gases leaving the absorber are scrubbed with process water recirculating through a heat exchanger. Most of the energy released in the combustion of fuel in the pyrohydrolysis reactor is expelled through the heat exchanger. The temperature of the scrubbing solution is controlled to condense the amount of water required for production of acid in the absorber. Gases leaving the scrubber are saturated with H_2O and contain only trace amounts of HCl.

Two parameters, relating to the material and energy balances, that have the most significant effect on the process economics, are discussed below.

Conversion Efficiency. The ratio of metal recovered as oxide to metal fed to the reactor as concentrated chloride solution is the reactor conversion efficiency. A conversion efficiency of 100 percent could be

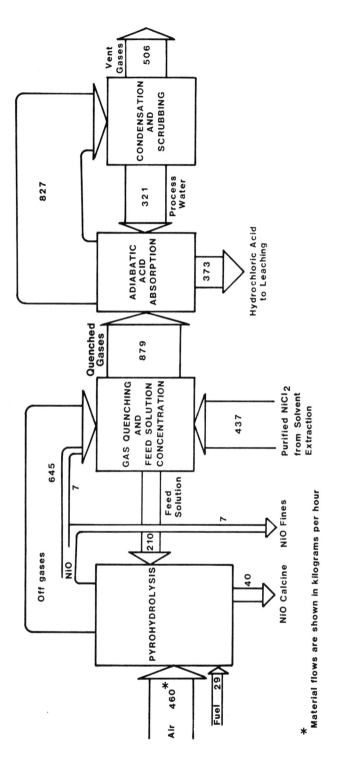

Figure 3. Material Balance for Nickel Chloride Pyrohydrolysis

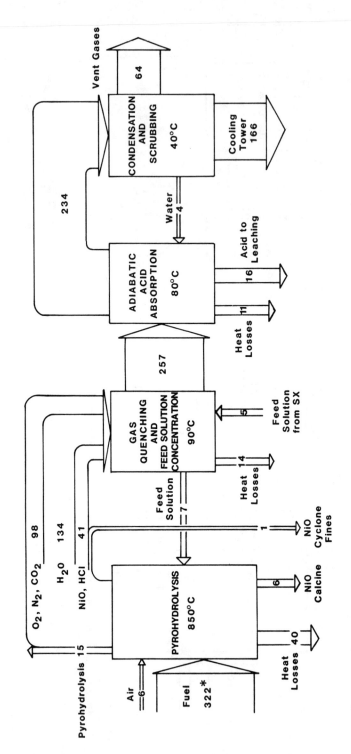

Figure 4. Energy Balance for Nickel Chloride Pyrohydrolysis

*Energy flows are shown in millions of calories per hour

achieved if all metal chloride in the feed solution was converted to oxide and none of the oxide escaped with off-gases and redissolved in the quenching circuit. While thermodynamics would indicate the possibility of attaining above 97 percent conversion efficiency, kinetic limitations based on the reactor design and performance may result in conversion efficiencies of 80 to 95 percent. Higher conversion efficiency results in lower fuel consumption rates and higher metal throughput capacity.

Fuel/Metal Throughput Ratio. Fuel efficiency of the pyrohydrolysis process is expressed as the mass ratio of fuel consumed to metal processed. The fuel/metal throughput ratio is lowered by increased reactor conversion efficiency, increased reactor feed concentration, decreased reactor heat losses, decreased fines recycle rate, and decreased operating temperature.

Design of the Fluid Bed Pyrohydrolysis Pilot Plant

The pyrohydrolysis reactor is the central component in the conversion process. The choice of reactor configuration therefore had to satisfy the major process development objectives including simplicity and reliability of operation, ease of control, and desired product characteristics. The design for the pyrohydrolysis section of the AMAX Chloride Refinery Pilot Plant was based on an initial decision to use a fluid bed reactor as opposed to a spray roaster such as those developed by Ruthner (15) or Woodall-Duckham (16) for conversion of metal chlorides. The fluid bed reactor was chosen for several reasons listed below.

1. The reaction product is deposited on existing oxide particles in the bed, producing a granular, dust-free and easily-handled product.

2. The product size distribution can be controlled over a wide range depending upon the type and rate of fines recycle.

3. Excellent temperature control can be achieved.

4. No separate fuel combustion unit need be provided since the fluid bed can act as a burner as well as the pyrohydrolysis reactor.

Following the selection of a fluid bed reactor, the pilot plant for development and demonstration of the pyrohydrolysis step of the refining process was designed to meet the following set of objectives.

1. The pilot plant must be of sufficient size to provide reliable data for process scale-up.

2. The plant must be designed and equipped to allow a reasonably wide variation in the key process variables and thus allow a thorough and systematic search for maximum conversion efficiency and maximum fuel utilization.

3. The plant must be equipped with sufficient instrumentation for the acquisition of comprehensive data.

4. The plant must be constructed of materials compatible with process fluids and environments and suitable for use in the construction of a commercial unit.

5. The plant must be designed to provide reliable, continuous operation and simplicity of control.

6. The plant must be designed to assure the safety of operating personnel.

7. The plant must be designed to produce an oxide product of desirable physical and chemical characteristics.

Following the choice of reactor configuration, a process flowsheet was developed for a pilot plant rated at one metric ton of nickel oxide per day. It is summarized in the schematic diagram shown in Figure 5. In addition to major pieces of equipment, the diagram also shows points of measurement of key process parameters as instrument labels. The temperatures, flow rates, pressures, and fluid densities so indicated were critical for process control and evaluation. Each major processing unit, starting with the fluid bed reactor, is described in the remainder of this section. The descriptions are of successful designs which in some cases involved significant modifications of the original designs. These are discussed in a latter section of this paper.

Fluid Bed Pyrohydrolysis Unit (Hydrolyzer)

A 76-cm (30-inch) diameter fluid bed reactor was deemed sufficiently large to simulate the fluidization behavior of a larger commercial unit. The fluid bed reactor consisted of a steel cylinder 244 cms (96 inches) in diameter, 427 cms (168 inches) in height, and 1.3 cms (0.5 inches) in wall thickness. The shell was first lined with a layer of insulating brick and then a layer of 70 percent alumina firebrick to give an inside diameter of 76 cms in the lower section and 107 cms (42 inches) in the upper or freeboard section. The reason for the larger freeboard diameter was to provide a better opportunity for disengagement of solids. This is in contrast to Keramchemie/Lurgi fluid bed hydrolyzers used by pickle liquor plants(17). The lower section containing the fluidized bed was 153 cms (60 inches) in height and the freeboard section was 244 cms (96 inches) in height. A 23-cm (9-inch) flared transition section joined the upper and lower sections of the reactor. The types and thicknesses of refractory materials used in the reactor were specified such that the steel shell temperature would remain above the condensation temperature of gases containing HCl and H_2O when the fluid bed temperature was maintained at $800^\circ C$.

The reactor was equipped with numerous thermocouples placed in the fluid bed, in the freeboard, and on the steel shell. The lower section of the reactor was equipped with several ports for insertion of a starting burner, introduction of recycle fines, withdrawal of product, direct cyclone discharge return, and injection of feed solution. The bottom of the reactor was a removable plug of cast refractory containing ten symmetrically arranged nozzles for mixing air and oil, as well as vertical openings for withdrawal of product and for drainage of the bed. A cross-section of the reactor showing the location of ports and the structure of the refractory lining is shown in Figure 6. For illustrative purposes, all ports are shown in the same plane in the diagram.

Combustion System

As discussed earlier, the hydrolysis reaction is endothermic and energy has to be supplied by the combustion of fuel. Although the use of

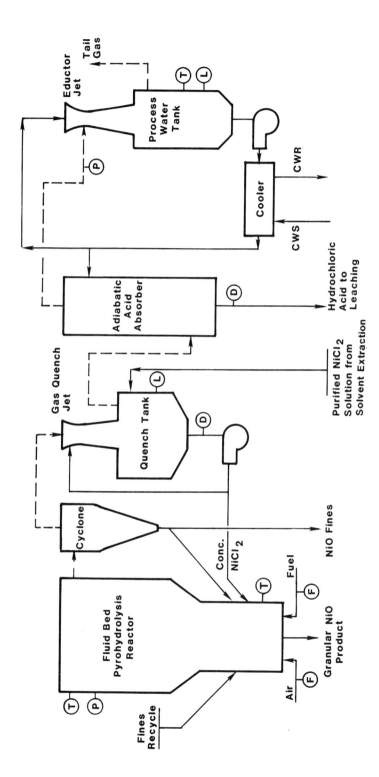

Figure 5. Process Flow Diagram for the Pyrohydrolysis Pilot Plant

natural gas is convenient as demonstrated by Falconbridge for hydrolysis of ferrous chloride solutions (18), fuel oil (No. 2) was selected for testing by AMAX because of its cheaper cost. A still cheaper energy source, coal, has been used for the hydrolysis of ferrous chloride solutions in Australia(19), but it cannot be considered for nickel because of the product purity considerations.

The combustion system was designed to deliver fuel and combustion air at flow rates that would result in a superficial space velocity of about 1 m/sec in the lower section of the reactor at 800°C. This flow rate of combustion gases is sufficient to fluidize a bed of coarse, granular nickel oxide (sp. gr. of 6.5). The combustion system consisted of a 25-horsepower air blower, fuel storage tank, fuel gauge tank, fuel pump, and ten nozzles symmetrically arranged in the bottom plug of the reactor. Air flow and fuel flow were metered and then evenly distributed to each of the nozzles. Each nozzle was constructed of 310 stainless steel using a no-weld design and had 6 horizontally-directed orifices. The nozzles were mounted in the bottom plug of the reactor such that the orifices were 3.8 cms above the reactor hearth. Mixing of fuel and air occurred in each nozzle and combustion occurred in the lower portion of the fluid bed.

Feed Injection System

Concentrated nickel chloride solution was injected into the pyrohydrolysis reactor through a feed gun fabricated from thick-walled (2.1 cm O.D., 0.6 cm I.D.) titanium pipe placed in a protective sheath consisting of a larger diameter 310 stainless steel pipe. High temperature corrosion of the titanium pipe by HCl was prevented by purging the annular space between the two pipes with air. A nozzle consisting of a 10-cm long titanium cylinder with a single 0.3 cm diameter bore was welded to the titanium pipe to produce a pencil-like stream of solution with a 5 m/sec linear velocity. The gun was mounted in a port in the reactor wall drilled at an angle 30 degrees from the vertical axis and positioned such that the stream of solution would impinge on the center of the unfluidized bed. Upon fluidization, the nozzle tip was submerged just below the surface of the expanded bed. The nozzle tip protruded beyond the inner wall of the reactor about 20 cms. Solution was supplied to the feed gun via a TFE tube tapped into the recirculating quench solution line. A solid Kynar control valve was used for flow control.

Dust Collection and Recycle

Nickel oxide particles not disengaged in the reactor freeboard were collected in a 41-cm (16-inch) diameter cyclone constructed using 310 stainless steel. The cyclone was insulated and operated at temperatures above 700°C. Two different arrangements for discharge of the cyclone underflow were used depending on whether the cyclone was used in the open or closed circuit mode. For open-circuit operation, the dust collection system consisted of a gate valve and a dust bin mounted below the barrel of the cyclone. The gate valve was the same diameter as the cyclone barrel thereby preventing any retention of fine dust by a conical cyclone bottom section. This dust collection system was also fabricated of 310 stainless steel.

For closed-circuit operation, the cyclone underflow was returned directly to the lower section of the reactor via a dipleg arrangement. The solids return line consisted of a 4-inch 310 stainless steel pipe

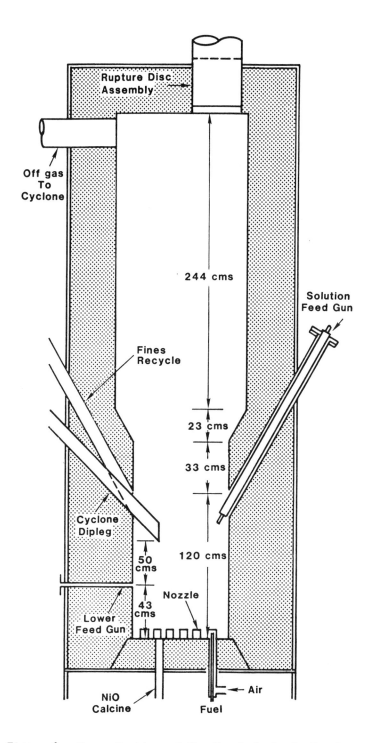

Figure 6. Cross-Section of the Pyrohydrolysis Reactor

equipped with a rod-out device. For this mode of operation, a conical section was used for the bottom section of the cyclone.

Off-Gas Quenching System

The function of the quenching system was to utilize the sensible heat available in the reactor off-gases to evaporate water and HCl from the dilute feed solution (from solvent extraction) prior to pyrohydrolysis. Since evaporation of water represents the major thermal load on the pyrohydrolysis reactor, it was highly desirable to control the quench system to produce a feed solution of as high a nickel chloride concentration as could be tolerated for smooth operation. This concentration was essentially the solubility limit of nickel chloride. Control of solution concentration was attained by closely monitoring the quench solution density and making adjustments by addition of dilute feed solution and hydrochloric acid.

Quenching was accomplished by contacting the pyrohydrolysis reactor off-gases with nickel chloride solution in a venturi-type jet constructed of titanium. In the quenching process, solution was recirculated through the jet, entering the side of the jet, overflowing a circular weir, and falling through the 5-inch jet throat. Hot gases entered the top of the jet and flowed downward where intimate mixing of gas and solution occurred. The quench jet also served to scrub fine nickel oxide dust that escaped the reactor cyclone. Most of the dust subsequently dissolved in the acidic quench solution to reform nickel chloride.

After leaving the jet, the two-phase mixture of gas and solution impacted directly on the surface of concentrated nickel chloride solution held in the quench tank. Disengagement of gas and liquid occurred at this point and in the tank freeboard before HCl-laden gases passed on to the acid absorber.

Acid Absorption System

Gases leaving the quenching system contain essentially all the chlorine introduced into the pyrohydrolysis system from the solvent extraction section. An adiabatic absorber was used to reclaim this chlorine as azeotropic hydrochloric acid for recycling to the sulfide leaching section of the refinery. The absorber consisted of a vertical 66-cm diameter cylinder constructed of Hetron 197 fiberglass reinforced plastic (FRP) and packed with 3-cm polypropylene Pall rings. Gases leaving the quenching system flowed through the column while process water from the scrubbing system flowed countercurrently to the HCl-laden gases (see Figure 5). The process water flow rate was controlled based on product acid strength and discharge of acid from the absorber was based on absorber sump level.

Gas Scrubbing and Condensation System

Gases discharged from the absorbing system were contacted with a large recirculating flow of process water in a venturi eductor. The eductor was constructed of FRP and provided the necessary draft to move process gases through the plant as well as scrubbing of residual HCl leaving the absorber. A graphite heat exchanger in the process water circuit was used to remove heat and maintain a water temperature of about $40^{\circ}C$, thus effecting the condensation of water from process gases for use in producing acid. System draft was controlled by bleeding sufficient air into the process upstream from the eductor so that a nearly neutral pressure was maintained in the freeboard of the pyrohydrolysis reactor. Process water temperature was controlled by adjustment of the cooling water flow rate through the heat exchanger.

Instrumentation and Control

The pyrohydrolysis pilot plant was adequately instrumented for effective process control and comprehensive data acquisition. In addition to conventional instrumentation for temperature, flow rate, and level control, an on-line digital computer was programmed and used to manipulate some of the critical process control loops. The computer was also used as the primary data acquisition and recording tool.

Control of the process was effected according to the following scheme:

1. Combustion of air flow rate was controlled at a fixed value established by a predetermined superficial space velocity through the fluid bed reactor.

2. Fuel flow rate was controlled at a fixed value established by combustion stoichiometry and a predetermined excess of combustion air.

3. Concentrated nickel chloride feed solution flow rate was regulated to maintain a predetermined fluid bed temperature. The computer was used for fluid bed temperature control, allowing convenient incorporation of alarm conditions and selection of the control thermocouple.

4. Inventory of concentrated feed solution in the quenching system was maintained at a constant value via addition of dilute nickel chloride solution from the solvent extraction section and hydrochloric acid from acid storage. The purpose of adding the acid was to control the density of the solution in the quench loop, at a point where nickel chloride would not crystallize. Solution inventory was determined using a nuclear density gauge to monitor solution concentration and a tank level measurement to monitor solution volume.

5. Hydrochloric acid strength was controlled by adjusting the water flow rate to the absorber according to density measurements of acid in the absorber sump.

6. Process water inventory in the scrubbing circuit was controlled by adjusting the cooling water flow rate to the process water heat exchanger, thereby controlling the rate of water condensation.

Operation of the Fluid Bed Pyrohydrolysis Pilot Plant

The pyrohydrolysis pilot plant was designed, constructed, and run to demonstrate the operability of the process and to obtain metallurgical performance data. In this section, pilot plant operating experience is described in terms of several of the major design objectives presented earlier in the paper. Briefly, the objectives addressed in this section are: a) reliable, continuous operation, b) adequate and simple process control, c) suitable materials of construction, and d) assured safety of operating personnel.

Reliability

After design and construction were completed, the AMAX Chloride Refinery Pilot Plant was operated for a total of 14 months resulting in completion of 23 campaigns. During the initial four campaigns, necessary modifications to the original plant design were made and operating personnel were trained. After this period, straightforward startup and planned shutdown of campaigns were readily achieved. For later campaigns, operating periods of one or two weeks were obtained without interruption of feed solution flow to the reactor.

Two systems in the process proved to be especially problematic and modifications made to achieve reliable and trouble-free operation were not successful until later campaigns. First of all, proper design of the fuel combustion system proved crucial to long-term operation. In early campaigns, when a windbox was used, unequal distribution of fuel and air to the nozzles in the fluid bed reactor resulted in localized reducing conditions in lower parts of the bed. Agglomerates of nickel oxide granules and nickel metal were formed, resulting in erratic fluid bed temperatures, local overheating, and eventual defluidization. When the air and fuel distribution system described previously under 'Combustion Systems' subheading was installed prior to the twelfth campaign, the agglomeration problem was essentially eliminated. When smaller nozzle orifices placed close to the reactor hearth were used along with a solids discharge line through the hearth instead of the reactor wall, the agglomeration problem disappeared completely.

Secondly, the off-gas cyclone, when operated in the open-circuit mode (no direct recycle of fines to the fluid bed), plugged frequently due to adherence and packing of the very fine nickel oxide dust on the cyclone walls. The problem was eliminated in later campaigns by removing the conical bottom section of the cyclone and installing a large gate valve and dust collection bin as described earlier. When the off-gas cyclone was operated in the closed-circuit mode and the mean particle size of the bed became smaller due to recycle of all of the oxide fines, surges of fine solids entered the quenching system via the cyclone dipleg and off-gas piping. The presence of undissolved solids in the quenching system produced erroneous solution density measurements and resulted in poor control of the feed solution concentration. This problem was not completely eliminated until recycle of fines was lowered to only a fraction of fines production, thus allowing the fluid bed to increase in particle size.

Process Control

The central scheme outlined earlier for the operation of the pilot plant proved to be satisfactory as stated. Experiences with the more critical control loops are worthy of mention. First of all, regulation of the reactor temperature using computer control of the feed solution flow rate proved to be very successful. Fluid bed temperature was controlled within one degree C of the set point value. Control of the feed solution concentration in the quenching system was more difficult because deviations from the operating set point (maximum $NiCl_2$ concentration without exceeding the $NiCl_2$ solubility limit) often resulted in crystallization of $NiCl_2$ and plugging of solution piping. False solution density measurements by the nuclear density gauge were the source of most of the problems. When excessive suspended solids or entrained gas bubbles were eliminated from the quench solution inventory at the point of measurement, control of the density at the set point value was readily achieved.

Materials of Construction

Most of the materials of construction selected in the original pilot plant design performed well. The high-alumina refractory brick used in the pyrohydrolysis reactor showed no appreciable signs of erosion or cracking.

The performance of 310 stainless steel in the lower section of the reactor was also noteworthy. All nozzles were constructed of this material and maintained their original dimensions during operation. The only major problem encountered with nozzle construction was the fracture of welds. Several types of nozzles, all fabricated from 310 stainless steel, were used. The first type was of a two-piece, welded construction. The weld encircled the nozzle, fastening the cap to the body. Upon defluidization, many of the welds cracked. The fracturing is believed to be caused primarily by thermal shock brought about by defluidization of the hot bed and the subsequent sudden heating of the tuyeres. To remedy this problem, a fabricated nozzle without welds was installed and used successfully.

The direct cyclone discharge return system, installed prior to Campaign 13, utilized a section of 310 stainless steel schedule 80 pipe for the portion of the dipleg that protruded from the hydrolyzer wall into the fluidized bed. This pipe section performed remarkably well with no sign of deformation or erosion.

Titanium was used to construct both the venturi quench jet and the solution feed gun. The performance of the material in this application was excellent as long as nickel chloride solutions containing oxidizing cations were fed to the quenching system. However, when nickel chloride raffinate from the solvent extraction section was used as feed to the pyrohydrolysis section, the quench jet quickly corroded and began to leak at the welds. The internal surfaces of the jet were also badly pitted. The apparent reason for the corrosion with the pure solution was the dissolution of the passive titanium dioxide layer on the metal in the absence of oxidizing cations such as Fe^{+3} and Cu^{+2}. Dissolving chlorine gas in the dilute feed solution to increase its oxidizing potential (up to +800 mv vs. saturated calomel electrode) did not solve the problem since the chlorine was immediately stripped from the concentrated solution in the quenching system. It was found necessary to dope the raffinate with $FeCl_3$ and $CuCl_2$ in order to continue operation of the pyrohydrolysis section. In a commercial plant, acid brick lining will be the appropriate choice.

For transport of hot nickel chloride solutions, Kynar-lined steel piping and teflon (TFE) tubing performed well. For many less severe applications chlorinated polyvinyl chloride (CPVC) piping proved satisfactory.

Control valves and orifice plates fabricated from Hastelloy B or C (depending upon pure HCl or pure $NiCl_2$ streams) used in the original design of the plant had to be replaced by solid Kynar control valves and TFE orifice plates. Replacement of many other metallic components with Kynar and TFE materials occurred during the operation of the plant.

Safety

The original design of the pyrohydrolysis pilot plant was adequate in terms of assuring the safety of operating personnel. However, several of the safety systems were improved to avoid problems encountered with the operation of the plant. The pressure relief system for the reactor was most noteworthy in this respect. A TFE rupture disk, 24 inches in diameter (protected from reactor temperatures and corrosive gases by a heat shield and air purge) replaced the removable reactor lid (see Figure 6). Ingress of trapped air and poor control of the reactor freeboard pressure were eliminated with this design change.

Performance of the Fluid Bed Pyrohydrolysis Pilot Plant

In addition to demonstrating the operability of the nickel chloride pyrohydrolysis process, the pilot plant also provided the metallurgical performance data for technical and economic feasibility studies and for design and scale-up to a production plant. In this section, the performance of the pilot plant is discussed in terms of three major design objectives. They are a) achievement of maximum fuel efficiency, b) achievement of maximum conversion efficiency, and c) production of nickel oxide of desirable physical and chemical characteristics. These objectives are reviewed in terms of the effect of important process variables such as feed solution composition, fluid bed temperature, bed particle size distribution, solution feeding system, and fines recycle system.

Effect of Feed Solution Composition

The nickel chloride concentration of the feed solution to the pyrohydrolysis reactor had the most pronounced effect on reactor performance in comparison to any of the process variables. A major portion of the heat supplied to the reactor is consumed in evaporating the water in the feed solution. Thus, it was expected that higher nickel chloride concentrations would result in higher nickel flow rates to the hydrolyzer under otherwise similar operating conditions.

In early campaigns, the nickel chloride concentration was maintained below 38% to avoid plugging problems. However, after the quench jet was modified and the piping between the quench loop and the feed gun was improved, the nickel chloride concentration was increased to 42% in Campaign No. 7 and finally to 45% in Campaign No. 8. It was demonstrated in Campaign No. 11 that the upper practical limit was about 46% $NiCl_2$. To avoid plugging under temporary upset conditions, about 45% $NiCl_2$ was the preferred operating maximum.

In Figure 7, fuel consumption ratios are plotted vs. reactor feed solution concentration for various runs made with an air flow rate of 440-460 kg/hr, an oil flow rate of 28-29 kg/hr, and a bed temperature of 850°C. These conditions encompassed runs from several campaigns. Included with the data are calculated curves for various conversion efficiencies. These curves were plotted by using a heat balance around the reactor to calculate feed flow rates for various feed solution concentrations. Heat losses from the reactor were determined experimentally using only water as the feed.

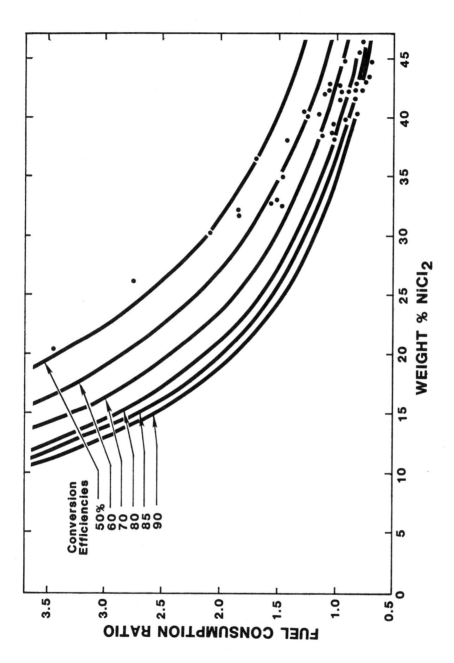

Figure 7. Effect of Feed Solution Concentration on Fuel Consumption

Effect of Fluid Bed Temperature

The effect of temperature on the reactor performance is twofold. First of all, an energy balance indicated that an increase of 25°C in the bed temperature will lower the feed flow rate by about 3%. In general, the data presented in Table I support this trend.

The second effect of temperature is analyzed in terms of conversion efficiency. The thermodynamics of the hydrolysis reaction discussed earlier suggest that, at low temperatures, the driving force and the vapor pressure of $NiCl_2$ are low. Thus, there is a chance the reaction may not reach completion in the bed, leading to low conversion efficiencies. On the other hand, a very high temperature may greatly improve the kinetics to the point where nickel oxide dust is formed by homogeneous nucleation again resulting in low conversion efficiencies.

In view of the above, an optimum temperature was sought. The data presented in Table I indicate that the temperature did not affect the conversion efficiency and if anything, 850°C appeared to be the optimum temperature. This temperature was therefore used in most of the runs to evaluate other process variables.

Table I. Effect of the Bed Temperature on the Performance of the Hydrolyzer

Run	Temperature, °C	Nickel Flow to Hydrolyzer, kg/hr	Feed Solution Concentration Wt.% $NiCl_2$	Conversion Efficiency, %	Oil Consumption kg/kg of Ni
6B	825	37.5	37.8	76	0.89
6D	850	40.5	38.2	86	0.87
6E	875	36.7	38.2	85	0.90
8D	850	46.8	44.9	87	0.70
8G	900	42.7	45.4	88	0.75
9E	800	42.8	44.6	73	0.92
9C	850	40.8	44.4	75	0.95
9D	875	37.9	44.5	75	1.02

Effect of Solution Feeding System Design

Feed solution was injected into the fluid bed reactor through two feed gun ports shown earlier in Figure 6. The lower feed port location was 43 cms above the reactor hearth and allowed injection of solution directly into the fluid bed. The upper feed position permitted solution injection onto the top surface of the bed.

Comparative test runs were made during Campaigns 9 and 10 to evaluate the performance of the two gun locations and the results are presented in Table II. The actual bed weight was about 1500 kg in all cases giving a static bed height of about 80 cms. The windbox was replaced with an air manifold beginning with Campaign No. 10. It is thought that this resulted in more complete combustion of the fuel within the bed and thus lower freeboard temperature.

Table II. Effect of Feed Gun Location on the Performance of the Hydrolyzer

Run	9A	9B	10A	10C
Feed gun location, cms above the bottom plug	120	43	120	43
Bed Temperature, °C	852	852	848	850
Freeboard Temperature, °C	817	843	802	822
Mean Particle Diameter, micrometers	530	605	419	475
Nickel Flow to Hydrolyzer, kg/hr	40.3	45.9	42.0	43.1
Conversion Efficiency, %	77	69	87	82
Oil Consumption, kg/kg of Ni	0.94	0.91	0.79	0.82

It is interesting to note that the freeboard temperature was higher in both campaigns when the lower gun was used. This is attributed to a more expanded bed due to the evolution of water vapor from the feed solution near the bottom of the bed. A bubbling bed model(20) was used to characterize the fluidization of the bed. The presence of large gas volumes, as bubbles, could lead to higher dust formation rates by homogeneous reaction and therefore to lower conversion efficiencies.

Results presented in Table II support this argument. The dust produced with the two gun locations and collected in the cyclone was analyzed for complete size distribution and the results are presented in Figure 8. It is again seen that the lower gun location, producing more bubbles, resulted in the formation of much finer dust (93% below 5 micrometers against only 18% below 5 micrometers).

To determine what effect the type of solution spray produced by the feed gun nozzle had on reactor performance, three comparative tests were performed in Campaign No. 12 and the resuls are presented in Table III. Two air sparge rates were used to create two types of solution dispersion. With no air sparge, a pencil-like stream was produced. Other experimental conditions were similar and normal. Apparently the high air sparge rates created a very fine mist which led to excessive dust formation as reflected by the highest cyclone discharge rate and lowest conversion efficiency. The results presented in Table III suggest that the fine mist is definitely undesirable.

Table III. Effect of Air Atomizing the Feed Solution on the Performance of the Hydrolyzer

Run	12C	12A	12B
Air Sparge Rate, SCFM	0	0.5	1.5
Nickel Flow to Hydrolyzer, kg/hr	46.7	44.9	49.2
Cyclone Discharge Rate, kg NiO/hr	5.0	6.4	7.4
Conversion Efficiency, %	85	86	70
Oil Consumption, kg/kg of Ni	0.72	0.75	0.84

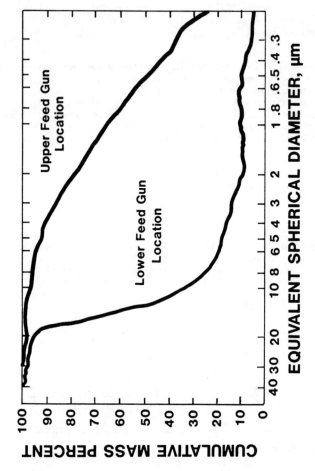

Figure 8. Effect of Feed Gun Location on Particle Size Distribution of Cyclone Fines

Effect of Fluid Bed Particle Size Distribution and Fines Recycle Rate

The particle size distribution of the fluid bed affected the performance of the pyrohydrolysis reactor in several ways. According to the two-phase or bubbling bed model for gas-solid fluidization systems, a larger portion of the fluidizing gas passes through the bed as bubbles as the mean particle diameter of the bed material decreases (20). This in turn results in a more expanded bed and increased transport of material into the freeboard or disengaging section of the reactor.

During Campaign No. 8, the starting bed was made entirely of minus 28 mesh particles. No fines were returned throughout the campaign and the solids were discharged at a rate that would maintain the bed at about 2200 kg. The lack of fines return led to a gradual coarsening of the bed as is illustrated by the data presented in Table IV. Again, the bubbling bed model suggests that with increasing particle size, the quantity of bubbles and the height of the expanded bed will decrease. This of course will result in increasing the difference between the bed and the freeboard temperatures and a lowering of cyclone discharge rates. As the particle size increases and the bubble size and quantity decrease, fewer of the coarse granules will escape to the cyclone and the bulk density of the cyclone discharge will approach that of the fine dust. Data presented in Table IV definitely support this line of reasoning.

Table IV. Effect of Stopping Fines-Return on the Hydrolyzer Bed and Cyclone Discharge Characteristics

Run	% +28M in Calcine	ΔT Between Bed and Freeboard °C	Cyclone Discharge Rate, kg NiO/hr	Bulk Density of Cyclone Discharge kg/L
8A	-	26	10.4	2.5
8B	12	27	10.7	2.0
8C	35	29	7.9	1.6
8D	39	30	7.2	1.8
8E	48	34	-	-
8F	57	45	5.3	0.9
8G	64	41	3.6	0.6

After improving the combustion system design and establishing that a small bed (about 80 cm static height) with the upper feed gun (about 120 cms from the bottom) location gave excellent results, the effect of fines return rate was studied in greater detail in Campaign No. 12. The dust collected in the calcine was mixed with the minus 28 mesh fraction of the calcine discharge and the fines mixture was fed to the reactor.

The particle size distribution of the bed was then varied from one extreme to the other by controlling the return rate of the fines. The results obtained during three runs are presented in Table V. They roughly represent a fine, a medium and a coarse size bed. The bed weight was the same in all cases.

Table V. Effect of Bed Particle Size Distribution on the Performance of the Hydrolyzer

Run	12G	12D	12E
% of +28M in Calcine	19	74	89
Temperature at the bottom of the bed, °C	854	868	873
Bed Temperature, °C	850	850	851
Freeboard Temperature, °C	833	786	758
Cyclone Discharge Rate, kg NiO/hr	13.4	7.8	4.0
Bulk Density of Calcine Discharge, kg/L	1.6	0.6	0.5
Fines Return Rate, kg NiO/hr	38.9	31.7	31.6
Nickel Flow Rate to Hydrolyzer, kg/hr	39.5	47.3	48.3
Conversion Efficiency, %	91	79	82
Oil Consumption kg/kg of Ni	0.81	0.77	0.73

The temperature profile clearly establishes that the bed expanded considerably with increasing fineness of the bed. The expanded bed and larger bubbles in the case of the fine bed also led to higher cyclone discharge rates and greater carryover of bed particles resulting in a higher bulk density of cyclone discharge material. Results presented in Table V also confirm this.

The lower nickel flow rates to the reactor for the finer bed are the result of a high fines recycle rate needed to maintain the small mean particle diameter of the bed. The recycle of fines represents a thermal load on the reactor and results in decreased nickel throughput.

Open Circuit vs. Closed Circuit Operation of the Off-Gas Cyclone

An ideal process for the pyrohydrolysis of nickel chloride solutions in a fluid bed reactor would deposit all nickel fed to the reactor on existing oxide particles in the bed. All nickel oxide produced would be granular and dust-free, conversion efficiency would be close to that predicted by thermodynamics, and fuel consumption would be relatively low. On the other hand, pyrohydrolysis by spray roasting would produce a fine nickel oxide powder and no granular material.

The fluid bed pyrohydrolysis reactor in the AMAX Chloride Refinery Pilot Plant converted about 85 to 90 percent of the nickel chloride solution from the solvent extraction section to granular nickel oxide and the remainder to fine nickel oxide powder discharged from the off-gas cyclone. This distribution of oxide product was typical of operation with the open circuit cyclone. Both oxide products met chemical composition specifications and the fine powder was easily briquetted or pelletized rendering it suitable for subsequent hydrogen reduction. Operation with the open circuit cyclone also required recycle of oxide fines to control the fluid bed particle size distribution and provide new nuclei for growth of oxide granules. Ground granular oxide or a mixture of cyclone discharge and ground oxide were successfully recycled to the reactor.

Closed circuit operation of the cyclone was studied in four of the latter campaigns. This mode of operation afforded a means to eliminate the separate fines recycling operation and to produce only a granular product. Operation in this mode produced a very fine, easily handled oxide product of acceptable chemical composition but problems with control of the reactor were encountered. Direct recycle of all fines discharged from the cyclone reduced the mean particle diameter of the fluid bed to 100 micrometers and lower, resulting in a highly expanded bed. Essentially, no temperature differential between the fluid bed and freeboard zones of the reactor was observed. By comparison, operation with a coarse bed produced up to a $90°C$ temperature difference between the fluid bed and freeboard as shown previously in Tables IV and V. In addition, short-circuiting of fine solids through the cyclone dipleg and into the quenching system made control of the feed solution concentration difficult due to a huge quantity of suspended solids. The overall effect of the direct recycle of all cyclone discharge was to increase fuel consumption in comparison to open circuit operation. Successful use of direct fine recycle could probably be obtained by cycling the cyclone operation betweeen open and closed circuit mode. This procedure would gain control of the fluid bed particle size distribution and still eliminate the need for a separate fines recycling step.

Nickel Oxide Product Characteristics

One of the primary reasons for selecting a fluid bed reactor instead of a spray reactor or other type of high temperature reactor was production of dust-free granular nickel oxide. This objective was met in all pilot plant campaigns and oxide granules ranging from about 400 micrometers to 600 micrometers mean particle diameter were consistently produced.

The bulk density of the calcine was measured and found to be between 4.0 and 4.2 kg/L in all cases. The particle density was found to be a very consistent 6.34 kg/L. The color of the calcine was usually greyish black, but at times it was green. No correlations could be found for the color change.

The nickel content of the calcine was slightly higher than the theoretical 78.6% in early campaigns, indicating some metallization. However, as the % excess air was increased and a better combustion system was installed, metallization was completely eliminated. Figure 9 shows a photomicrograph of a metallized particle, along with that of others with no metal ring. Typical onion skin-type growth is seen in each of the particles.

Samples of the granular oxide were regularly submitted for chemical analyses. Results from several runs are presented in Table VI. In general, the chlorine and sulfur content of the granular oxide product was below 100 ppm, while the carbon content was below 10 ppm. For all cases, the levels of carbon, sulfur, and chlorine were entirely acceptable in terms of product quality. (ASTM Class I nickel specifications are 100 and 300 ppm for sulfur and carbon, respectively.) For some of the critical trace metal impurities, such as zinc and lead, levels present in the samples were below the detection limit for the analytical instrument. Because utility water was sometimes used for emergency cooling, levels for magnesium and calcium were quite high. Exclusive use of de-ionized water would most likely lower these values. Copper and iron levels were high due to the doping procedure necessary to passivate the titanium quench jet. For commercial operation, quenching equipment constructed of acid brick or SiC should eliminate the need for doping and the nickel oxide product should contain

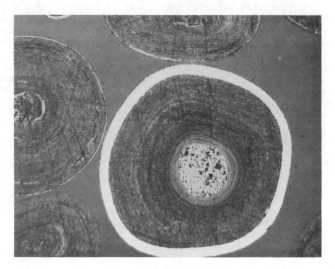

Figure 9. Photomicrograph of Particles Showing Onion Skin Growth and Occasional Presence of Metallic Rims

about 75 ppm iron and 15 ppm copper. These values correspond directly to iron and copper levels in the nickel chloride solution generated in the solvent extraction section.

Of particular interest is the nickel-to-cobalt ratio of the nickel oxide product. This ratio closely followed the nickel-to-cobalt ratio in the feed solution (raffinate from the solvent extraction circuit), confirming that any desired ratio can be obtained by properly operating the solvent extraction circuit. The observation also indicated that the cobalt chloride solution can be converted to a pure cobalt oxide product in a separate reactor at efficiencies similar to those obtained in nickel operation.

Table VI. Chemical Analyses of Nickel Oxide Produced During Various Runs

Run	19G	19H	19I	20C
Analysis, Weight %				
Nickel	77.95	78.07	78.39	77.38
Cobalt	0.101	0.111	0.126	0.560
Chlorine	0.01	0.01	0.01	0.01
Sulfur	0.01	0.01	0.01	0.01
Carbon	0.014	0.007	0.007	0.007
Analysis, ppm				
Chromium	90	157	20	135
Titanium	ND	ND	ND	273
Magnesium	160	140	140	59
Calcium	981	1000	1000	1000
Lead	ND	ND	ND	ND
Zinc	ND	ND	ND	ND

ND - Not Detected

Summary and Conclusions

The main findings of the nickel chloride pyrohydrolysis development program are summarized below:

1. The fluid bed pyrohydrolysis unit of the AMAX Chloride Refining Process was successfully demonstrated on a pilot scale. This conversion process can simply and efficiently convert concentrated nickel chloride solutions to low-chlorine, low-sulfur granular nickel oxide and recycle the chloride component of the feed solution as hydrochloric acid of azeotropic composition.

2. Nickel chloride concentration in the feed solution to the reactor was the most important process parameter affecting reactor performance. When feed solutions containing 45-46 weight percent $NiCl_2$ were used, fuel/nickel throughput ratios as low as 0.7 were obtained.

3. The nickel oxide product reflected the chemical purity of the solvent extraction raffinate used as dilute feed solution. Nickel oxide product of excellent physical characteristics and with a Ni:Co ratio of at least 500:1 was readily produced. Carbon content was below 10 ppm, while the sulfur and chlorine contents were below 100 ppm. Nickel content was close to the stoichiometric 78.6%.

4. A quenching system utilizing a venturi jet and disengagement tank performed well in cooling reactor off-gases, concentrating nickel chloride solution, and scrubbing fine NiO dust that escaped the off-gas cyclone.

5. In the temperature range investigated (800 to 900°C), the temperature had no major effect on reactor performance. The energy balance on the process as well as the thermodynamics of nickel chloride pyrohydrolysis support this conclusion.

6. When the reactor was operated with an open-circuited off-gas cyclone, about 10 to 15 percent of the nickel oxide product was collected as a fine dust. Recycle of ground granular oxide product was necessary for control of fluid bed particle size.

7. When the hydrolyzer was operated using the direct cyclone discharge return system, it was very likely that the large number of new nickel oxide particles produced in the reactor forced the fluidized bed size distribution into a range of particle sizes much smaller than those encountered when running the open-circuited cyclone. As a result of the fine particle size distribution of the fluidized bed, a large bed expansion was indicated by the temperature profiles observed in the reactor. The bubbling bed theory for gas-solid fluidized systems support these observations.

8. A properly designed combustion system was of extreme importance. It affected not only the process fuel consumption but also process reliability. Improper fuel and combustion air distribution resulted in localized reducing zones, agglomeration of bed material, and defluidization.

9. The best pyrohydrolysis reactor performance was obtained by injecting feed solution as a pencil-like stream near the top of a fluid bed composed of 400 to 600 micrometer particles.

10. The performance of 310 stainless steel in all applications in the lower section of the reactor was excellent both in terms of resistance to corrosion and to deformation. The high-alumina refractory used in the reactor also performed well as no replacement of brick was necessary through the entire time of operation. Kynar, TFE, CPVC, polypropylene, and Hetron 197 based FRP performed well in the low temperature sections of the plant.

11. Use of a process computer greatly enhanced data acquisition efforts as well as the manipulation of critical control loops in the process. Temperature control of the reactor using the computer was especially effective.

Acknowledgments

The authors would like to thank the AMAX management for their permission to present and publish this paper. We would also like to acknowledge that there were many individuals other than the authors who made substantial contributions to the results reported here. Our thanks are particularly directed to them.

References

1. H. Kay and E.J. Michal, "The AMAX Acid Leach Process for Oxide Nickel Ores", TMS paper no. A78-36, presented at the 107th AIME Annual Meeting, Denver, 1978.

2. W.P.C. Duyvesteyn, G.R. Wicker, and R.E. Doane, "An Omnivorous Process for Laterite Deposits", Chapter 28 in International Laterite Symposium, D.J.I. Evans, R.S. Shoemaker, and H. Veltman (editors), Society of Mining Engineers of AIME, New York 1979, pp. 553-570.

3. M.C. Jha, G.A. Meyer, and G.R. Wicker, "An Improved Process for Precipitating Nickel Sulfide from Acidic Laterite Leach Liquors", Journal of Metals, November, 1981, pp. 48-53.

4. E.J. Michal, S.O. Fekete, and H.J. Roorda, "Hydrometallurgical Refining of Nickeliferous Sulfides," United States Patent 4,214,901, July 29, 1980.

5. M.C. Jha and G.A. Meyer, "AMAX Chloride Refining Process for Recovery of Nickel and Cobalt from Mixed Sulfide Precipitates", HYDROMETALLURGY Research, Development and Plant Practice, K. Osseo-Asare and J.D. Miller (editors), The Metallurgical Society of AIME, New York, 1982, pp. 903-924.

6. M.C. Jha, J.R. Carlberg, and G.A. Meyer, "Hydrochloric Acid Leaching of Nickel Sulfide Precipitates", Hydrometallurgy, vol. 9, 1983, pp 349-369.

7. E.C. Chou, B.E. Bonnema, and G.A. Meyer, "Solvent Extraction in AMAX Chloride Refining Process", TMS paper no. A84-62, presented at the 113th AIME Annual Meeting, Los Angeles, 1984.

8. G.A. Meyer, M.A. Peters, and H.S. Leaver, "Purification of Nickel Chloride Solutions," United States Patent 4,221,765, September 9, 1980.

9. P.G. Thornhill, E. Wigstol, and G. Van Weert, "The Falconbridge Matte Leach Process", Journal of Metals, July, 1971, pp. 13-18.

10. S. Nishimura, "Novel Hydrometallurgical Processing of Nickel and Cobalt Mixed Sulfide in Japan, "Extraction Metallurgy '81", The Institution of Mining and Metallurgy, London, 1981, pp. 404-412.

11. A. Setsuma, N. Ono, T. Iio and K. Yamada, "Sumitomo's New Process for Nickel and Cobalt Recovery from Sulfide Mixture", TMS paper selection A80-2, presented at the 109th AIME Annual Meeting, Las Vegas, 1980.

12. J.M. Demarthe, "Elaboration de Nickel Electrolytique a Partir de Matte de Nickel", in Chloride Hydrometallurgy Proceedings, an Inernational Symposium organized by Benelux Metallurgie, Brussels, September 26-28, 1977, pp. 231-248.

13. A.D. Mah and L.B. Pankratz, Contributions to the Data on Theoretical Metallurgy XVI. Thermodynamic Properties of Nickel and Its Inorganic Compounds, Bulletin 668, United States Bureau of Mines, 1976.

14. O. Kubaschewski and C.B. Alcock, Metallurgical Thermochemistry, 5th Edition, Pergamon Press, Oxford, 1979.

15. H. Jedlicka, "New Applications of the Spray Roasting Process in the Chloride Hydrometallurgy", Chloride Hydrometallurgy Proceedings, an International Symposium organized by Benelux Metallurgie, Brussels, September 26-28, 1977, pp. 154-181.

16. A. Conners, "Hydrochloric Acid Regeneration as Applied to the Steel and Mineral Processing Industries", CIM Bulletin, February, 1975, pp. 75-81.

17. G.H. Rupay and C.J. Jewell, "The Regeneration of Hydrochloric Acid from Waste Pickel Liquor Using the Keramchemie/Lurgi Fluidized-Bed Reactor System", CIM Bulletin, February 1975, pp. 89-93.

18. G. Van Weert, E.C. Robertson, and J.H. Christiansen, "Treatment of Ferrous Chloride Liquors in the Faclonbridge Fluid-Bed Hydrolyzer", CIM Bulletin, January, 1975, pp.87-95.

19. H.N. Sinha and J.R. Tuffley, "Recovery of Hydrochloric Acid from Ferrous Chloride Liquor in a Coal-Fired Fluidized Bed", in Extraction Metallurgy '81, the Institute of Mining and Metallurgy, London, 1981, pp. 421-429.

20. D. Kunii and O. Levenspiel, Fluidization Engineering, John Wiley and Sons, Inc., New York, 1969.

CONTRIBUTION TO CHLORIDE HYDROMETALLURGY

René WINAND
Université Libre de Bruxelles - CP 165
Department Metallurgy-Electrochemistry
50 , avenue F.D.Roosevelt - B 1050 - Brussels - Belgium

ABSTRACT

The paper gives a synthesis of research work performed in the department during the last ten years and dealing with cobalt , copper and manganese electrodeposition in chloride aqueous solutions. Special emphasis is given to electrocrystallization. For copper, properties of the electrolyte such as solubility limits, electrical conductivity and complex ions stability are discussed in connection with leaching, purification of the electrolyte, and metal recovery. When possible, kinetic parameters and probable electrode mechanisms are given. Problems that are still to be solved are also listed.

INTRODUCTION

Work performed in the department is characterized by an intimate relationship between fundamental and applied research. However, the guideline followed is usually more fundamental , and applications are derived from it. In the specific case of chloride hydrometallurgy, the guideline is electrocrystallization.
The author had the opportunity to present a synthesis on the subject at a recent meeting in Victoria - B.C. - Canada. This work is published now(1)(2) sothat only a short summary of the most important conclusions will be given here.
Making use of FISCHER'S classification of the most important types of deposits(3) (FI: field orientated isolated crystals;BR:basis-orientated reproduction type;FT:field orientated texture type;UD: unorientated dispersion type), the author proposed some years ago (4)-(8) to show the fields of stability of them in a schematic diagram (figure 1),as function of two main

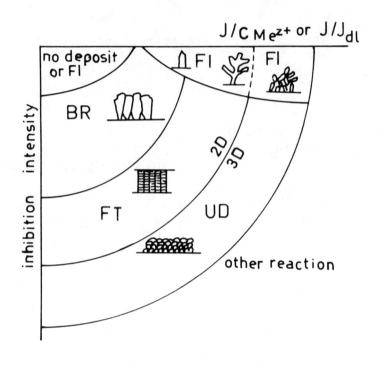

Figure 1
Schematic diagram showing different possible types of polycrystalline electrodeposits as a function of $J/c_{Me^{z+}}$ or J/J_{dl} and inhibition intensity
FI : field orientated isolated crystals(whiskers,dendrites, or crystalline powder; uncoherent deposit).
BR : basis reproduction (large crystals, coherent deposit, rather high surface roughness)
FT : field orientated texture type (elongated crystals;coherent deposit;rather low surface roughness)
UD : unorientated dispersion type (small crystals;coherent deposit; surface roughness function of experimental conditions; may end in metal powder.)
All types of deposit are obtained by two-dimensionnal nucleation(2D) except FI when crystalline powder is obtained and UD that are obtained by threedimensionnal nucleation (3D). To simplify the diagram, Z type of deposit (twinning intermediate type) was dropped: it appears only seldom, and lies between BR and FT

parameters : - the ratio $J/c_{Me^{z+}}$ of the current density to the bulk
concentration of metal ions to be discharged, or more
generally , if the electrolyte flow rate is to be changed
drastically , the ratio J/J_{dl} of the current density to the
diffusion limiting current density. This parameter charac-
terizes mass transfer to the electrode.

- the inhibition intensity. This parameter is not very easy to
define. According to FISCHER(3)(9), inhibition is due to the
presence at the surface of the electrode, in the double
layer or in the the diffusion layer, of substances (molecules, atoms, or ions) different from Me^{z+} or the corresponding adatom. These substances hinder the cathodic process
and are called inhibitors. They usually do not cover completely the cathode surface and favour active sites. They are
physically or chemically adsorbed. Their main effects result
in changes of the metallographic structure and of the crystallographic texture of the deposit and/or in values of the
overvoltage.

These effects depend on the nature of the metal(normal metals: low sensi
tivity to inhibition; intermediate metals: medium sensitivity; inert metals:
high sensitivity), of the anion (inhibiting anion like ClO_4^- ; intermediate
anion like $SO_4^=$; activating anion like Cl^-), and of the inhibitor. It
should be noted that although most inhibitors are organic additives, when
no such additive is present in solution, the electrochemical behaviour of
the metal in presence of a given mineral solution already defines a certain
inhibition intensity. From experiments performed with silver in nitrate solutions,copper in sulphate and chloride solutions, and cobalt in sulphate
and chloride solutions,in each case without organic additive, the author
found the inhibition intensity to be linked to the order of magnitude of
the exchange current density J_o : the higher J_o, the lower the inhibition
intensity. This relationship is only qualitative up to now because the
values of exchange current density are not very reproducible and depend on
the measurement method. Moreover, when organic additives are present, the
problem is much more complicated, although cases can be found where inhibition intensity in the diagram increases with the concentration of the organics in solution.

After this short fundamental introduction, let us now consider successi-
vely cobalt, copper, and manganese electrodeposition in chloride aqueous
solutions. Experimental results will be discussed mainly in connexion with
industrial practice.

Cobalt electrodeposition from acid chloride solutions
Cobalt, like iron and nickel, is generally considered as an "inert metal",
showing high electrodeposition overvoltage, low exchange current density,
high surface tension, low hydrogen overvoltage and low surface diffusion
coefficient for adatoms. It shows high tendency to adsorb inhibitors, and
"secondary inhibition", due to an adsorbed layer of CoOH or of another
intermediate specie is usually observed, even in the absence of any added
inhibitor to the electrolyte. Accordingly, the replacement of the sulphate
anion, commonly used in most industrial electrowinning plants (10)(16), by
the chloride activating anion is likely to be beneficial.
A fundamental research was thus made in the laboratory in order to obtain data concerning cobalt electrocrystallization in pure chloride solution (17)(18), with a soluble cobalt anode. Standard conditions for all the
experiments were as follows :
$[Co^{2+}]$: 50 g/l; temperature: 25°C; charge 125 mAh; cathodic area: 2.5 cm^2;
agitation: constant, by magnetic stirrer and by nitrogen bubbling;

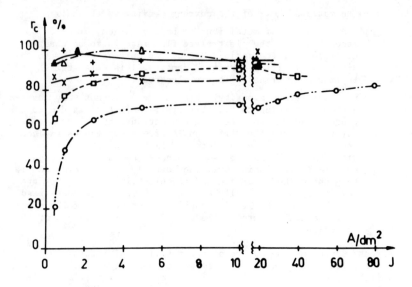

Figure 2 : Cobalt electrodeposition from pure $CoCl_2$-HCl aqueous solutions at 25°C. Current efficiency r_c against current density J at different pH values :
pH 4.5 : + ——— ; 3.5 : △ — . — . — ; 2.5 : X — — — ;
1.5 : ☐ --------- ; 1.0 : ○ — .. —.,—

no inhibitor added; pH subject to precise automatic control by HCl additions.

The most interesting results can be summarized as follows.

Figure 2 gives the current efficiency r_c against current density J for different pH values. It can be noticed that current efficiency is higher than 90% in the whole range of investigated current densities when pH is higher than 3. Rather low values are obtained at low pH due to hydrogen evolution. Better results are found when current density is increased. For instance, at pH = 1 and 80 A/dm^2, current efficiency is higher than 80%. It should be noted that this result is obtained without drastic stirring, and that the cathode deposit is of good quality. Figure 3 gives a cut at 2 A/dm^2 in figure 2, to allow a comparison with results obtained by FENEAU and BRECKPOT (11) in pure sulphate solutions. As can be seen, current efficiencies at 25°C in acid chloride solutions are already better than those achieved at 60°C in sulphate solutions. These results should further be improved by increasing the temperature . This is confirmed by SUMITOMO's operating results (19) , given in the same diagram for comparison .GRAVEY and PEYRON (20) also claim 90% current efficiency in pure chloride solutions at 60°C, 3.5 A/dm^2 and 20-30 g/l Co^{++}; however , the pH value is not available.

Figure 4 gives the galvanostatic cathodic potential as a function of pH for various current densities in pure chloride solutions. The curves are characterized by two minima (in fact , two maxima in overvoltage): a first one around pH = 1.5, and a second one around pH = 2.5-3; and also by a sudden decrease in overvoltage between pH 3.5 and 4. For comparison, a curve derived from FENEAU and BRECKPOT (11) in pure sulphate solution is given in the same graph. It can be seen that overvoltages at low pH are higher in sulphate solutions than in chloride solutions.

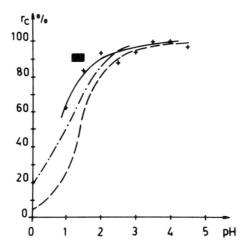

Figure 3 - Comparison between sulphate and chloride solutions for cobalt electrodeposition. Current efficiency against pH at 2 A/dm^2;50 g/l Co^{++}
+ ─────SCOYER and WINAND(18)-Chloride solution-25°C;── ── ──FENEAU and BRECKPOT (11)-Sulphate solution-20°C;──·──·──FENEAU and BRECKPOT (11)-Sulphate solution-60°C; ▨ SUMITOMO's operating conditions(19):all chloride solutions:pH:1.2-1.5;Co^{++}:55-65g/l;J:2.3 A/dm^2;temperature:55-60°C

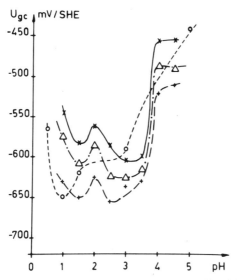

Figure 4 - Galvanostatic cathodic potentials measured against SHE in function of pH in pure chloride solutions,25°C;
X─────J=2.5 A/dm^2⎤ after SCOYER
△──·──J=5   A/dm^2⎥ and WINAND(18)
+── ──J=10 A/dm^2⎦

O ──────Results for pure sulphate solution, at 20°C, and 4 A/dm^2, after FENEAU and BRECKPOT (11)

Despite the fact that points are lacking in sulphate solutions to allow for more detailed discussion, the same important decrease in overvoltage like in chloride solutions seems to occur between pH 3 and 5.

After drawing the Tafel plots for chloride solutions, exchange current densities for stationnary conditions were obtained for various values of pH. They are listed in Table I.

TABLE I

Exchange current densities at 25°C; 50 g/l Co^{++}
for pure chloride solutions (18)

pH	J_{o_2} A/dm^2
1	6.10^{-5}
1.5	$0.14.10^{-5}$
2	2.10^{-5}
2.5	50.10^{-5}
3	13.10^{-5}
3.5	$2.4.10^{-5}$
4	$1.6.10^{-2}$
4.5	$2.8.10^{-2}$

In pure acid sulphate solutions, values between 10^{-6} and 10^{-8} A/dm^2 are commonly cited, sothat the change from sulphate to chloride results in an important increase in exchange current densities in acid solutions.

The crystallographic characteristics of the deposits were also investigated (18). Both α (close-packed hexagonal) and β (face-centered cubic) allotropic forms of cobalt were found in the deposits, the α phase (thermodynamically stable below 420°C) being predominant. The percentage in α phase is maximum near pH=3-3.5, where it is almost alone (Fig.5). This is surprising, because the overvoltage is maximum at this pH. The textures of both α and β phases were determined.

Figure 6 shows the results. The α phase is textured along the planes (11$\bar{2}$0) and/or (10$\bar{1}$0) whereas the β phase is textured along the planes (220) and/or (422). The textures of both phases are correlated : the β (220) and β (422) textures are respectively associated with the α (11$\bar{2}$0) and α (10$\bar{1}$0) textures. It should be noticed however that the β phase textures are weaker and much broader than for the α phase.

Comparison with pure sulphate solutions is difficult, because FENEAU and BRECKPOT (11) investigated mainly temperature instead of current density, and also because they did not detect any β phase. Nevertheless, at 25°C, it can be derived from their paper that, at current densities between 2 and 10 A/dm^2 and at pH between 0 and 3-3.5, the α phase is mainly textured (11$\bar{2}$0) whereas at pH higher than 4, it is textured (0001). These results are completely different from those obtained in chloride solutions. It is interesting to note that FENEAU and BRECKPOT observed at low pH (lower than 1.5 at 80°C and than 0.5 at 50°C) and higher temperature than in our experiments a (10$\bar{1}$0) texture for the α phase, i.e. a texture that we obtained at 25°C in the same range of current densities but at higher pH values. Another important remark is that we did not observe the orientation

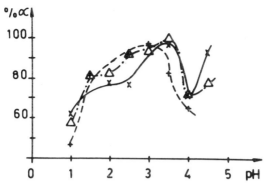

Figure 5 : Percentage of α phase in cobalt electrodeposits obtained in pure chloride solutions (18) at 25°C and 50 g/l Co^{2+}, for various current densities. X ─── 2.5 A/dm^2; Δ ─── • ─── 5 A/dm^2; + ─ ─ ─ 10 A/dm^2

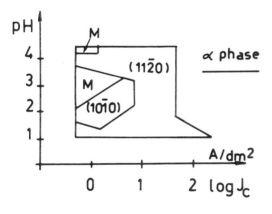

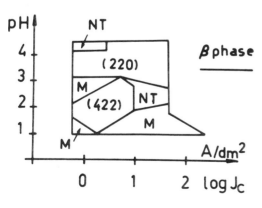

Figure 6 : Textures of α and β phases of cobalt electrodeposits from pure chloride solutions, at 25°C and 50 g/l Co^{++}. For α phase, M denotes mixed texture $(11\bar{2}0) + (10\bar{1}0)$. For β phase, M denotes mixed texture (220) + (422), and NT means no texture

(0001), usually considered as brittle, while the other orientations show some ductility.

Concerning the morphology of the deposits, the results for chloride solutions can be summarized as follows :
appearance : deposits showing a($11\bar{2}0$) texture are satin-like pale grey; those textured ($10\bar{1}0$) and the mixed texture produced at low current density and at pH between 2 and 3.5 are dull, dark grey to black; the mixed texture at pH 4.5 is pale-grey satin-like. Bright deposits (mirror-like) have also been produced at pH 4 from 2.5 to 20 A/dm² and at pH 3.5 at 10 A/dm².

Figure 7 shows electron micrographs of the surface of the most characteristic types of deposits.
Metallographic structure : Figure 8 shows the most important types of structure observed. Whereas in sulphuric solutions, very fine grained FT-UD structure with a nodular tendency is always observed, it can be seen in the picture that, at pH 4-4.5, BR and FT types can be obtained, even showing epitaxy with the cobalt substrate. It is only at rather high current densities (40 A/dm²) and low pH that similar structures to those obtained in sulphuric solutions are observed.

In conclusion, it can be said that replacing the sulphate anions by the chloride ones results effectively in activating the electrochemical behaviour of cobalt in pure aqueous solutions, despite the complexation effect of Cl^- on Co^{++} ;
- current efficiency are improved in acidic solutions
- overvoltages are lowered in acidic solutions
- exchange current densities are increased at least by three powers of ten
- metallographic structure types pertaining to lower inhibition are
observed. Crystallographic textures observed at higher temperatures in sulphuric solutions are obtained at room temperature in chloride solutions.

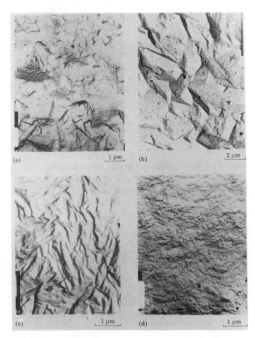

Figure 7 : Electron microscopy (carbon replicas) : characteristic appearance of the surface of cobalt deposits with varying textures. (a): ($11\bar{2}0$) texture satin-like type; (b) : mixed texture at pH 4.5; (c) : ($10\bar{1}0$) and mixed principal textures; (d): ($11\bar{2}0$) texture bright deposits

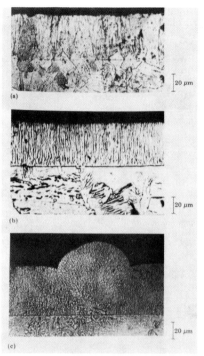

Figure 8 : Metallographic structure of the main types of cobalt deposits observed in chloride solutions.
(a) pH 4.5 −0.5 A/dm^2 BR type showing epitaxy with the substrate
(b) pH 4− 0.5 A/dm^2 − FT type showing some epitaxy with the substrate
(c) pH 1.5− 40 A/dm^2 −FT-UD nodular type (similar to most deposits obtained in sulphuric solutions under industrial conditions)

In industrial electrowinning practice, this results in the possibility of making in chloride solutions cobalt starting sheets and rather thick coherent deposits without nodules(19), which is not possible in sulphuric solutions. Our experimental results show also that high current densities (at least 80 A/dm^2) could be achieved easily in chloride solutions, even without drastic stirring.

For electroplating purposes, bright cobalt deposits can be obtained without organic additive in chloride solutions. This was investigated in the laboratory (21)(22), and very good results were achieved on steel sheets electroplated with a thin layer of copper.

For instance, figure 9 illustrates adherence and reflectivity of such a deposit.

Although very good papers have already been published on cobalt electrochemistry (see the literature survey in (17)(18) and (23)), further research on the subject should be concentrated on reaction mechanisms (what are the adsorbed species responsible for the overvoltage; are those species responsible for the drastic change in overvoltage observed in a short range of pH; is crystallization overvoltage an important part of the total overvoltage); on deposit morphology (how could crystallographic texture and metallographic structure of the deposits be related to reaction mechanism); and on deposit properties like hydrogen content , internal stresses, adherence to the substrate and surface roughness.

Copper electrodeposition from acid chloride solutions

Interest for copper electrowinning in chloride solutions in the laboratory arose mainly from its industrial potentialities, sothat fundamental research in the field was not as systematic as it was for cobalt. However, enough information was gathered to derive sound conclusions.

The process is very attractive at first sight for the following main reasons : it could help reducing air pollution by SO_2 since the leaching

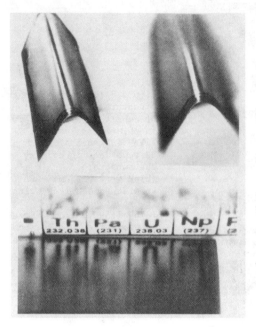

Figure 9 : Photographs showing the adherence and reflectivity of cobalt deposit on copper electroplated steel. Steel roughness amplitude is 5 to 10 micrometers. Copper plating (2 micrometers thickness) is achieved from a 50 g/l Cu^{++}, 100 g/l H_2SO_4 sulphuric solution at 4 A/dm^2, 25°C. Subsequantly, cobalt (5 micrometers thickness) is electroplated from a 50 g/l Co^{++}, 2 M in Cl^- (as $CoCl_2$ + NaCl; pH adjusted near to 3.7 by HCl) solution without organic additive, at 14 A/dm^2, 25°C.

of sulfide ores by chloride solutions leaves sulfur in the elemental state; it could also decrease by a factor of 2 to 4 the energy required for the electrowinning itself as compared with the conventional sulfuric process because the copper ions discharged at the cathode are monovalent instead of bivalent, and because the anodic oxygen evolution reaction is replaced by the oxidation of copper ions from the monovalent to the bivalent state at a less positive reversible potential and with a much smaller overvoltage; thus the overall voltage drop in the electrolytic cell is much lower than in sulfuric solutions, even taking into account the voltage drop in the diaphragm, for a productivity increased by a factor of 2 at the same current density.

However, most of the papers published on the subject report the high tendency of the deposit to grow in nodules, dendrites or powders(24)-(29). In fact, industrial processes were designed for powder production, the most striking one being the DUVAL's CLEAR process(30) in which advantage was taken of this type of deposit to make the electrolysis continuous.

In the laboratory, it was considered that full advantage of the chloride route could only be obtained if coherent cathodes of the same quality level as those of the conventional electrorefining process could be made, in order to avoid any further expense. So, again electrocrystallization is to be considered, and also the necessary purification level of the electrolyte before the electrowinning step.

The first experiments performed in the laboratory with a GOKHALE type electrolyte (31) (CuCl 0.7 N; NaCl 4N; HCl 0.5 N) in a diaphragm electrolytic cell confirmed both the potential advantages and the drawbacks of the process. Figure 10 gives the current efficiency and figure 11 the overall voltage drop in function of current density for three different temperatures, resulting in the specific energy consumption shown in figure 12. It is evident that electrowinning can be achieved with approximately 500kWh/T copper instead of the 2000 currently reported for the sulphate route.

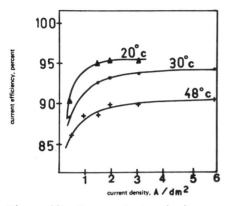

Figure 10 - Copper electrowinning in a pure aqueous chloride solution. Current efficiency in function of current density.

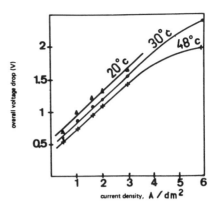

Figure 11 - Copper electrowinning in a pure aqueous chloride solution. Overall voltage drop in function of current density.

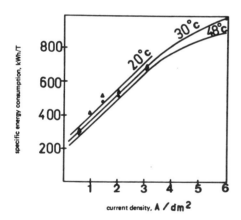

Figure 12 - Copper electrowinning in a pure aqueous chloride solution. Specific energy consumption in function of current density.

Figure 13 - Optical micrograph of a copper deposit from a GOKHALE pure chloride solution after 24 h.- 2.5 A/dm^2 -30°C (magnification 77 times)-10s ammoniacal attack-small initial BR layer, ending in FI

However, despite a rather thin initial layer of BR type coherent deposit, the deposit ends in isolated crystals pertaining to the FI type (figure 13), sothat an organic additive is necessary to avoid contamination of the deposit by entrapped electrolyte. CATHRO(32) had already obtained a dense cathode from cuprous chloride at 2 g/l gelatine, sothat rather high concentrations of inhibitors were considered.

Three industrial inhibitors were tried (33)(34). Jaguar between 25 and 250 mg/l, was found to have only a small influence on the deposit morphology. It reduces grain size and changes the type of crystallization but has no levelling effect and causes the formation of nodules at high concentration. Thiourea between 1 and 100 mg/l was found to have no effect at all on the metallographic structure of the deposits. Finally, gelatine A.S.F. (X) was studied between 50 and 1000 mg/l. In the small electrolytic cell used for the experiments and with the type of agitation used at that time, 300 mg/l gelatine were found necessary to obtain a porefree 3 mm thick deposit at 2 A/dm^2. Figure 14 shows the metallographic structure of the deposit, which seems to be at the limit between FT and UD. Later, in a large diaphragm cell allowing deposits on a 1 meter high 0.2 m wide copper starting sheet with a uniform flow of electrolyte at 0.35 m/s, 125 mg/l gelatine A.S.F. were enough at 30°C to achieve a 5 mm thick dense and smooth deposit on the whole cathode, however at 1 A/dm^2. However, the structure of the deposit was found to be very sensitive to local changes in electrolyte flow and/or current density, sothat with the same gelatine concentration but at 2 A/dm^2, a few localized dendrites appeared already in a 3 mm thick deposit. These results could certainly be improved by making use of an electrolytic cell designed to operate at a higher electrolyte flow rate, and also by increasing the acidity of the solution :gelatine is a protein that becomes cationic in acid solution (2).

The behaviour of the main impurities encountered in complex sulfide ores was also studied(33). Figure 15 shows the molar equilibrium potentials measured in the laboratory at 30°C in a 4 M NaCl and 0.5 M HCl solution. It can be seen that silver and bismuth are the two most probable contaminants of the copper deposits. However, if silver is effectively codeposited with copper, bismuth is not, even when its concentration reaches 5 g/l in solution. Finally, in a small laboratory cell, a 3 mm thick smooth and dense copper deposit was made at 300 mg/l gelatine ASF and 2 A/dm^2 in a solution containing at least 5 to 10 g/l of each impurity : zinc, cobalt, lead, iron, nickel and bismuth. Chemical analysis of the deposit, made in an independant laboratory was :

Bi < 2 ppm		Co < 1.5 ppm	
Ni < 1 ppm		Pb < 5 ppm	
Zn < 3 ppm		Fe < 2 ppm	

It may thus be concluded that satisfactory copper cathodes can be produced provided silver purification of the electrolyte is achieved prior to electrolysis and that an adequate combination can be found and controlled through the process between electrolyte flow rate, current density, duration of the electrolysis, gelatine concentration, and pH.

Figure 16 shows that cathodic overvoltages are low, even with 300 mg/l gelatine (34). They are certainly much lower than in sulphate solutions, especially taking into account that,in Cu^o production, current density in chloride solution should be multiplied by 2 against sulphate solutions. Making use of a method described by BESSON and GUITON(35), exchange current densities were determined by the double impulse galvanostatic method. Values in the order of 24 A/dm^2 were obtained without organic additive while values in the range of 0.2 to 2.0 A/dm^2 are currently recorded in pure sulphate solutions,except at very high current densities (36).

(X) atomisée, soluble à froid,i.e. atomized,soluble at room temperature

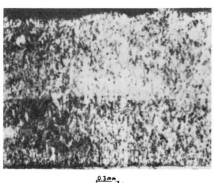

Figure 14 - Optical micrograph of a dense and smooth copper deposit obtained in pure chloride solution with 300 mg/l gelatine after 24 hrs - 2 A/dm^2 - 30°C

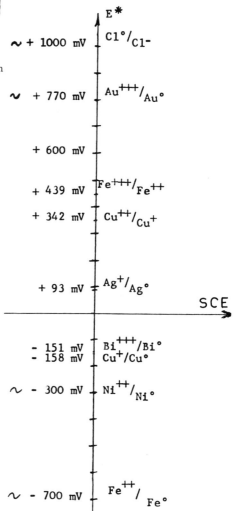

Figure 15 - Molar equilibrium potentials measured against SCE at 30°C in a 4 M NaCl - 0.5 M HCL solution.

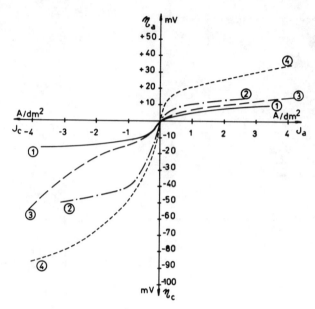

Figure 16 - Direct current steady state polarization curves for the reduction of Cu^I to Cu^0 from chloride solutions - Electrolyte composition:4M NaCl;0.5M HCl;1M CuCl:temperature:25°C;Curve ① ——— ① : without gelatine; curve ② — · —② with 300 mg/l gelatine. For comparison , reduction of Cu^{2+} to Cu^0 from sulphate solutions. Electrolyte composition :37g/l Cu^{++},192g/l H_2SO_4-Temperature:61°C Curve ③ — — —③ without gelatine Curve ④ -------- ④ with 10 mg/l gelatine .

If their values are recalculated for a CuCl concentration of 1 M, WINTER, COVINGTON and MUIR (31) found on a rotating disk copper electrode an exchange current density of 5 A/dm^2 and KARABINIS and DUBY (38) values between 46 and 50 A/dm^2 by an entirely different method. Anyway, the reaction is certainly highly reversible, sothat the sensitivity of the metal to inhibition is very low as was shown by the type of deposits and the difficulty to find a suitable organic additive. It was shown also (34) that the same Cu^I specie reacts at the cathode to be reduced to Cu^0 and at the anode to be oxidized to Cu^{II}, with a diffusion coefficient at 25°C of

$$D_{Cu^I} = (6.4 \pm 1).10^{-10} \text{ m}^2/s$$

This value is in agreement with a value given by WINTER, COVINGTON and MUIR(37):

$$D_{Cu^I} = 7.3 . 10^{-10} \text{ m}^2/s,$$ not too far away from a value found by KARABINIS and DUBY (38) :

$$D_{Cu^I} = 9-9.3 .10^{-10} \text{ m}^2/s.$$ Anyway, the order of magnitude is the same as for sulphate solutions, however for a monovalent ion instead of a bivalent one sothat under equivalent hydrodynamic conditions, the diffusion limiting current density would be reduced by a factor of two. Adequate stirring will thus be required in both anodic and cathodic half cells. It was also found that the anodic oxidation of Cu(I) to Cu(II) is highly reversible, with an exchange current density of at least 8 A/dm^2.

Concerning the species reacting at the electrodes, a spectrophotometric study of cuprous chloride solutions was performed, at high CuCl concentrations(39). It was found that the LAMBERT-BEER's law is satisfied for all concentrations of dissolved copper up to quasi saturation when the Cl^- (Na^+ + H^+) concentration does not exceed 4 M, and that the results could thus be interpreted in terms of $CuCl_2^-$ and $CuCl_3^=$ ions; the first ones being in excess to the second ones when (NaCl + HCl) is lower than 3.5 M. However when the solutions are 5 M in Cl^- and more than 0.6 M in CuCl, the LAMBERT-BEER's law is no more satisfied, and a third complex ion, probably $Cu_2Cl_4^{2-}$ appears. Figure 17 summarizes those results. In a 4 M NaCl, 0.5 M HCl, 1 M CuCl solution, the different cuprous species would thus be about 0.42 M $CuCl_2^-$; 0.55 M $CuCl_3^=$ and 0.03 M $Cu_2Cl_4^=$.

In an industrial electrowinning plant, the material balance of the global flowsheet plays a very important role in determining the possible electrolyte composition range. As the process was considered for complex sulfide ores, solubilities , densities and electrical conductivities of aqueous copper (I) and copper(II)chlorides in solutions containing other chlorides such as iron, zinc, sodium and hydrogen chlorides were measured(40). It was found that increasing $FeCl_2$ concentration decreases less the solubility of CuCl than an increase in $ZnCl_2$ concentration does. Moreover, increasing the total Cl^- concentration increases the CuCl solubility if Cl^- is added as $FeCl_2$ while it is decreased if Cl^- is added as $ZnCl_2$. NaCl solubility remains unchanged by $FeCl_2$ additions , and is increased by $ZnCl_2$ additions (Figure 18). For Cu(II) chloride, $FeCl_3$ additions decrease $CuCl_2$ solubility in a more drastic manner than $ZnCl_2$ additions; however, NaCl additions have a much higher influence (Figure 19). Concerning electrical conductivity, $FeCl_2$ additions slightly decrease the conductivity of Cu(I) solutions, but this can easily be compensated by a slight increase in temperature or in acidity. In the normal range of concentrations and temperature, specific electrical conductivity in the electrolytic cell is in the range of 0.25 to 0.35 $\Omega^{-1}cm^{-1}$, in good agreement with values obtained by WINTER, COVINGTON and MUIR(37).

In conclusion, copper electrowinning as pure dense coherent cathodes in

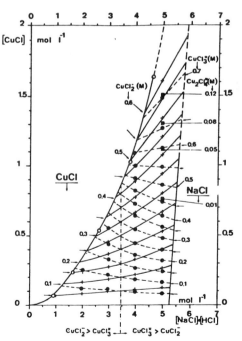

Figure 17 - Occurrence of complex species $CuCl_2^-$ (+); $CuCl_3^{2-}$ (•) and $Cu_2Cl_4^{2-}$ (■) in CuCl-NaCl-HCl solutions at 25°C as a function of the copper and chloride concentrations (pH=0.0) (39)

223

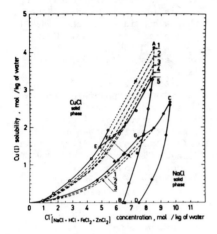
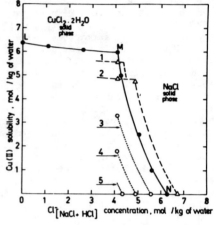

Figure 18-Cu(I) solubility in the CuCl-FeCl$_2$-ZnCl$_2$-NaCl-HCl system at 30°C. Continuous lines: Zn(II) iso-concentration; broken lines: Fe(II) iso-concentration. Zn(II) concentration, 0: zero; ●: 0.64m; ▲: 0.5m; Fe(II) concentration, 1: zero; 2: 0.25m; 3: 0.5m; 4: 0.75m; 5: ≥ 1m. Data of PETERS and al(41): ⊙

Figure 19-Solubility data in complex Cu(II) chloride solution at 50°C ● CuCl$_2$-NaCl-HCl-H$_2$O system. ▲ CuCl$_2$-ZnCl$_2$-NaCl-HCl-H$_2$O system; 1: [Zn(II)] =0.5m; 2: [Zn(II)] =1.5m; 0: CuCl$_2$-FeCl$_2$-NaCl-HCl-H$_2$O system; 3: [Fe(III)] =0.5m; 4: [Fe(III)] =1.5m; 5 [Fe(III)] =1.5m

aqueous chloride solutions can be considered as a potential alternate route to known processes for the treatment of complex sulphide ores. However, the following important factors are to be taken in consideration while making a decision :

  - leaching must occur at rather high temperature and in a multistage equipment in order to achieve high copper recoveries: three stages with approximately 1 hour residence time at 90°C in each are commonly needed to achieve more than 96% copper recovery. Zinc and especially lead are very readily dissolved. However, an important part of iron is dissolved at the same time, resulting in an increased flow of solution and in an increase in volume of the goethite precipitation equipment.

  A pretreatment of the concentrate with sulphur at 400-430°C was recently studied with some success in the laboratory, but the results need confirmation.

  - silver must be removed from the electrolyte prior to the electrolysis. A process based on the recovery of silver on a copper amalgam proved to be satisfactory at laboratory scale, but seems to be difficult to extrapolate at industrial scale.

  - the electrolytic cell including a diaphragm, the structure of the cathode must be under very good control so that no local dendritic growth could result in destroying the diaphragm while pulling out the cathodes.

  It is only if satisfactory solutions are found to those three main problems that the process could become competitive.

<u>Manganese electrodeposition from chloride solutions.</u>

  As the complete paper appeared recently(42), only the main conclusions will be recalled here.

The study was made in a diaphragm cell, allowing high electrolyte flow rates in both compartments. A comparison was made between pure sulphate, pure chloride, and mixed chloride-sulphate solutions.

Cathode current efficiencies are better in chloride solutions than in sulphate ones: values better than 80% are achieved up to 100 A/dm^2(figure 20);these results are further improved by increasing the electrolyte speed.

The main part of the overall voltage drop $U_{A/C}$ is ohmic, sothat straight lines are obtained for $U_{A/C}$ in function of current density. Figure 21 gives accordingly the specific power consumption for this cell, with a distance of 1.2cm between anode and cathode. It is evident from the figure that a pure chloride electrolyte is advantageous. Moreover, type A deposit refers to a dense pure γ-manganese deposit,usually with rather large crystals, probably pertaining to the BR type (figure 22)while type B deposit refers to a less textured mixedα and γ-manganese deposit,with smaller rounded crystals (figure 23).

Due to considerable hydrogen evolution at the cathode, only an order of magnitude of exchange current density could be determined from high current density measurements. Values between 4 and 9 A/dm^2 were obtained in pure chloride solutions against 0.5 to 2.2 A/dm^2 in pure sulphate solutions. These values and the structure of the cathode deposit show again the activating role of the chloride ions.

From a more practical point of view, it should however be kept in mind that if only a minor amount of chlorine is evolved at the anode even in pure chloride solutions, it is replaced by nitrogen evolution according to the following reaction :

$$2 NH_4^+ + 3 Cl_2 \longrightarrow N_2 + 8 H^+ + 6 Cl^-$$

sothat ammonium replenishment of the solution would be a drawback for manganese electrowinning from chloride solutions. Our results show that manganese electroplating could be satisfactorily achieved up to high current densities even without selenious acid addition to the electrolyte.

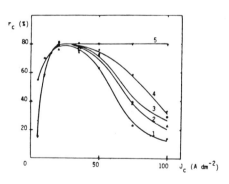

Figure 20-Cathodic current efficiency for manganese electrowinning-Electrolyte speed:0.5m/s-50°C.All solutions are 30g/l Mn^{++} and 50g/l NH$_4^+$.X_{Cl^-} = $\frac{Cl^-}{Cl^- + SO_4^=}$ molar fraction is as follows:0 for curve 1;0.33 for curve 2; 0.5 for curve 3;0.67 for curve 4 and 1 for curve 5.

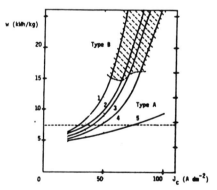

Figure 21 - Specific power consumption versus current density for manganese electrowinning-same conditions as in figure 20.Type A and type B refers to the two main types of deposits. The horizontal broken line refers to the value obtained in the conventional sulphate process,at 5 A/dm^2

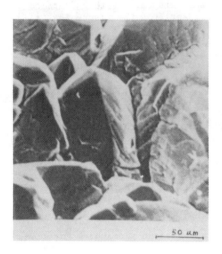
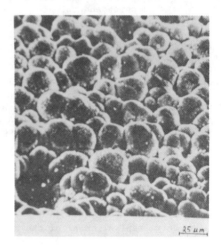

Figure 22-Surface micrograph (SEM)- Type A deposit. Amount of electricity 1.15 Ah on 0.1dm^2; current density: 50 A/dm^2; electrolyte speed: 4m/s; $X_{Cl^-} = 1$

Figure 23-Surface micrograph(SEM)- Type B deposit. Amount of electricity: 1.15 Ah on 0.1dm^2; current density: 100 A/dm^2; electrolyte speed: 0.5m/s; $X_{Cl^-} = 0.5$

General conclusions

Cobalt, copper and manganese electrodeposition was studied in the laboratory from pure chloride solutions. Comparing with sulphate solutions, the activating character of the chloride ion results in every case in shifting the type of metallographic structure of the deposits in the direction of a lower inhibition intensity. This is favourable to cobalt electrowinning, because cobalt behaves in sulphate solutions as an inert metal and is moved in chloride solutions to an almost intermediate metal behaviour. For copper however, it is moved from intermediate to normal, and the sensitivity of the metal to organic additives, usually necessary in this case to keep a smooth surface and avoid dendritic growth, is decreased. Sulphate solutions would thus be better in this case. Other potential advantages, mainly connected to specific energy consumption, justify the efforts devoted to this type of electrolyte. For manganese also, an important move in the diagram of figure 1 is observed in the same direction at high current densities. The advantage of a lower specific energy consumption for electrowinning in chloride solutions is partially lost, due to nitrogen evolution at the anode. However high current density electroplating in chloride solutions is to be considered.

LITERATURE

1. WINAND R. - Electrocrystallization - in Application of polarization measurements in the control of metal deposition - I.H.WARREN - Editor - Process Metallurgy 3 -(New-York, Elsevier,1984)p47 to 83

2. WINAND R., TROCH.G.,DEGREZ M., and HARLET Ph. - Industrial electrodeposition of copper. Problems connected to the behaviour of organic additions- ibid.-p 133 to 145

3. FISCHER H. - Elektrolytische Abscheidung und Elektrokristallisation von Metallen (Berlin - Springer Verlag, 1954), 729

4. WINAND R. - Contribution à l'étude de l'électrocristallisation en sels fondus et des phénomènes connexes - Application au cas particulier du zirconium - Ph.D.Thesis - Université Libre de Bruxelles - 1960

5. WINAND R. - same tittle- Mém.Scient.Revue Mét. - $\underline{58}$, 1961,25-35

6. WINAND R. - Procédés électrolytiques de recouvrement - Dépôts cathodiques et anodisation - Revue ATB Mét. - $\underline{10}$, 1970,53-61

7. WINAND R. - Electrochimie - Notes de cours (Bruxelles, The University, 1973)

8. WINAND R. - Electrocrystallization of copper - Trans.Sect.C. - IMM-$\underline{84}$, 1975, 67-75

9. FISCHER H. - Wiskungen der Inhibitoren bei der Elektrokristallisation Electrochim.Acta $\underline{2}$, 1960, 50-96

10. THOUMSIN Fr. - Recherche d'un procédé de fabrication de cobalt électrolytique marchand par électroextraction. Problèmes posés - Mise au point par l'Union Minière du Procédé de Luilu - Revue ATB Mét. - $\underline{9}$ (1969)33-42

11. FENEAU C. and BRECKPOT R. - Optimalisation des conditions d'électroextraction du cobalt - Revue ATB Mét. - $\underline{9}$ (1969) 115-125

12. BRUMMIT R.N. - A review of cobalt production in Central Africa - Proceedings of Cobalt 80 - 10th Annual Hydrometallurgical Meeting of CIMM - Edmonton - October 26-28 , 1980

13. AIRD J., CELMER R.S. and MAY A.V. - New cobalt production from RMC's Chambishi roast-leach-electrowin process - ibid.

14. LENOIR P., VAN PETEGHEM A. and FENEAU C. - Extractive metallurgy of cobalt - Proceedings of the International Conference on Cobalt : Metallurgy and Uses - Benelux Metallurgie - Brussels, November 10-13, 1981 -Published by Revue ATB Met. Vol.1 - p. 51-62

15. SHUNGU T. and CHARLES Ph. - Recent improvements in metallurgical processing at the metallurgical plant of Shituru (GCM Zaïre) - ibid.-Vol.1- p. 73-78

16. PEARSON D. - Recovery of cobalt from metallurgical wastes - ibid.-Vol.1 p 139-148

17. SCOYER J. - Contribution à l'étude de l'électrocristallisation du cobalt en solutions de chlorures - Ph.D.Thesis - Université Libre de Bruxelles - 1977

18. SCOYER J. and WINAND R. - Electrocrystallization of cobalt from acid chloride solution - Surface Technology - $\underline{5}$ (1977)169-204

19. FUJIMORI M., ONO N., TAMURA N. and KOHGA T. - Electrowinning from aqueous chloride in SMM's nickel and cobalt refining process - Chloride electrometallurgy - P.D.PARKER - Editor - Proceedings of a symposium sponsored by the Electrolytic Process Committee of TMS - AIME - 111th AIME

Annual Meeting - Dallas - February 15-16, 1982 - p.155-166

20. GRAVEY G. and PEYRON F. - A French producer of cobalt : Métaux spéciaux S.A. - same Proceedings as 14.

21. DEGREZ M., HUBEAU V. and WINAND R. - Cobaltage brillant sans inhibiteur Revue ATB Mét. - XX(1980)95-101

22. DEGREZ M. , HUBEAU V. and WINAND R. - Further progress in bright cobalt electroplating in chloride solution without levelling agent -Poster session Benelux Metallurgie - Same Proceedings as 14 - Abstract in Vol.2 of the Proceedi ngs, p 317-318

23. SCOYER J. and WINAND R. - Electrocrystallization of cobalt in chloride solutions - Proceedings of Chloride Hydrometallurgy , an International Symposium organized by Benelux Metallurgie, Brussels, September 26-28,1977, p 294-318

24. GOKHALE S.D. - J.Sci.Ind.Res.10 B (1951)316

25. MITTER G.C. , ROSE B.K., DIGHE S.G., GOKHALE Y.W. and CHOUDHURY B.P.- J.Sci.Ind.Res. - 20 D (1961)114

26. ARAMUTHAN V. and SRINIVASAN R. - J.Sci.Ind.Res. - 21D (1962)381

27. KURESI P.R. ,- (Cymet Process) - U.S.Pat. 3,673,061 (1972)

28. ATWOOD G.E. and CURTIS C.H. - U.S.Pat. 3,785,944 (1972)

29. MUSSLER R.E., OLSEN R.S. and CAMPBELL T.T. - US Bureau of Mines - Report of Investigation 8076 (1975)

30. SCHWEITZER F.W. and LIVINGSTONE R.W. - Duval's Clear Hydrometallurgical Process - in Chloride Hydrometallurgy - Same Proceedings as 19, pp.221-227

31. ANDRIANNE P.A. , DUBOIS J.P. and WINAND R. - Electrocrystallization of copper in chloride aqueous solutions - Met.Trans. B, $\underline{8}$ (1977)315-321

32. CATHRO K.J. - Extractive Metallurgy of Copper - A symposium held during the 105th AIME Annual Meeting - Las Vegas, February 1976, p. 776

33. ALBERT L. and WINAND R. - Characterization of the behaviour, in chloride solutions, of copper and of the most important impurities contained in a sulphide concentrate, in order to recover the copper - same Proceedings as 14 - p. 319-335

34. ALBERT L. and WINAND R. - Copper electrowinning in chloride aqueous solutions - Chloride electrometallurgy - Same Proceedings as 19 p. 189-202

35. BESSON J. and GUITON J. - Manipulations d'Electrochimie - Masson et Cie Paris (1972)

36. DEGREZ M. and WINAND R. - Détermination des paramètres cinétiques de l'électrodéposition du cuivre à haute densité de courant. Cas des solutions sulfuriques sans inhibiteur. Electrochimica Acta - $\underline{29}$ (1984)365-372

37. WINTER D.G. ; COVINGTON J.W. and MUIR D.M. - Studies related to the electrowinning of copper from chloride solutions - Chloride Electrometallurgy - Same Proceedings as 19 - p.167-188

38. KARABINIS V.D. and DUBY P. - Chronopotentiometric studies of copper deposition from chloride electrolytes - ibid. - p.203-220

39. FONTANA A., VAN MUYLDER J.and WINAND R.- Etude spectrophotométrique de solutions aqueuses chlorurées de chlorure cuivreux, à concentrations élevées - Hydrometallurgy $\underline{11}$ (1983)297-314

40. BERGER J.M. and WINAND R. - Solubilities, densities and electrical conductivities of aqueous copper (I) and copper (II) chlorides in solutions containing other chlorides such as iron, zinc, sodium and hydrogen chlorides - Hydrometallurgy 12 (1984) 61-81

41. PETERS M.A. and JOHNSON R.K. - U.S.Pat. 4,101,315 (1978)

42. PARISSIS G. and WINAND R. - Manganese electrodeposition from sulphate, chloride and mixed solutions, including high current densities - Chloride Electrometallurgy - Same Proceedings as 19 - p 131-154

EXTRACTION OF SILVER THROUGH COMPLEXATION IN THE VAPOR PHASE

J. P. Hager
Department of Metallurgical Engineering
Colorado School of Mines
Golden, Colorado 80401

M. C. Rupert
Handy and Harmon
1770 Kings Highway
Fairfield, Connecticut 08430

W. A. May
AMAX Research Center
1600 Huron Parkway
Ann Arbor, Michigan 48106

The formation of the vapor complex $AgFeCl_4(g)$ has been identified and its thermodynamic properties determined. The vapor pressure of silver chloride can be enhanced by a factor of 15-20 due to the formation of the vapor complex.

The vapor pressure of both $AgCl(g)$ and $AgFeCl_4(g)$ were measured using a double-zone transpiration technique. Gibbs free energies have been determined for the vapor transport reactions.

Laboratory kiln tests were conducted to confirm the enhanced recovery of silver through the formation of the $AgFeCl_4(g)$ vapor complex and to indicate the opportunities for process development using this new vaporization chemistry.

Introduction

In 1966 Kellogg (1) reviewed the use of vapor-state transport of metals in extractive metallurgy and concluded that through the expansion of knowledge of vaporization chemistry many new opportunities seemed possible. The enhanced volatilization of gold and silver through the formation of vapor complexes has been an area of research at Colorado School of Mines for several years. In an earlier publication (2) the formation of the vapor complex $AuFeCl_6(g)$ was identified by means of the reaction

$$2Au(s) + Fe_2Cl_6(g) + 3Cl_2(g) = 2AuFeCl_6(g) \qquad (1)$$

The vapor pressure of $AuFeCl_6$ was found to be a factor of 25 greater than the vapor pressure of Au_2Cl_6 formed by the reaction

$$2Au(s) + 3Cl_2(g) = Au_2Cl_6(g) \qquad (2)$$

In a subsequent paper (3) the enhanced volatility of AgCl was noted in the presence of either $Cu_3Cl_3(g)$ or $FeCl_3(g)$ due, presumably, to the formation of the complexes $AgCu_2Cl_3(g)$ and $AgFeCl_4(g)$.

The importance of the formation of chloride complexes of silver can be inferred from the early work of Varley et al (4), who reported an enhanced volatility of AgCl in the presence of pyrite. Likewise, Rampacek et al (5) and Pollandt and Pease (6) reported that silver volatilized with copper chloride in the copper segregation process. The complexing action of $AlCl_3$ and $FeCl_3$ on AgCl has also been reported by Schafer (7,8) with the species $AgAlCl_4(g)$ being identified by mass spectrometry.

The purpose of this study was to investigate the specific vaporization chemistry that gives rise to an enhanced volatility of AgCl in the presence of $FeCl_3$ and to apply the results to the vapor phase extraction of silver from an ore.

Experimental

To measure the vapor pressure of the Ag-Fe-Cl vapor complex the double-zone transpiration system described by James and Hager (3) was used. The vapor pressure of pure AgCl was first measured to provide a baseline for the subsequent measurements on the vapor pressure of the Ag-Fe-Cl complex.

To examine the feasibility of vaporizing silver from an ore, as a Ag-Fe-Cl complex, extraction studies were carried out in a laboratory-size rotary kiln. Conditions for these runs were determined from the results of the fundamental studies conducted with the transpiration apparatus.

Transpiration Studies

Apparatus. A schematic diagram of the transpiration apparatus is shown in Figure 1 with an overall view given in Figure 2. The unit is similar to transpiration systems described in the literature (9, 10); however, there is one significant difference in that two furnaces were placed end to end. In this way, the carrier gas could be doped with $FeCl_3(g)$ before equilibrating with the $AgCl(l)$ sample.

Two Marshall resistance furnaces with multiple taps on the furnace windings were used to provide for constant temperature zones at each sample equilibration point and for an increase in temperature from the $FeCl_3(g)$ saturation zone to the $AgCl(l)$ equilibration zone. The reaction tube, the sample carrier tube, the condenser sheath, and the sample boats were all made of high purity recrystallized alumina.

The sample carrier tube provided a convenient means of rapidly inserting and removing the samples. The water cooled cold-finger placed inside the condenser sheath provided for the condensation of all volatile species in the gas mixture leaving the AgCl equilibration zone.

The carrier gas was either purified chlorine or a mixture of purified chlorine and purified argon. Control of the composition and flow rate was achieved through the use of capillary flow meters.

Procedure. Samples of high purity AgCl and $FeCl_3$ obtained from a chemical supply house were further purified before use in the transpiration apparatus. The chlorides were heated in a stream of dry chlorine at a temperature just below their melting point for a period of 12 hours. The purified salts were loaded into the sample boats under the protective atmosphere of a dry box and then transferred to the transpiration apparatus in an atmosphere of dry nitrogen.

For experiments designed to measure the vapor pressure of pure AgCl, the $FeCl_3$ sample boats were filled with inert Al_2O_3. For experiments designed to measure the vapor pressure of the Ag-Fe-Cl complex, iron chloride was added to these sample boats and the temperature adjusted to give the desired partial pressure of $FeCl_3(g)$ in the silver chloride equilibration zone.

The furnaces were preheated to the desired set point and the sample carrier tube inserted into the reactor tube under a flow of the carrier gas mixture. After an equilibration period of 120-180 minutes, the sample carrier tube was withdrawn and the carrier gas switched to argon. The condenser assembly was then removed and the deposited material dissolved in a strip solution. An atomic absorption analysis for silver or for silver and iron was performed on the strip solutions. For experiments where $FeCl_3(g)$ was added to the carrier gas, samples of the equilibrated silver chloride were submitted for analysis to determine the degree of contamination of the molten silver chloride with iron chloride.

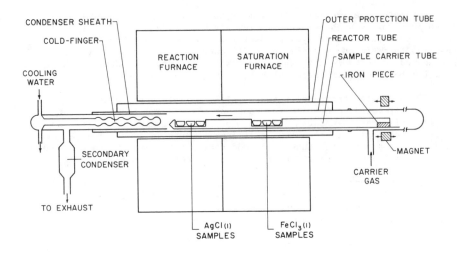

Figure 1 - Schematic diagram of the transpiration apparatus.

Figure 2 - Overall view of the transpiration system.

Rotary Kiln Studies

Apparatus. A flow schematic for the rotary kiln system is shown in Figure 3. Figure 4 is an overall view of the kiln reactor system. The system was constructed by Hazen Research of Golden, Colorado and then modified to provide for the insertion of the same condenser assembly as used in the transpiration reactor.

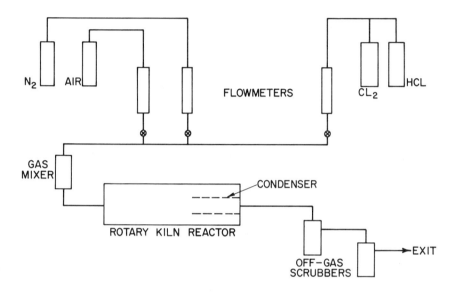

Figure 3 - Flow schematic for rotary kiln reactor

The split shell furnace was fitted with a fused quartz reactor fabricated with ground-glass ball joints at each end. The ball joints thus provided for a gas-tight seal as the reactor was rotated. The quartz reactor was dimpled so that the charge would be lifted and mixed during rotation. A sprocket was fastened to one end of the reactor with asbestos tape and set screws and connected by a machine chain to a variable speed gearmotor. The matching ball joint sections on each end of the reactor were supported on wooden bearing blocks which provided for the support and alignment of the reactor.

The quartz reactor had a reaction zone volume of 650 ml and could handle charges of up to 175 grams. For this study, 160 gram charges were used. A gas mixing train was constructed to prepare sweep gas mixtures of N_2, O_2, HCl, and Cl_2. For this study, however, only chlorine or nitrogen at a flow of 750 ml/min was used.

Procedure. To minimize the potential side reactions that could occur with some of the metals present in natural silver ores, a synthetic ore was prepared and then doped with Fe_2O_3 to different levels. The synthetic ore was prepared from chemically cleaned silica sand that was mixed with a silver nitrate solution, vacuum filtered, and then calcined

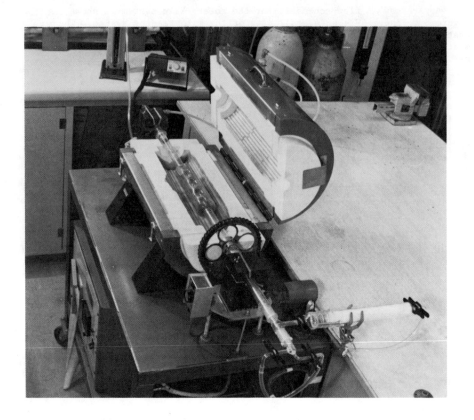

Figure 4 - Overall view of rotary kiln reactor system.

in air at 550 C to decompose the silver nitrate to metallic silver. The targeted head assay was 9.8 troy ounces per ton in order to give approximately 50 mg of silver available for extraction in each kiln charge.

After charging the ore, the furnace was turned on and the charge brought up to temperature in a sweep stream of nitrogen. The sweep gas was then changed from nitrogen to chlorine. After a 30, 45, or 60 minute extraction period, the sweep gas was switched from chlorine back to nitrogen, the furnace power was turned off, and the furnace was opened to provide for a rapid cooling of the charge. The condenser assembly was then removed and the deposited material dissolved in a strip solution for subsequent assay on the atomic absorption spectrophotometer. The processed charge was also submitted for assay.

A temperature of 700° C was selected for these tests based on the extraction enhancement calculations made at the completion of the fundamental studies. Also, calculations showed that the in situ generation of the $FeCl_3(g)$ complexing agent might best be controlled by the reaction of a Cl_2 sweep gas with Fe_2O_3 (which could be added to the ore).

Results

Ag-Cl Vapor Pressure Measurements

The principal vapor species in equilibrium with molten AgCl has been established (11-15) as $AgCl(g)$. Thus, at chlorine pressures high enough to prevent the decompositon of the $AgCl(l)$ to metallic silver, the vaporization reaction is

$$AgCl(l) = AgCl(g) . \qquad (3)$$

The vapor pressure of $AgCl(g)$ can be calculated from the measurement of the moles of silver transported, the moles of carrier gas species, and the total pressure, according to the equation

$$P_{AgCl} = M_{AgCl} P_T / (M_{AgCl} + M_{Cl_2} + M_{Ar}) \qquad (4)$$

where

M_{AgCl} = moles of AgCl transported,
M_{Cl_2} = moles of Cl_2 in the carrier gas,
M_{Ar} = moles of Ar in the carrier gas, and
P_T = total pressure in the reactor.

The experimental results obtained for reaction (3) are given in Figure 5 along with the results of Von Wartenburg and Bosse (16) (1528-1715 K using a boiling point technique), Maier (17) (1127-1429 K using a static technique), and Bloom et al (18) (1301-1533 K using a boiling point method). Since the temperature range of this study (930-1072 K) is considerably below those of the previous investigations, the lines shown for the previous studies are based on an extrapolation of their data using a second law analysis. The line for Bloom et al (18) is an exception and is based on the vapor pressure equation given in their paper.

To determine that equilibrium saturation was established for reaction (3), the effect of flow rate of a pure chlorine carrier gas on the vapor pressure was studied at constant temperature (see Figure 6). To test for the presence of higher valence state gaseous silver chlorides, experiments were conducted with different partial pressures of chlorine in the carrier gas (see Figure 7).

The thermodynamic consistency of the results obtained for reaction (3) was examined using second and third-law analysis techniques (19-21). The sigma function plot for reaction (3) is shown in Figure 8 (for determination of the second-law heat of reaction at 298 K) and a plot of

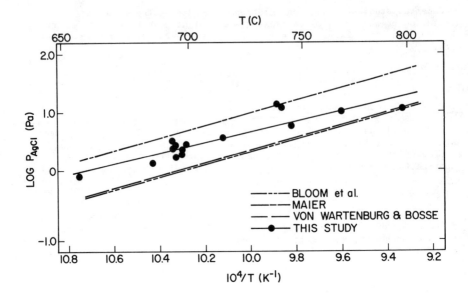

Figure 5 - Vapor pressure as a function of temperature for the reaction AgCl(l) = AgCl(g)

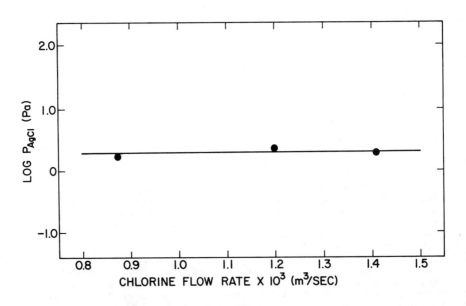

Figure 6 - Vapor pressure of AgCl(l) at 973 K as a function of carrier gas flow rate.

the heats of reaction at 298 K determined by the third-law method is given in Figure 9. A tabular comparison of the results is given in Table I.

Table I. Heat of Reaction at 298 K for AgCl(l) = AgCl(g)

Investigator	ΔH°_R (298 K)			
	Second Law		Third Law	
	Kcal·Mole^{-1}	(KJ·Mole^{-1})	Kcal·Mole^{-1}	(KJ·Mole^{-1})
This Study	46.903	(196.242)	47.271	(197.782)
Maier	44.648	(186.807)	43.017	(179.983)
Von Wartenburg & Bosse	54.208	(226.806)	43.855	(183.489)

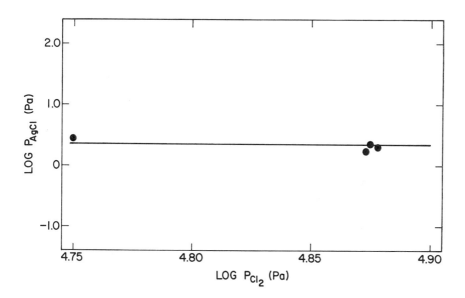

Figure 7 - Vapor pressure of AgCl(l) at 973 K as a function of chlorine partial pressure in the carrier gas.

Results are not given for Bloom et al (18) since detailed experimental data were not included in their paper. The C_p^0 data used in the above analyses were taken from (22) and the S° data were taken from (23).

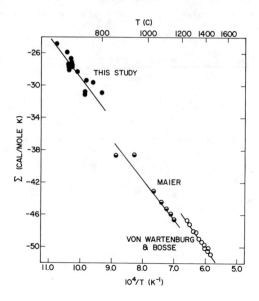

Figure 8 - Sigma function for the reaction AgCl(l) = AgCl(g) as a function of temperature.

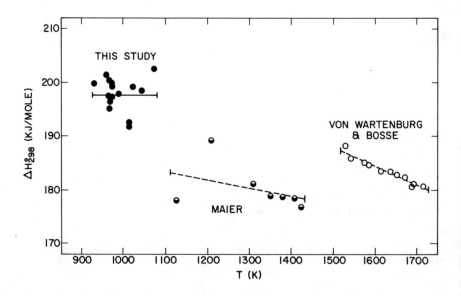

Figure 9 - Third-law heat of reaction at 298 K for the reaction AgCl(l) = AgCl(g) as a function of temperature.

The thermodynamic properties of reaction (3) can be calculated from the sigma function plot (Figure 8). The results are given in Table II. The standard state for AgCl(g) is one atmosphere pressure ideal gas.

Table II - Thermodynamic Results for the Reaction AgCl(l) = AgCl(g)

(930-1072 K)

$\Delta H°_T$ = $-7.07T + 0.34 \times 10^5 T^{-1} + 4.89 \times 10^4$ (cal·mole^{-1})
= $-29.58T + 1.42 \times 10^5 T^{-1} + 2.05 \times 10^5$ (J·mole^{-1})

$\Delta G°_T$ = $7.07T \ln T - 77.86T + 1.70 \times 10^4 T^{-1} + 4.89 \times 10^4$ (cal·mole^{-1})
= $29.58T \ln T - 3.26 \times 10^2 T + 7.11 \times 10^4 T^{-1} + 2.05 \times 10^5$ (J·mole^{-1})

Log P_{AgCl} = $-1.5458 \ln T - 3.7169 \times 10^3 T^{-2} - 1.0682 \times 10^4 T^{-1} + 17.0155$ (ATM)
= $-1.5458 \ln T - 3.7169 \times 10^3 T^{-2} - 1.0682 \times 10^4 T^{-1} + 22.0208$ (Pa)

(1 atm = 101325 Pa)
(1 cal = 4.184J)

Ag-Fe-Cl Vapor Complex Measurements

With the vapor pressure of AgCl(g) according to reaction (3) established, the formation of the Ag-Fe-Cl vapor complex could be addressed. If the vapor complex species is AgFeCl$_4$ as proposed by (3), the following set of simultaneous reactions will occur in the liquid silver chloride equilibration zone

$$AgCl(l) = AgCl(g) \quad (3)$$

$$AgCl(l) + FeCl_3(g) = AgFeCl_4(g) \quad (5)$$

The partial pressure equation for reaction (3) can be combined with mass balance equations for silver, iron, and the total gaseous species to yield the following set of four simultaneous equations in four unknowns.

$$P_{AgCl} = (M_{AgCl}/M_T) P_T \quad (6)$$

$$M_{Fe} = M_{FeCl_3} + M_{AgFeCl_4} \quad (7)$$

$$M_{Ag} = M_{AgCl} + M_{AgFeCl_4} \quad (8)$$

$$M_T = M_{AgCl} + M_{FeCl_3} + M_{AgFeCl_4} + M_{Cl_2} + M_{Ar} \quad (9)$$

where

M_{Fe} = moles of iron in the condensate,
M_{Ag} = moles of silver in the condensate, and
M_T = total moles of gaseous species.

However, the value of P_{AgCl} in equation (6) can only be calculated (using the results in Table II) if the activity of the liquid silver chloride can be established. The analytical results given in Table III show that the degree of contamination of the liquid AgCl with $FeCl_3$ was slight and that to a good approximation the activity of AgCl(l) can be taken as unity (in this composition range a_{AgCl} = mole fraction of AgCl).

Table III - Results of Commercial Analysis Performed on $FeCl_3$ Contaminated AgCl Samples

SAMPLE NUMBER	Wt % $FeCl_3$	Wt % AgCl	Mole % $FeCl_3$	Mole % AgCl	Comments
AgFeCl-14	0.639	99.361*	0.565	99.435*	Analysis Performed by
AgFeCl-15	0.581	99.419*	0.514	99.486*	Colo. Analytical Labs.
AgFeCl-16	0.378	99.622*	0.334	99.666*	Brighton, Colorado
AgFeCl-17	0.407	99.593*	0.359	99.641*	* -- by difference
AgFeCl-18	0.367	96.456	0.337	99.663	Analysis Performed by
AgFeCl-19	0.421	98.715	0.376	99.624	Hazen Research Inc.,
AgFeCl-20	0.084	98.316	0.076	99.924	Golden, Colorado

Assuming that $AgFeCl_4(g)$ is the vapor complex, the equilibrium constant for reaction (5) was then calculated. The results are given in Figures 10 and 11.

To help established the molecularity of the complex, a series of experiments were run at a constant equilibration temperature for the liquid silver chloride but with different levels of $FeCl_3(g)$ saturation in the iron chloride equilibration zone. If the complexation reaction is written in the form

$$X AgCl(l) + Y FeCl_3(g) = Ag_X Fe_Y Cl_{(X+3Y)} (g) \quad (10)$$

and the activity of the liquid silver chloride is taken as unity, the equilibrium constant can be written in the following form

$$LOG\ P_{Ag_X Fe_Y Cl_{(X+3Y)}} = Y\ LOG\ P_{FeCl_3} + LOG\ K \quad (11)$$

where Y can be obtained from a plot of LOG $P_{COMPLEX}$ versus LOG P_{FeCl_3}. Such a plot in shown in Figure 12 with the result that the value of Y is unity.

Although mass spectrometer experiments were conducted to help determine the value of X in reaction (10) no meaningful results could be obtained because of difficulties arising from the large difference in vapor pressure for the two salts. Given that Y has been determined to be

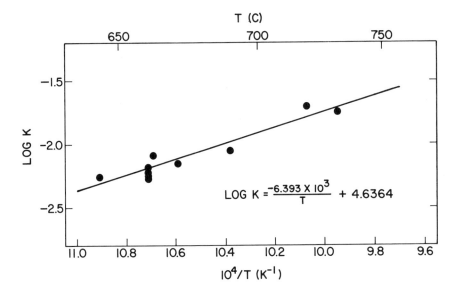

Figure 10 - Equilibrium constant for the reaction $AgCl(l) + FeCl_3(g) = AgFeCl_4(g)$ as a function of temperature.

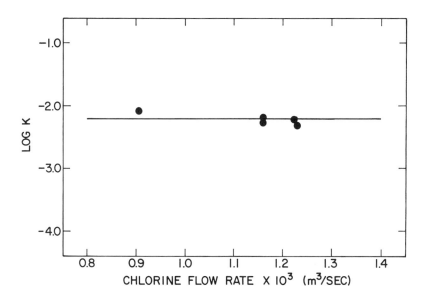

Figure 11 - Equilibrium constant for the reaction $AgCl(l) + FeCl_3(g) = AgFeCl_4(g)$ at 933 K as a function of carrier gas flow rate.

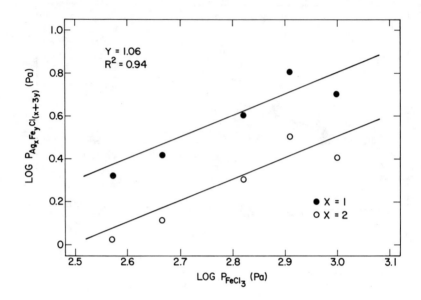

Figure 12 - Determination of Y in the reaction $XAgCl(l) + YFeCl_3(g) = Ag_XFe_YCl_{(X+3Y)}(g)$ at 933 K.

unity and that for a similar system (Ag-Al-Cl) the species $AgAlCl_4(g)$ has been determined mass spectrometrically (8) the composition of the Ag-Fe-Cl complex was taken as $AgFeCl_4$.

The thermodynamic properties of $AgFeCl_4(g)$ were calculated from the results given in Figure 10 and are tabulated in Table IV.

Table IV - Thermodynamic Results for the Reaction $AgCl(l) + FeCl_3(g) = AgFeCl_4(g)$
(917 - 1000 K)

$\Delta H°$ = 29,256 (cal·mole^{-1})
 = 1.22 x 10^5 (J·mole^{-1})
$\Delta G°_T$ = 29,256 - 21.21T (cal·mole^{-1})
 = 1.22 x 10^5 - 88.75T (J·mole^{-1})
$\Delta S°$ = 21.21 (cal·mole^{-1} °K^{-1})
 = 88.74 (J·mole^{-1} °K^{-1})

In the absence of C_p^0 data for $AgFeCl_4(g)$ the value of ΔC_p^0 for reaction (5) was taken as zero. The standard state for $AgFeCl_4(g)$ is one atmosphere pressure ideal gas.

Rotary Kiln Extraction Tests

To identify the proper experimental conditions for the rotary kiln tests, the enhancement of the vapor pressure of silver chloride was calculated for two different temperatures for the use of Fe_2O_3 with Cl_2 as the sweep stream. The results are given in Table V.

Table V - Enhancement of the Vapor Pressure of Silver Chloride Due to The Formation of the Complex $AgFeCl_4(g)$

Temperature °C	P_{AgCl} (mm Hg)	P_{FeCl_3} (1) (mm Hg)	P_{AgFeCl_4} (mm Hg)	Apparent % Increase P_{AgCl}
727	0.03	27	0.49	1629
827	0.23	68	4.56	1980

(1) equilibrium partial pressure for the reaction
$2Fe_2O_3(s) + 6Cl_2(g) = 4FeCl_3(g) + 3O_2(g)$

Based on these results a temperature of 700 C was selected as providing a realistic test of the ability of the complex $AgFeCl_4(g)$ to achieve high levels of extraction in a reasonable period of time.

The results for the kiln extraction tests are given in Figures 13 and 14. In Figure 13 all kiln charges contained in excess of 2% Fe_2O_3 based on the results shown in Figure 14. The % Ag extraction was based on a head versus tails assay of the charge. However, the silver mass balance closure averaged 90% based on the head and tails assays and the silver collected in the condenser.

Discussion of Results and Conclusions

The vapor pressure measurements for the reaction

$AgCl(l) = AgCl(g)$ (3)

are seen to be in reasonable agreement with earlier measurements (16-18). However, the results of this study give much better agreement between the second-law and third law $\Delta H°$ (298K) values. Furthermore, there is no tendency for the third-law $\Delta H°$ (298K) values to be dependent on the measurement temperature, which is not the case for the data of (16) and (17). While the results of this study do show more scatter than the results of the other studies, this is no doubt due to the very low vapor pressures being measured (0.02 MM Hg at 700 C) in contrast to the other studies. The results of this study have also been shown to be independent of the carrier gas flow rate (measure of equilibrium saturation) and do not depend on the partial pressure of chlorine in the carrier gas (indication of no higher valence state gaseous silver chlorides). The results of this study appear to give a more accurate measure of the vapor pressure of AgCl(l) in the temperature range 650-800C that has been available previously.

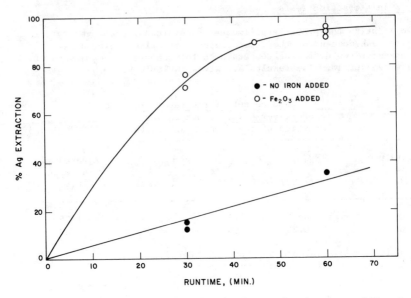

Figure 13 - Silver extraction from synthetic ore in the rotary kiln at 700 C using a chlorine sweep stream.

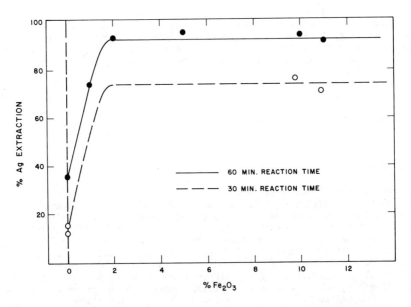

Figure 14 - Silver extraction from synthetic ore in the rotary kiln at 700 C as a function of the % Fe_2O_3 in the charge (chlorine sweep stream).

The results of this study show that in the presence of $FeCl_3(g)$ the vapor complex species $AgFeCl_4(g)$ can be formed according to the reaction

$$AgCl(l) + FeCl_3(g) = AgFeCl_4(g) \qquad (5)$$

While there is some uncertainty in the value of X in the formula $Ag_XFeCl_{4+X}(g)$ the value is almost surely unity based on results in analagous systems ($AgAlCl_4$). The second law plot for reaction (5) (see Figure 10) shows very little scatter and allows for the accurate calculation of the thermodynamic results given in Table IV. The results shown in Figure 11 were taken as evidence that equilibrium saturation of the carrier gas with $AgFeCl_4(g)$ was achieved. The measured mole fraction of silver chloride in the equilibrated molten silver chloride samples was certainly large enough (average = 0.996) to accurately define the activity of AgCl in these samples. The thermodynamic results given in Table IV will now allow for an accurate calculation of the vapor pressure of the complex $AgFeCl_4(g)$ under different processing conditions.

An enhancement in the volatility of silver chloride by a factor of 15-20, or higher, is clearly indicated by the results of this study. The silver extraction results from the rotary kiln experiments show that it is possible to achieve better than 90% extraction at 700 C with only an hour of contact time. Further, the required minimum concentration of Fe_2O_3 appears to be only 2% at 700 C. At higher temperatures the rate of extraction can be expected to be even higher. These results should provide the basis for process development efforts for new extraction processes for silver.

Acknowledgements

The authors are pleased to acknowledge the support of ASARCO, Incorporated through a research grant and through the sponsorship of a graduate fellowship in extractive metallurgy. The help and support of Drs. Roe and Kudryk at ASARCO's central research department are especially appreciated.

References

1. H.H. Kellogg, "Vaporization Chemistry in Extractive Metallurgy," Transactions of the Metallurgical Society of the AIME, 236 (1966) p. 602.

2. J.P. Hager and R.B. Hill, "Thermodynamic Properties of the Vapor Transport Reactions in the Au-Cl System by a Transpiration - Mass Spectrometric Technique," Metallurgical Transactions, 1 (1970) p. 2723-31.

3. S.E. James and J.P. Hager, "High Temperature Vaporization Chemistry in the Gold-Chlorine System Including Formation of Vapor Complex Series of Gold and Silver with Copper and Iron," Metallurgical Transactions, 9B (1978) p. 501-508.

4. T. Varley, et al., "The Chloride Volatilization Process of Ore Treatment," U.S. Bureau of Mines Bulletin 211, 1923.

5. C. Rampacek, W.A., and P.T. Waddleton, "Treating Oxidized and Mixed Oxide-Sulfide Copper Ores by the Segregation Process," U.S. Bureau of Mines Report of Investigations 5501, 1959.

6. F. Pollandt and M.E. Pease, "Extraction of Copper and Silver by the Segregation Process in Peru," Institute of Mining and Metallurgy Bulletin, 69 (1960) pp. 687-97.

7. H. Schafer, "Gas Complexes with $AlCl_3$, AlI_3 or $FeCl_3$, Their Abundance and Stability," Z. Allorg. Allg. Chem., (1975), p. 151.

8. H. Schafer, "Gaseous Chloride Complexes with Halogen Bridges - Homo-Complexes and Hetero-Complexes," Ang. Chem. (International Edition in English), 15(12) (1976) p. 713.

9. U. Merten and W.E. Bell, "The Transpiration Method," p. 91 in The Characterization of High Temperature Vapors, J.L. Margrave, ed; John Wiley and Sons, New York, N.Y., 1967.

10. J.H. Norman and P. Winchell, "Vapor Pressure Methods," p. 131 in Physiochemical Methods in Metals Research, R.A. Rapp, ed; IV, Part 1, Interscience Publishers, New York, N.Y., 1970.

11. C. Svedburg, "The Determination of Activities in Molten Salt Solutions by Mass Spectrometry," Ph.D. Dissertation, University of Pennsylvania, 1966.

12. A. Visnapuu and J.W. Jensen, "Composition and Properties of Vapors Over Molten Silver Chloride," Journal of Less Common Metals, 20 (1970) p. 141.

13. M. Binneweis, K. Rinke, and H. Schafer, "The Vaporization Equilibrium of Silver Chloride," Z. Anorg. Allg. Chem., 395 (1973) p. 50.

14. P. Graber and K.G. Weil, "Mass Spectrometric Investigations of Silver Halides I: Mass Spectrum, Appearance Potentials, and Fragmentation Scheme of Silver Chloride," Ber. Bunsenges Phsik Chem., 76 (1972) p. 410.

15. S.K. Chang and J.M. Toguri, "Effect of Temperature and Electron Energy on the Mass Spectrum of Silver Chloride," Journal of Chemical Thermodynamics, 7 (1975) p. 423.

16. H. Von Wartenburg and O. Bosse, "Der Dampfdruck Einiger Salze III," Zeit. Fur Elek., 28 (1922) p. 384.

17. C.E. Maier, "Vapor Pressures of the Common Metallic Chlorides and a Static Method for High Temperatures," U.S. Bureau of Mines Technical Paper 360, 1925.

18. H. Bloom, J. Bockris, N.E. Richards, and R.G. Taylor, "Vapor Pressure and Heat of Vaporization of Some Simple Molten Electrolytes," Journal of the ACS, 80 (1958).

19. G.N. Lewis and M. Randall, Thermodynamics, Second Edition; McGraw-Hill, Inc., New York, N.Y., 1961.

20. L.S. Darken and R.W. Gurry, Physical Chemistry of Metals, McGraw-Hill, Inc., New York, N.Y., 1953.

21. E.D. Cater, "Measurement of the Gross Vaporization Rate (Knudsen Methods)," pp. 21-93 in Physiochemical Measurements in Metals Research, R.A. Rapp, ed; Interscience Publishers, New York, N.Y., 1970.

22. K.K. Kelley, "Contributions to the Data on Theoretical Metallurgy: XIII High-Temperature Heat-Content, Heat Capacity, and Entropy Data for the Elements and Inorganic Compounds," U.S. Bureau of Mines Bulletin 584, 1960.

23. NBS Technical Note 270-4, U.S. Department of Commerce, National Bureau of Standards, 1969.

RECOVERY OF FLUORITE AND BYPRODUCTS FROM THE FISH CREEK DEPOSIT,

EUREKA COUNTY, NEVADA

Donald G. Foot, Jr., Freddy W. Benn, Jerry L. Huiatt

Bureau of Mines, U.S. Department of the Interior
Salt Lake City Research Center
729 Arapeen Drive
Salt Lake City, UT 84108
USA

Summary

The Bureau of Mines, U.S. Department of the Interior, investigated flotation methods for recovering fluorite and byproduct beryllium from the Fish Creek deposit of Nevada. The preferred method includes (1) fluorite rougher and cleaner flotation, (2) desliming at 20 micrometers, (3) muscovite flotation, (4) silicate flotation, (5) hypochlorite wash of silicate concentrate, and (6) beryllium rougher and cleaner flotation from the silicate concentrate. Laboratory results showed that over 94 pct of the fluorite was recovered in a concentrate containing 96 pct CaF_2. Beryllium flotation produced concentrates containing in excess of 5 pct BeO with recoveries over 70 pct. A 100-lb/h continuous flotation unit was constructed to demonstrate this procedure.

Introduction

Fluorspar is the commercial name for an aggregate of rock and mineral matter containing a sufficient amount of the mineral fluorite (CaF_2) to qualify as a marketable commodity. Pure fluorite is calcium fluoride (CaF_2), which contains 51.3 pct calcium and 48.7 pct fluorine. Fluorite is the principal source of fluorine, an element for which there are no adequate substitutes in any of its uses. The principal uses of fluorite are as a source of fluorine for making hydrofluoric acid (HF), as a flux in metallurgy, and as a raw material in the manufacture of glass and enamels.

Fluorite is marketed in three grades that have different physical and chemical specifications. They are—

o Metallurgical Grade.- This is also known as "metspar" or "lump spar" and is sold on the basis of effective calcium fluoride content rather than the actual calcium fluoride content. The effective CaF_2 content is derived by subtracting 2.5 times the percentage of the silica content from the percentage of calcium fluoride content. "Metspar" users require 60 to 70 pct effective CaF_2, limit silica to 5 or 6 pct, limit sulfides to under 1/2 pct, and limit lead to under 0.25 pct. Metallurgical grade fluorite is used as a flux in steel and as electrolyte in aluminum smelting, metal welding, procelain enameling and in glazing.

o Ceramic-Grade.- This is also known as "glass" and "enamel" grade and must not contain less than 95 pct CaF_2, with a maximum of 2.5 pct SiO_2 and 0.12 pct Fe_2O_3. It must be finer grained for its uses in the manufacture of opaque glass, flint glass, as an ingredient in welding and coatings, in making white and buff colored clay bricks, and in vitreous enamels for coating metal articles and appliances.

o Acid-Grade.- This is used in the manufacture of hydrofluoric acid and should contain a minimum of 97 pct CaF_2 and not over 1.1 pct silica with a low content of $CaCO_3$ and sulfur. Acid-grade fluorite is the highest quality marketed and, therefore, commands the highest price. Major uses include the manufacture of fluorine chemicals, synthetic cryolite (essential for aluminum production), preservatives, insecticides, aerosols, and in metal coatings.

At projected consumption rates, known fluorite deposits of the world are expected to be depleted by the end of the century. The long-term U.S. fluorite requirements, of which 85 pct is imported, will have to be derived from secondary sources, submarginal domestic ores, or multicomponent ores. In accordance with this, the Bureau of Mines Salt Lake City Research Center investigated the recovery of fluorite and byproduct beryllium and muscovite from the Fish Creek deposit in Nevada.

Domestic shipments of finished fluorite declined for the fourth consecutive year in 1980 as shown by Morse (8). Fluorite output failed to exceed 100,000 tons for the first time since 1938. Mexico remained the major supplier of metallurgical- and acid-grade fluorite (62 pct) with South Africa the second largest supplier (23 pct). Minor amounts were received from the People's Republic of China, Italy, Spain, and Kenya.

A long-term downward effect on the demand for fluorite and fluorine compounds continues as steelmakers consume less fluorite per unit of output, and primary aluminum smelters adopt and refine technology that recovers and recycles more fluorine. New and expanded uses for fluorine chemicals will offset these decreases in long-term demand. From a 1981 base, consumption of fluorite equivalent is expected to increase at an annual average rate of 2.3 pct through 1990, Pelham (9).

Most fluorite in the United States is produced in the Midwest, with Illinois accounting for well over 90 pct of all U.S. shipments. In the West, production was reported from small mines in Nevada and Texas. Small unreported amounts of fluorite were produced in Utah, Idaho, and New Mexico, as indicated by Morse (8). Nevada fluorspar deposits were described by Horton (7) and Holmes (6). Approximately 90 deposits were reported, which contained 5 to 50 fluorite with byproducts of barite, beryllium, scheelite, zinc, gold, and silver. Byproduct recovery is necessary to enable economic processing of submarginal domestic reserves.

Traditional methods of concentrating fluorite use gravity separation followed by flotation. Clemmer and Clemmons (2) patented a procedure using oleic acid as a flotation collector. This procedure recovered acid-grade fluorite from an Illinois jig tailings that contained 44 pct CaF_2. Bloom, et al (1) depressed barite with lignin sulfonate while floating fluorite with oleic acid from a Yuma County, AZ, fluorspar ore. This paper describes the characterization and beneficiation of fluorite and beryllium from the Fish Creek deposit of Nevada.

Experimental Materials

A sample was obtained from the Fish Creek fluorspar deposit near Eureka, NV, which was known to contain beryllium values. The sample was crushed through 10 mesh, thoroughly blended, and samples split for chemical and petrographic analysis. The chemical analyses are shown in Table I.

TABLE I - Chemical Analysis of Fish Creek Fluorspar-Beryllium Ore

Constituents	Chemical composition, pct
CaF_2	10.5
SiO_2	82.6
$CaCO_3$	0.8
BeO	0.54
Zn	0.3
Total Fe	1.2

Petrographic analysis of the fluorspar-beryllium ore indicated that it contained major amounts of quartz (SiO_2), fluorite (CaF_2), and muscovite (light colored mica, $KAl_3Si_3O_{10}(OH)_2$). Minor amounts of hematite (Fe_2O_3), limonite (hydrous iron oxides), sphalerite (ZnS), potassium-feldspar ($KAlSi_3O_8$), beryl ($Be_3Al_2Si_6O_{18}$), calcite ($CaCO_3$) and chlorite (hydrous silicates of aluminum, ferrous iron, and magnesium) were observed. Liberation size was 35 mesh.

Minus 35 mesh material was prepared for bench scale testing by stage crushing through a jaw and roll crusher and pulverizing. The ore was blended and split into 500-gram samples. The screen analysis of the ground ore is shown in Table II.

TABLE II - Screen Analysis of Fluorspar-Beryllium Ore

Size, mesh	Wt pct	Analyses, pct		Distribution, pct	
		CaF_2	BeO	CaF_2	BeO
-35 + 48	15.4	8.3	0.58	12.2	16.4
-48 + 65	18.8	9.7	0.61	17.4	21.1
-65 + 100	24.0	15.2	0.65	34.7	28.8
-100 + 200	19.1	10.2	0.49	18.7	17.2
-200 + 325	10.3	8.3	0.43	8.4	8.1
-325 + 400	7.5	7.6	0.38	5.4	5.2
-400	4.9	6.8	0.36	3.2	3.2
Composite	100.0	10.51	0.54	100.0	100.0

Reserves of this deposit are estimated in excess of 250 million tons with a content of 14 pct CaF_2. Currently, no beryllium analysis or reserve estimates are available.

Experimental Procedures and Results

Conventional gravity separation techniques, such as tabling, jigging, and heavy medium-separation, were investigated to concentrate the fluorite. Some upgrading was achieved, but these methods did not produce concentrates of sufficient grades or recoveries. Froth flotation proved to be the most successful. A Denver D-1 Laboratory Flotation machine[2] was used in the subsequent flotation testwork.
_____
[2]Reference to specific trade names does not imply endorsement by the Bureau of Mines.

Fluorite Flotation
_____

Flotation testing was performed to determine the effects of (1) collector addition and conditioning time, (2) flotation pH variation, (3) sodium silicate addition and conditioning time, and (4) sodium carbonate addition and conditioning time on fluorite rougher flotation.

Collector addition. Oleic acid, as a 5.0-pct solution in ethanol was used as the fluorite collector. The ethanol was used only as a diluent for the oleic acid for ease of handling during testing. The oleic acid was varied from 0.5 to 3.0 lb/ton with 3.0 lb/ton of sodium carbonate, 2.0 lb/ton sodium silicate at flotation pH of 9.0. The fluorite rougher concentrate grades ranged from 62.1 to 75.1 pct CaF_2, with recoveries ranging from 93.4 to 97.0 pct (Figure 1). The optimum oleic acid addition was 2.0 lb/ton, where 97.0 pct of the fluorite was recovered with a grade of 73.5 pct CaF_2.

Oleic acid conditioning was varied from 3 to 30 minutes with 3.0 lb/ton of sodium carbonate, 3.0 lb/ton of sodium silicate, and 2.0 lb/ton of oleic acid. The fluorite rougher concentrate grades ranged from 64.7 to 72.8 pct CaF_2 with recoveries ranging from 95.2 to 98.0 pct. The optimum conditioning time for oleic acid was 15 minutes, where 97.0 pct of the fluorite was recovered with a grade of 72.8 pct CaF_2.

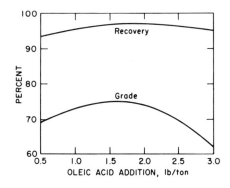

Figure 1 - Effect of oleic acid addition on fluorite grade and recovery.

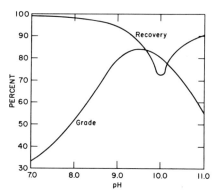

Figure 2 - Effect of pH variation on fluorite grade and recovery.

Flotation pH variation. Prior investigators, Hiskey (5) and Fuerstenau (3), reported the point-of-zero-charge for fluorite was pH 10. Optimum pH for fluorite flotation using an anionic collector should be below pH 10 because of the positive charge on the surface. The pH was varied from 7.0 to 11.0 with sodium hydroxide or hydrochloric acid and allowed to condition for 3 minutes with 3.0 lb/ton of sodium carbonate, 2.0 lb/ton of sodium silicate, and 2 lb/ton of oleic acid. The fluorite rougher concentrate grades ranged from 33.3 to 79.3 pct CaF_2 with recoveries ranging from 72.3 to 96.5 pct (Figure 2). The optimum pH was 9.0, where 94.6 pct of the fluorite was recovered with a rougher grade of 79.3 pct CaF_2.

Sodium Silicate. Sodium silicate was used as a quartz depressant and a dispersant. The sodium silicate addition was varied from 0 to 3.0 lb/ton with 3.0 lb/ton of sodium carbonate and 2.0 lb/ton of oleic acid at pH 9.0. The fluorite rougher concentrate grades ranged from 38.0 to 82.0 pct CaF_2, with recoveries ranging from 95.2 to 98.7 pct (Figure 3). Optimum flotation occurred using 3.0 lb/ton of sodium silicate, where 96.2 pct of the fluorite was recovered with a grade of 82.2 pct CaF_2.

The conditioning time for the sodium silicate was varied from 3 to 30 minutes, with 3.0 lb/ton of sodium carbonate, 2.0 lb/ton of sodium silicate and 2.0 lb/ton of oleic acid at pH 9.0. The fluorite rougher concentrate grades ranged from 65.5 to 80.4 pct CaF_2, with recoveries ranging from 95.2 to 96.5 pct. Optimum flotation occurred using a 15-minute conditioning time, where 96.2 pct of the fluorite was recovered with a grade of 80.4 pct CaF_2.

Sodium Carbonate. Sodium carbonate was used as a water conditioner and to aid in gangue depression. The sodium carbonate addition was varied from 0.5 to 4.0 lb/ton, with 2.0 lb/ton of sodium silicate and 2.0 lb/ton of oleic acid as the fluorite collector at a pH of 9.0. The fluorite rougher concentrate grade ranged from 42.2 to 73.2 pct CaF_2 with recoveries ranging from 96.2 to 98.7 pct (Figure 4). The optimum results were with 3.0 lb/ton of sodium carbonate, where 97.5 pct of the fluorite was recovered at a grade of 73.2 pct CaF_2.

Sodium carbonate conditioning time was then varied from 3 to 30 minutes with 3.0 lb/ton of sodium carbonate, 2.0 lb of sodium silicate

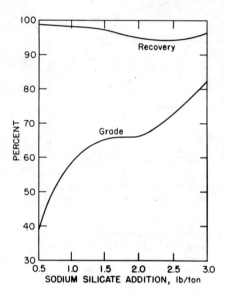

Figure 3 - Effect of sodium silicate addition on fluorite grade and recovery.

and 2.0 lb/ton of oleic acid at pH 9.0. The fluorite rougher concentrate grades ranged from 78.5 to 82.0 pct CaF_2 with recoveries ranging from 96.8 to 97.8 pct. Optimum flotation occurred using 15 minutes of conditioning time, where 97.8 pct of the fluorite was recovered at a grade of 81.7 pct CaF_2.

Flotation Under Optimum Conditions. Under optimum conditions (Table III), a fluorite rougher concentrate was produced that contained 96.1 pct of the fluorite with a grade of 83.0 pct CaF_2. The rougher concentrate was cleaned two times; no additional reagents were used except those needed to maintain a pH of 9. The second cleaner concentrate contained 98.1 pct fluorite with a recovery of 94.0 pct.

TABLE III. Optimum Reagent Dosages and Conditioning Times For Fluorite Flotation, pH 9.0

Reagent	lb/ton	Conditioning time, minutes
Sodium carbonate	3.0	15
Sodium silicate	3.0	15
Oleic acid	2.0	15

Locked-Cycle Flotation Testing. Locked-cycle flotation testing was performed to determine the effects of the recycling of the cleaner tailings products and the recycle water on fluorite flotation, and to obtain data for scale-up planning. The cleaner tailings and recleaner tailings were combined and scavenged, with the scavenger concentrate being recycled to the head of the next cycle. The fluorite rougher tailings were combined with the scavenger tailings to make the final tailings. Five complete cycles were performed to assure that the system was in equilibrium. Results are shown in Table IV. Ninety-four percent of the fluorite was recovered at a grade of 96 pct CaF_2. The final tailings contained

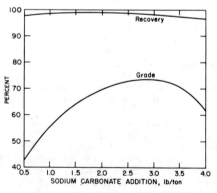

Figure 4 - Effect of sodium carbonate addition on fluorite grade and recovery.

over 90 pct of the weight with a grade of 0.62 pct CaF$_2$, representing a loss of 5.78 pct of the fluorite. The concentrate contained 0.01 pct BeO, while the tailings product contained over 99 pct of the BeO values with a grade of 0.60 pct BeO.

Table IV. Analyses and Distribution of Fluorite in Locked Cycle Flotation

Product	Analyses, pct		Distribution, pct	
	CaF$_2$	BeO	CaF$_2$	BeO
Final CaF$_2$ product............	96.00	0.01	94.22	0.17
Final CaF$_2$ tailings...........	0.62	.60	5.78	99.83

Byproduct Recovery

Various procedures were investigated to concentrate the beryl. The preferred method consisted of (1) fluorite flotation using oleic acid, (2) desliming at 20 micrometers, (3) muscovite flotation using sulfuric acid and tallow amine acetate, (4) silicate minerals flotation using hydrofluoric acid and tallow amine acetate, (5) collector removal from the silicate concentrate using hypochlorite, (6) water washing of the silicate concentrate, and (7) beryl flotation from the silicate concentrate using oleic acid, as indicated by Fuerstenau (4). The flowsheet is shown in Figure 5.

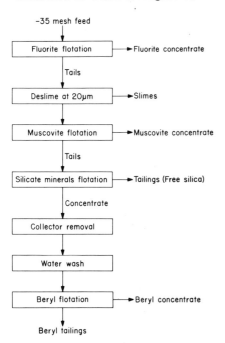

Figure 5 - Flowsheet for concentrating beryl from CaF$_2$ tailings product.

Muscovite Removal. Flotation testing was performed on the muscovite flotation circuit to determine the effects of (1) acid addition, and (2) collector addition on the flotation of the muscovite. The muscovite posed a problem with all the previous tests, because it would concentrate with the beryl, thus lowering the beryl concentrate grade. No reliable analytical method was available for determining muscovite in the flotation products. Optimum flotation reagent dosages were based on (1) the percent BeO values in the muscovite flotation products, (2) weight percent removed from the deslimed fluorite tailings products, and (3) the BeO grade of the muscovite flotation products.

Acid Addition. Sulfuric acid was used as the pH modifier and as the beryl depressant. The acid was varied from 0 to 4.0 lb/ton. The collector (tallow amine acetate) was added at 0.3 lb/ton. The amount of beryl reporting to the muscovite rougher concentrate ranged from 4.5 to 6.4 pct. The grade of the muscovite rougher tailings products ranged from 0.62 to 0.75 pct BeO (Figure 6). The optimum sulfuric acid addition was determined to be 3.0 lb/ton at pH 3.0, where only 5.7 pct of the BeO values were lost. The most weight was removed in the muscovite rougher concentrate, and the grade of the muscovite rougher tailings product was 0.75 pct BeO.

Collector Addition. Tallow amine acetate was used as the muscovite collector. The sulfuric acid was held constant at 2.0 lb/ton and the tallow amine acetate was varied from 0.3 to 1.0 lb/ton. The amount of beryl reporting to the muscovite rougher concentrate ranged from 5.6 to 10.4 pct with the grades of the muscovite rougher tailings ranging from 0.63 to 0.73 pct BeO (Figure 7).

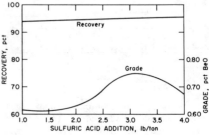

Figure 6 - Effect of sulfuric acid addition on BeO recovery and grade in muscovite rougher tails.

The optimum collector addition was determined to be 0.7 lb/ton of tallow amine acetate. At this collector addition (1) 8.9 pct of the beryl was lost to the muscovite concentrate, (2) the muscovite rougher tailings contained 0.73 pct BeO, and (3) the highest weight percent was removed in the muscovite concentrate.

Silicate Removal. Direct flotation of the beryl produced concentrates with grades less than 2.0 pct BeO, therefore, silicate flotation was used to depress the free silica and improve the grade prior to beryl flotation. The flotation tests were performed on the silicate flotation circuit to determine the effects of acid addition and collector addition on the flotation of the silicate (beryl and feldspar) minerals.

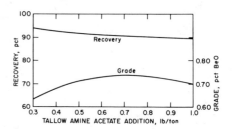

Figure 7 - Effect of tallow amine acetate addition on BeO recovery and grade in muscovite rougher tails.

Acid Addition. Hydrofluoric acid was used as an activator for the beryl, as a quartz depressant, and as a pH modifier. The hydrofluoric acid was varied from 0 to 10 lb/ton with the tallow amine acetate collector held constant at 1.0 lb/ton. The amount of beryl recovered in the silicate stage ranged from 24.0 to 89.0 pct (overall) with silicate concentrate grades ranging from 0.86 to 4.80-pct BeO (Figure 8). The optimum hydrofluoric acid addition was 5.0 lb/ton, at pH 2.75, where the grade was 4.80 pct BeO with an overall recovery of 89.0 pct of the beryl.

Collector Addition. The collector, tallow amine acetate in a 5-pct solution, was used to remove the silicate minerals from the free silica. The collector addition was varied from 0 to 5.0 lb/ton with the hydrofluoric acid held constant at 5.0 lb/ton. The amount of beryl recovered in the silicate stage ranged from 34.0 to 90.5 pct with the grade ranging from 1.19 to 7.07 pct BeO (Figure 9). The optimum amount of tallow amine acetate was determined to be 0.5 lb/ton, where 81.0 pct of the beryl was recovered with a grade of 4.55 pct BeO.

The tailings from the silicate flotation stage was a silica product containing 99 pct SiO$_2$; silica recovery was 90 pct.

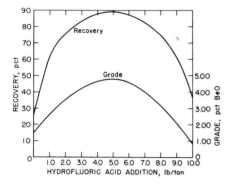

Figure 8 - Effect of hydrofluoric acid addition on BeO recovery and grade in silicate rougher concentrate.

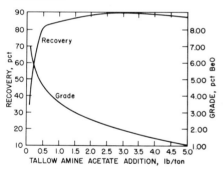

Figure 9 - Effect of tallow amine acetate addition on BeO recovery and grade in silicate rougher concentrate.

Collector Removal. Sodium hypochlorite was used to remove the collector (tallow amine acetate) from the mineral surfaces to gain selectivity in the beryl circuit. The residual collector would have promoted gangue flotation in the beryl circuit, thus lowering the grade. The pulp was washed with 2.0 lb/ton of sodium hypochlorite for 5 minutes and allowed to settle for 5 minutes. The excess liquid was decanted and solids repulped with fresh water. The pulp was conditioned for 5 minutes and allowed to settle for 5 minutes. The excess fresh water was decanted.

Beryl Flotation. The test results indicated that beryl selectivity could be improved by using a flotation pH of 11.0. The flotation testwork was performed to determine the optimum amount of collector needed for beryl flotation at this pH.

Collector Addition. Oleic acid was used as the collector for the beryl. The oleic acid was varied from 0.3 to 1.0 lb/ton with the pH constant at 11.0. The amount of beryl recovered ranged from 7.5 to 70.0 pct with beryl concentrate grades ranging from 3.11 to 4.15 pct BeO (Figure 10). The optimum oleic acid addition was determined to be 1.0 lb/ton where the grade was 3.7 pct BeO with an overall recovery of 70 pct. Grades lower than those achieved during silicate flotation optimization were due to overreagentizing of the silicate stage, thus promoting gangue. Higher beryl grades were achieved but with lowered recoveries.

Byproduct Flotation Under Optimum Conditions. Under optimum reagent dosages listed in Table V, a rougher beryl concentrate was produced that contained 75.3 pct of the beryl with a grade of 5.45 pct BeO. Using one cleaner, a concentrate grade was produced that contained 54 pct of the beryl with a grade of 6.11 pct BeO. A second cleaner upgraded the concentrate to 6.73 pct BeO with a beryl recovery of 37.3 pct. The acceptable grade for further processing is between 7 and 8 pct BeO. A material balance for fluorite and byproduct recovery is shown in Table VI.

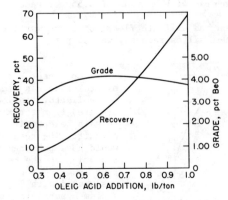

Figure 10 - Effect of oleic acid addition on BeO recovery and grade in beryl rougher concentrate.

TABLE V. Optimum Reagent Dosages for Byproduct Recovery

Flotation Stage	Reagent	lb/ton
Muscovite flotation	Sulfuric acid	3.0
	Tallow amine acetate	0.7
Silicate rougher flotation	Hydrofluoric acid	5.0
	Tallow amine acetate	0.5
Collector removal	Sodium hypochlorite	2.0
Beryl rougher	Oleic acid	1.0

Table VI. Material Balance for Fluorite and Byproduct Flotation

Product	Analysis, pct		Distribution, pct	
	CaF_2	BeO	CaF_2	BeO
CaF_2 cleaner concentrate	98.1	0.01	94.0	1.6
Slimes	3.4	0.20	2.7	1.4
Muscovite concentrate	2.4	0.12	1.5	2.2
Silicate tailings	0.1	0.02	0.9	3.8
Beryl rougher concentrate[1]	0.8	5.45	0.7	75.3
Beryl Tailings [2]	0.4	0.83	0.2	15.7

[1] To be cleaned.
[2] To be recycled to head of silicate circuit.

Continuous Flotation Unit (CFU)

Large Tonnage Sample

A 23.3-ton sample was obtained from the Fish Creek fluorspar deposit for use in a 100-lb/h continuous flotation unit (CFU). The chemical analyses of the head sample are shown in Table VII. A petrographic analysis showed major amounts of quartz, muscovite, and fluorite, with minor amounts of feldspar, calcite, hematite, and clays. The liberation size was determined to be 48 mesh. The ore for the CFU contained one-third the BeO and four times the calcite content of the bench-scale ore. Changes in the design and construction of the CFU may be necessary due to the co-flotation of calcite and fluorite, thus lowering fluorite concentrate grades and recoveries. Lower beryl grades and recoveries will be expected owing to the lowered grade of the feed material.

TABLE VII. Chemical Analysis of Large Tonnage Sample of Fish Creek Fluorspar Ore

Constituents	Chemical composition, pct
CaF_2	12.95
SiO_2	72.50
$CaCO_3$	3.26
BeO	.18
Zn	.20
Total Fe	1.10

Conclusions

Experimental results showed that fluorite can be readily floated from the Fish Creek ore using oleic acid at pH 9.0. Maximum grades were achieved in bench-scale flotation using 3.0 lb/ton sodium carbonate, 3.0 lb/ton sodium silicate, and 2.0 lb/ton oleic acid. Under these conditions, 94 pct of the fluorite was recovered in a recleaner concentrate that contained 98.1 pct CaF_2. A muscovite concentrate was obtained from the deslimed fluorite tailings, using 3.0 lb/ton sulfuric acid and 0.7 lb/ton tallow amine acetate. The silicate minerals were floated from the free silica using 5 lb/ton hydrofluoric acid and 1.0 lb/ton tallow amine acetate. The silicate tailings product contained 99 pct SiO_2 with a recovery of 90 pct. A beryl concentrate was produced containing 75.3 pct of the beryl with a grade of 5.45 pct BeO.

A 100-lb/h continuous flotation unit (CFU) was designed and constructed to demonstrate the bench-scale procedure. Changes in the flowsheet may be necessary owing to the amount of calcite, which will co-float with the fluorite, and the lowered beryl grade in the feed material. Operation of the CFU is now in progress.

References

1. P. A. Bloom, W. A. McKinney, and L. G. Evans, "Flotation Concentration of Complex Barite-Fluorite Ore," Bureau of Mines, U.S. Dept. of the Interior, Report of Investigations 6213, 1963, 16 pp.

2. J. B. Clemmer, and B. H. Clemmons, "Concentrating Fluorspar by Froth Flotation," U.S. Pat. 2,407,651, Sept. 17, 1946.

3. M. C. Fuerstenau, D. A. Elgillani, and G. Gutierrez, "The Influence of Sodium Silicate in Nonmetallic Flotation Systems," AIME Trans., 241, 1968, p.319.

4. D. W. Fuerstenau, editor, Froth Flotation- 50th Anniversary Volume, AIME, New York, 1962, pp. 439-440.

5. J. B. Hiskey, "Electrokinetics of CaF_2," Masters Thesis, Department of Metallurgy, University of Utah, Salt Lake City, UT, 1971.

6. George H. Holmes, Jr., "Beryllium Investigations in California and Nevada," Bureau of Mines, U.S. Dept. of the Interior, Information Circular 8158, 1963, 19 pp.

7. Robert C. Horton, "An Inventory of Fluorspar Occurances in Nevada," Nevada Bureau of Mines, McKay School of Mines, University of Nevada, Reno, NV, Rep 1, 1961.

8. David E. Morse, "Fluorspar," Ch. in Minerals Yearbook 1980, Bureau of Mines, U.S. Dept. of the Interior, 1981, vol. 1, pp. 321-334.

9. Lawerence Pelham, Mineral Commodity Summaries 1984, Bureau of Mines, U.S. Dept. of the Interior, 1983, pp. 50-51.

THE INFLUENCE OF LiF AND BATH RATIO ON

PROPERTIES OF HALL CELL ELECTROLYTES

Suzanne Young & R. O. Loutfy

ARCO Metals Company
Primary Metals R&D
Tucson, Arizona 85726
USA

Summary

The effect of LiF additions on liquidus temperature and electrical conductivity over a wide range of cryolite bath ratios (1.0 to 1.5 weight ratio) was investigated. Results did not agree with multi-regression equations reported by Dewing and Choudhary over the entire region. New regression equations were developed. Reasons are given for the differences. The models are used to discuss the economic implications of LiF additions to Hall cell electrolytes.

Introduction

The energy requirement for aluminum electrolysis can be decreased by the addition of lithium fluoride (LiF) to the Hall cell electrolyte. Among the most important properties affected are liquidus temperature and electrical conductivity. Previous studies (1-4) and internal company work have shown that LiF will decrease the melting point by 8°-12°C/1% LiF for concentrations up to 5% LiF, and increase the electrical conductivity by 8.8-10%/5% LiF.

The addition of aluminum fluoride (AlF_3), or decreasing the cryolite ratio ($NaF:AlF_3$) will also have a favorable effect on liquidus temperature, decreasing it by 1°C/1% AlF_3 for 0-15% AlF_3 (excess). However, AlF_3 decreases the electrical conductivity by 6.1%/15% AlF_3 (1).

The measurement of the effects of LiF and AlF_3 has been done primarily at cryolite weight ratios of 1.2 or greater. Few determinations have been made of the effect of LiF in more acidic melts, or in the presence of other standard components of Hall cell electrolytes, such as alumina (Al_2O_3) and calcium fluoride (CaF_2).

In this investigation, the effect of adding 1-5% LiF was studied for melts of cryolite weight ratios of 1.0 to 1.5. The melts also contained conventional concentrations of CaF_2 and Al_2O_3. Liquidus temperature and electrical conductivity were measured, and the results compared to models developed earlier by Dewing (5) and Choudhary (6). Since the correlation did not agree over the entire concentration region, a new mathematical model was developed. The results can be used to predict changes in electrolyte properties resulting from LiF additions to electrolytes in operating smelters.

Experimental

Materials and Apparatus

Nine melts were tested for liquidus temperature and conductivity. The cryolite weight ratios used were 1.00, 1.25, and 1.50, each with lithium fluoride concentrations of 1.0, 3.0, and 5.0 wt %. All melts contained 4.0 wt % CaF_2 and 3.5 wt % Al_2O_3. A 300-g charge was used.

The experimental cell can be seen in Figure 1. The melt was contained in a graphite crucible protected by coke. A chromel-alumel thermocouple in a fused alumina thermocouple well was used for temperature measurement. The dense graphite conductivity probe was encased in boron nitride (BN) for protection. The other lead to the conductivity bridge was a nickel rod screwed into the wall of the crucible. The furnace was electrically heated.

Temperature was recorded with a Molytek 2701 multichannel strip chart recorder. A GenRad Model GR 1688 Precision LC Digibridge was used to make conductivity measurements.

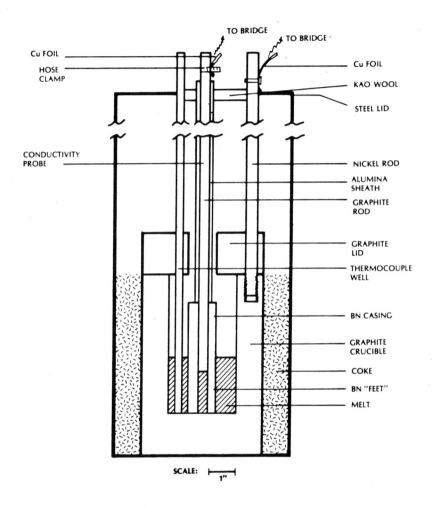

Figure 1 - Apparatus

Procedure

Liquidus Temperature. The method of cooling and heating curves was used. A typical cooling rate was 5.5°C/minute, and heating rate, 1.5°C/minute. Samples of the melt were taken before and after the curves were made.

Conductivity. The conductivity bridge measures resistance (R), which is related to conductivity (K) by the expression:

$$K = 1/R \ (l/A) \tag{1}$$

where l/A is the cell constant. When the bridge is balanced, the measured resistance (Rp) is related to the true solution resistance (Rs), interfacial capacitance (Cs), and frequency (f) by the following equation:

$$Rp = Rs + \frac{1}{RsCs^2 \ (2 \pi f)^2} \ (8) \tag{2}$$

Since resistance varies with the frequency of the generated sine wave (f), measurements of Rp and Cs were made and Rs calculated for several frequencies. The true Rs was found by extrapolating the curve to infinite frequency.

The cell constant was calculated by determining the resistance for a salt of known conductivity, sodium chloride.

Conductivity measurements were made for each melt at one point, 9°-42°C above the liquidus.

Results and Discussion

In developing models to describe the data, we had to account for changes in the composition of the melts over the duration of the experiments. Primarily, the cryolite ratio had increased because of the loss of AlF_3, and the Al_2O_3 concentration had increased from the dissolution of the alumina thermocouple sleeve and hydrolysis of the electrolyte. Final concentrations and experimental data can be found in Table 1.

From the corrected melt compositions, the following model for liquidus temperature was developed:

$$LT(°C) = 782.64 + 21.63 \ (CR_m) - 0.424 \ (Al_2O_3)$$
$$- 5.806 \ (LiF) + 21.62 \ (CaF_2) \tag{3}$$

Concentrations are in weight percent, and CR_m refers to the molar ratio $NaF:AlF_3$. An R-square correlation coefficient of 0.99 was obtained.

The model is illustrated in Figure 2 for melts of 9.0% Al_2O_3 and 6.0% CaF_2, which represent the averages of the melt concentrations. The magnitude of the effect of LiF addition is to lower the liquidus temperature by -6.0°C/1% LiF, and that of excess AlF_3 is to decrease it by -1.5°C/1% AlF_3.

Table I – Final Melt Compositions and Experimental Data

No.	Al_2O_3 (dissolved)	NaF	AlF_3	Cryolite Ratio	Excess AlF_3	CaF_2	LiF	T °C	Conductivity $\Omega^{-1}cm^{-1}$	ln K
1	13.16	51.84	28.02	1.85	-8.16	6.14	0.84	925.5	2.0	0.693
2	8.92	45.25	38.19	1.19	9.65	4.83	2.81	920.0	3.1	1.131
3	8.86	44.65	37.22	1.20	9.15	4.99	4.28	911.0	2.9	1.065
4	11.40	49.94	32.76	1.52	-.60	4.96	0.94	945.0	2.5	0.916
5	11.31	49.36	31.96	1.54	-1.16	4.81	2.56	934.5	3.1	1.131
6	7.99	46.36	35.77	1.30	5.97	5.38	4.50	927.0	3.2	1.163
7	9.36	52.36	32.19	1.63	-3.16	5.14	0.95	954.0	2.9	1.065
8	9.78	56.43	25.78	2.19	-14.36	5.50	2.51	937.0	2.8	1.030
9	12.68	52.73	26.37	2.00	-11.06	5.47	2.75	933.5	2.4	0.875

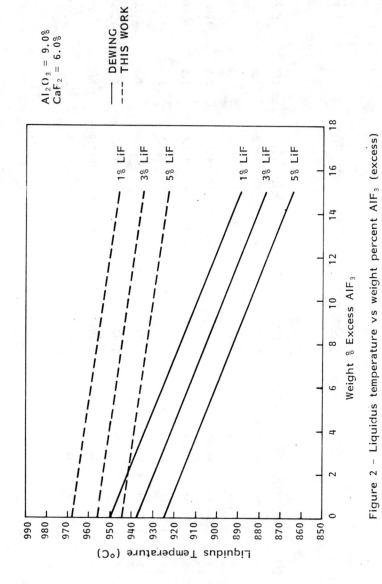

Figure 2 – Liquidus temperature vs weight percent AlF_3 (excess)

Also presented in Figure 2 are curves from Dewing's (5) model, which is expressed as:

$$LT(°C) = 793 + 60.83 \ (CR_m) - 2.75 \ (Al_2O_3) + 0.875 \ (CaF_2)$$
$$- 1.125 \ (MgF_2) - 6.19 \ (LiF) - 9.31 \ (MgCl_2) \qquad (4)$$

Again, concentrations are in weight percent, and CR_m is the cryolite molar ratio. According to Dewing, LiF lowers the liquidus temperature by 6.0°C/1% LiF, which is identical to the findings of this investigation. However, in his model AlF_3 decreases it by -1.7°C/% AlF_3 (excess), which is greater than the decrease found by the new model.

The experimental data is within 2% of the predictions of Dewing's model at a high bath ratio (low excess AlF_3), but at a low bath ratio, the deviations become substantial, thus confirming the need for the new model in this region.

The decrease in liquidus temperature associated with going to a low bath ratio is not as great as was anticipated based on the earlier model. This implies that when using LiF as an additive, the benefits of simultaneously lowering the ratio do not justify the added operational difficulties of using a large excess of AlF_3. These difficulties arise from its high vapor pressure, which makes frequent bath additions necessary to maintain the desired ratio, and increased loading of the off-gas system.

Again using the data in Table 1, the following conductivity model was developed:

$$\ln K = 4.976 - \frac{3277.9}{T°C} - 0.1135 \ (Al_2O_3) - 0.235 \ (CaF_2)$$
$$+ 0.67 \ (CR_w) + 0.0504 \ (LiF) \qquad (5)$$

Concentrations are in weight percent, and CR_w is the weight ratio of $NaF:AlF_3$. According to the model, LiF will increase conductivity by 0.12 $\Omega^{-1}cm^{-1}/1\%$ LiF, and excess AlF_3 will lower it by -0.059 $\Omega^{-1}cm^{-1}/1\%$ AlF_3.

The model is illustrated in Figure 3. Also shown is the model developed by Choudhary (6), which is expressed as:

$$\ln K = 2.0156 - \frac{2068.4}{T} - 0.0207 \ Al_2O_3 - 0.0050 \ (CaF_2)$$
$$+ 0.4349 \ (CR_w) + 0.0178 \ (LiF) + 0.0077 \ (Li_3AlF_6)$$
$$- 0.0166 \ (MgF_2) + 0.0063 \ (NaCl) \qquad (6)$$

At the lower bath ratios, the new model is much more closely in line with Choudhary's model than at higher bath ratios, where the magnitude of the experimental values exceeds the early model's predictions. It is evident then, that for a given percentage of added LiF, the greatest benefit in terms of conductivity will be realized if the bath ratio is kept high. These findings support the conclusions drawn from the new liquidus temperature model, where the low temperatures expected at low bath ratios on the basis of the earlier model were not achieved.

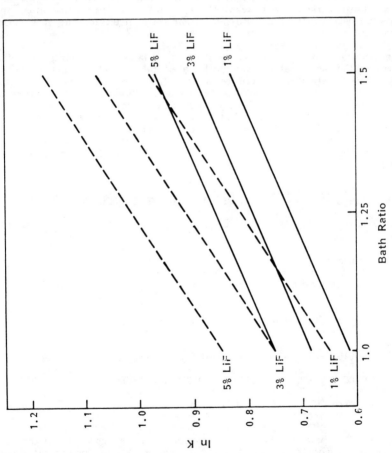

Figure 3 - ln K vs bath ratio

Addition of LiF will accomplish the desired change in properties of the electrolyte, in that it lowers the liquidus temperature and raises the conductivity. AlF_3 will also decrease the liquidus, but it has a negative impact on conductivity. LiF confers the added benefits of increased current efficiency and decreased fluoride emissions, but it has the disadvantage of depressing alumina solubility. From the experimental data presented here, it can be concluded that the electrolyte modifications sought will be gained with LiF added at bath ratios >~1.2.

These conclusions are supported by an analysis of several key economic factors. In Figure 4, the cost-benefit factor for three areas is plotted as a function of LiF concentration for two cryolite weight ratios. The three areas are: (1) the operating cost of lithium carbonate arising from losses to the aluminum and to the suction dust, (2) the changes in power cost, and (3) the cost of metal production from current efficiency improvements. Not considered in this analysis are the costs/benefits associated with reduced alumina solubility, the alterations in the heat balance of the cell, the increase in pot life, and the depression of fluoride emissions.

Curves a and b show the cost factor for Li_2CO_3. The bath loses Li_2CO_3 primarily to the suction dust and to the metal, and the expected amounts can be calculated from equations given by Kulikov (7). The expressions account for Li_2CO_3 requirements as a function of LiF concentration in the electrolyte and electrolyte temperature. It can be seen that the cost of Li_2CO_3 rises with the LiF concentration in the electrolyte and with the increase in bath ratio. The magnitude of each effect is low compared to that of the other factors studied.

Curves c and d show how the power cost factor changes with LiF concentration and bath ratio. Calculations were based on bath resistivity from our model, and on current efficiency, which was estimated from expressions developed by Qui and Fong (8). The curves show that the power cost factor declines substantially with the increase in LiF concentration. Both the increase in conductivity and the lower liquidus temperatures contribute to the trend. At a bath ratio of 1.5, from <1% LiF up, there is a benefit associated with LiF addition, but at a bath ratio of 1.0, that point does not occur until 5% LiF has been added. The greatest savings in power will therefore be achieved at a high bath ratio. The magnitude of both the effect of LiF concentration and of cryolite ratio is great.

Curves e and f show the variation in the production cost factor associated with changing the LiF concentration and the bath ratio. These costs are calculated solely on the basis of current efficiency, again from the expressions of Qui and Fong (8). An increase in LiF concentration lowers the production cost factor, and a lower bath ratio yields a lower cost factor. The magnitude of the production cost factor lies somewhere between that of the Li_2CO_3 cost factor and the power cost factor.

When all cost factors are combined (Figure 5), it is clear that costs can be significantly lowered by LiF addition. The position of the curves reveals that from 0.0-5.5% LiF, which is currently considered a reasonable operating range, a lower cost factor is associated with a higher bath ratio for the same LiF concentration. This is consistent with our experimental findings.

It can also be seen from the curves that the maximum cost benefit can be achieved with an electrolyte of low bath ratio at a relatively high LiF

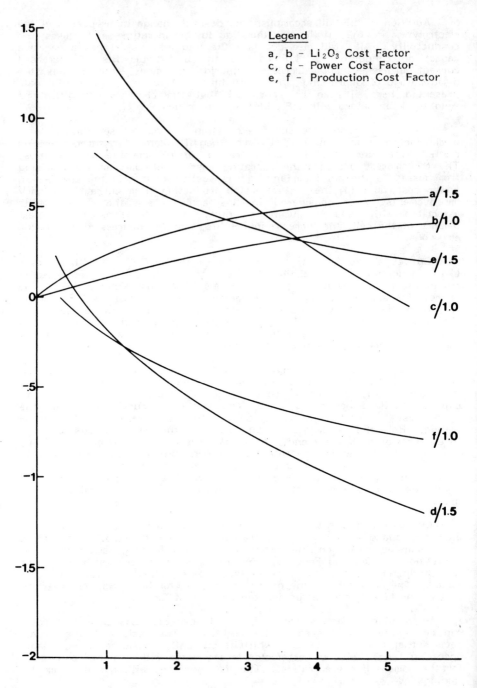

Figure 4 - Cost-benefit factors for three areas as a function of LiF concentration and bath ratio

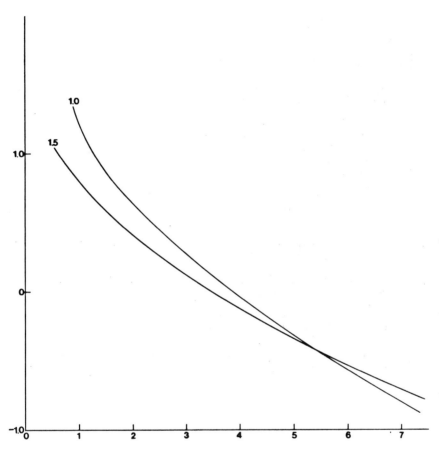

Figure 5 - The combined cost-benefit factor as a function of LiF concentration and bath ratio

concentration (>6%). It is expected that the alumina solubility of such a bath will be low and operation will be difficult, and this scheme is not considered feasible at the present time. With improved computer control and multi-point feeding of the bath, however, the maximum benefits of LiF addition may be able to be realized.

References

1. J. Adam and E. Balazs, "Investigations of the Physicochemical Properties of Cryolite Base Electrolyte Melts," Proceedings of the Research Institute for Non-Ferrous Metals, pp. 243-249 (Budapest: Research Institute for Non-Ferrous Metals, 1971).

2. K. Grjotheim, K. Matiasovsky, and M. Malinovsky, "Influence of NaCl and LiF on the Aluminum Electrolyte," Electrochimica Acta 15, (1970) 259-269.

3. K. Grjotheim, J. L. Holm, C. Krohn, and K. Matiasovsky, "Studies Connected with the Constitution, and the Physicochemical Properties of the Electrolyte for Aluminum Production," Presented at the 95th AIME Annual Meeting, New York, 1966.

4. K. Grjotheim, M. Malinovsky, and K. Matiasovsky, "The Effect of Different Additives on the Conductivity of Cryolite - Alumina Melts," Journal of Metals, January 1969, pp. 28-33.

5. E. W. Dewing, "Liquidus Curves for Aluminum Cell Electrolyte," J. Electrochem Soc., 117 (1970) 6, 780-781.

6. Halvor Kvande, "The Optimum Bath Composition in Aluminum Electrolysis - Does It Exist?," Erzmetall 35, No. 12, 601.

7. Yu V. Kulikov, "Testing Additives of Lithium Salts in Aluminum Electrolysis," Tsvet. Met. 44(3), 53-55, 1971.

8. Z. Qui and N. Fong, "Current Efficiency in Aluminum Electrolysis," Aluminum 59, Jahrg. 1983, 4.

Copper

RESIDENCE TIME PREDICTIONS DURING FLUID-BED ROASTING

C. A. Natalie, J. P. Hager, and T. Li

Department of Metallurgical Engineering
Colorado School of Mines, Golden, CO   80401, U.S.A.

Prediction of average residence times for different particle size intervals during sulfide roasting was investigated. Calculations were done to predict average residence times using a model which employs empirical correlations for elutriation constants. Experimental average residence times were determined using a continuous, six inch diameter, fluid bed roaster and compared with calculated residence time values. Comparison of calculated and experimental data showed that geometry effects may have been significant when attempting to apply elutriation data to this roaster.

Introduction

During the continuous fluid bed roasting of sulfides, feed material with a wide size distribution is fed to a bed and is subsequently discharged through an overflow pipe from the bed or entrained and carried out by the fluidizing gases. The material that is elutriated in the carryover stream is determined by its particle size, the gas velocity, gas and solid densities, and other characteristics as described by Stoke's law (1).

The average residence time of material of a given particle size in the bed area is more a function of the complex process by which particles disengage from the bubbling fluid bed and then move upward as dispersed solids. Extensive work has been done (2,3,4) to define the rate of elutriation from fluid beds by correlations which determine elutriation constants as a function of physical properties and operating parameters of fluid bed roasters. However, most of the experimental work that has been done to determine elutriation correlations has been done using rather coarse material with a somewhat narrow size range as typified by requirements in the petroleum industry.

When sulfide materials are roasted, they are usually the concentrated product of a prior flotation process. The flotation process can often yield material that is 70% finer than 325 mesh. Therefore, fluid bed roasting of sulfides requires treatment of very fine material with a broad size range. A concern when roasting sulfides is whether the elutriation correlations determined for coarser size materials can be used to predict elutriations phenomena in roasting of fine material. It is the purpose of this paper to attempt to predict average residence times of various particle size intervals during fluid bed roasting of sulfides based on present elutriation correlations.

Entrainment and Elutriation

The model used for prediction of residence times during fluid bed roasting comes directly from the work of Kunii and Levenspiel (5).

If the size distribution of material fed to a roaster is given by its size distribution function $p_o(R)$, it is desired to determine the bed size distribution function $p_b(R)$, size distribution function in the carryover stream, $p_2(R)$, and the mass flow rates of material which exit the reactor, F_1 and F_2. The calculations assume that the bed weight, feed properties, and elutriation constants are all known.

It is first assumed that there is sufficient back mix flow of solids so the size distribution of the overflow stream from the bed equals the size distribution in the bed:

$$p_b(R) = p_1(R) \tag{1}$$

The elutriation constant is a function of particle size and is defined as:

$$K(R) = \frac{F_2 p_2(R)}{W p_b(R)} \tag{2}$$

The overall mass balance on the reactor is given as:

$$F_o = F_1 + F_2 \tag{3}$$

The mass balance of any size interval between R and dR is:

$$F_o p_p(R)dR = F_1 p_1(R)dR + F_2 p_2(R)dR \tag{4}$$

For solids as a whole the mass average residence time is given by:

$$\bar{t} = \frac{W}{F_o} = \frac{W}{F_1 + F_2} \tag{5}$$

Due to extrainment, the fine material spends less time in the bed and the mean residence time for a particular size is given by:

$$\bar{t}(R) = \frac{\text{weight of a particular size in the bed}}{\text{flow rate of that size into bed}} \tag{6a}$$

$$\bar{t}(R) = \frac{W p_b(R)}{F_o p_o(R)} \tag{6b}$$

$$\bar{t}(R) = \frac{W p_b(R)}{F_1 p_1(R) + F_2 p_2(R)} \tag{6c}$$

Combining equation 6c with equations 1 and 2 gives:

$$\bar{t}(R) = \frac{1}{F_1/W + K(R)} \tag{7}$$

The composition of outflow stream is found by combining equations 1, 2, and 4:

$$F_1 p_1(R) = \frac{F_o p_o(R)}{1 + (W/F_1)K(R)} \tag{8}$$

upon integration yields:

$$\frac{F_1}{F_o} = 1 - \frac{F_2}{F_o} = \int_{R_1}^{R_2} \frac{p_o(R)dR}{1 + (W/F_1)K(R)} \tag{9}$$

A computer program was written which solves the above set of equations with input of specific operating conditions and physical properties of solids and gases. An important part of the program is the fact that it calculates the elutriation constant, $K(R)$, for each particle size based on the correlation of Yagi and Aochi (6)

Experimental Procedure

The experimentally determined residence times for various particle size intervals was obtained using the fluid-bed roaster pictured in figure 1 and diagramed in figure 2. The fluid-bed reactor body consists of a 6 inch I.D. stainless steel pipe with a 60° cone which contains the windbox and distributor plate assembly. The reactor bed overflow is cut into the side of the reactor at a height of 18 inches above the distributor plate. The reactor body is surrounded by an Applied Test Systems split-shell furnace.

An expanded freeboard is located above the body of the reactor. The freeboard consists of a 52 inch stainless steel pipe which is 8 inches in diameter. The off-gases exit the freeboard area and flow through two dust cyclones in series. The primary cyclone is approximately 6 inches in diameter at its widest point and the dimensions of the secondary cyclone are exactly half those of the primary cyclone. A complete wet-column scrubber system is used to clean the roaster gas.

Figure 1. Continuous six-inch diameter fluid-bed roaster.

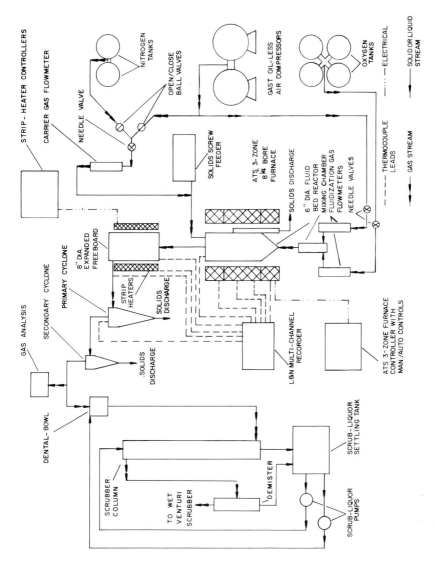

Figure 2 – Schematic of fluidized bed roasting laboratory.

A screw feeder system is used to feed the concentrate into the reactor. The feed is pneumatically carried from the end of the auger down a feed tube to a point approximately nine inches above the distributor plate. The feed tube was constructed of 1/2 inch, stainless steel tubing.

Operation of the roaster consisted of starting with silica as a bed material. The temperature of the reactor was brought up to experimental temperature and feeding of the concentrate began. The rate of concentrate feed was established at such a rate that the reaction between sulfide and oxygen provided the heat necessary to maintain experimental temperature. The reactor was run for many hours until it reached steady-state operation. At this time, overflow samples were saved from the bed overflow pipe and carryover samples were saved from the two cyclones.

The overflow and carryover material was screened using micro-screens to determine their size distribution. The screening was done wet and alcohol was used as the liquid to avoid dissolution of any water soluble oxide that was produced.

The experimental mean residence times for particle size intervals was determined using the equation below:

$$\bar{t}(R) = \frac{W X(R)_w}{F_o X(R)_F}$$

where: W – is the measured weight of the bed
$X(R)_w$ – is weight per cent of sieve fraction R in bed material.
F_o – is feed rate of concentrate expressed as cinder equivalent.
$X(R)_F$ – is the weight per cent of sieve fraction R in the feed.

The experiments run in this study were done with a copper sulfide concentrate. The size distribution of the concentrate used as feed material is given in figure 3. The average particle sizes used for calculation of mean residence times were the average screen opening between two screens which made up the size interval.

Results and Discussion

Figure 4 and figure 5 show the experimental residence times determined from experiments using the 6 inch roaster previously described. Results in figure 4 were determined for a feed rate of 79 gm/min, a space velocity of 27.3 cm/sec and a roasting bed temperature of 700°C. The experimental results are represented by the symbols and are for three roasting campaigns run at the above listed conditions. The value $\bar{t}_{ave}$ is the mass average residence time determined for this feed rate from equation 5.

The solid line in figure 4 are the results of a computer run which calculated the mean residence times of particle size intervals based on an elutriation correlation (6) and the equations 1-4 and 6-9.

Figure 5 shows similar results for a set of different roasting conditions. The symbols represent the four roaster campaigns at these conditions. The computer calculated residence times are also shown by the solid line.

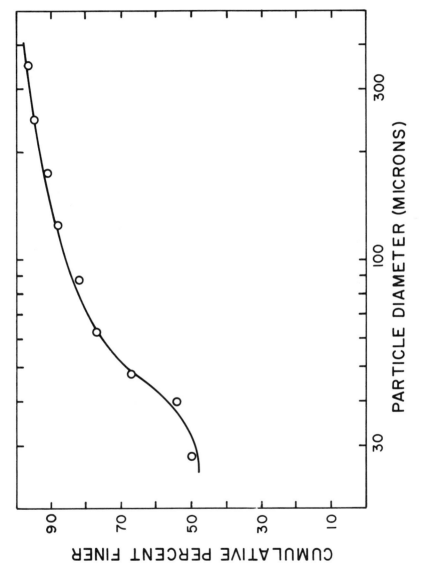

Figure 3 - Particle size analysis of concentrate.

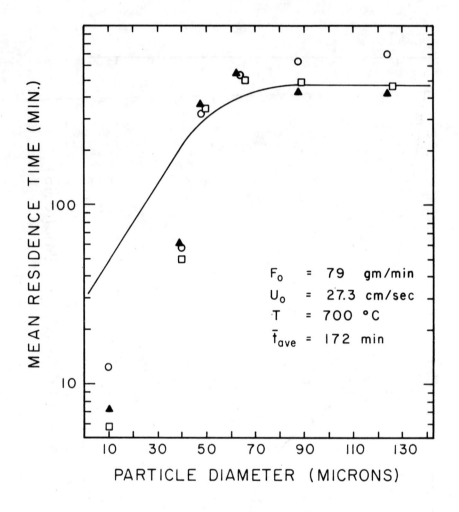

Figure 4 - Experimental and calculated particle residence times.

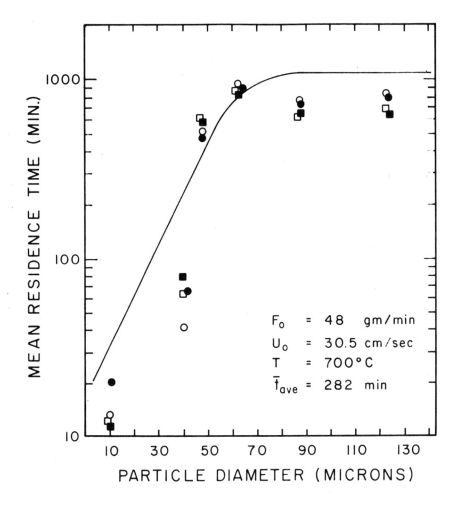

Figure 5 - Experimental and calculated particle residence times.

As can be seen in both figures 4 and 5, the results predicted by computer calculations do not agree closely with the experimental results. The disagreement is more evident for smaller particle sizes and shorter residence times.

The disagreement between calculated results and experimental results is due to the fact that the computer calculations rely on an empirical correlation for the elutriation constant. The correlation takes into account only the physical properties of the particles, the physical properties of the fluidizing gas, and the total weight of the bed material.

For example, Tanaka and Yoshimura (7) found that elutriation constants could be successfully predicted from the Yagi and Aochi correlation for a 14 foot diameter coal drier. However, they also determined that there was an affect on the elutriation constant caused by changing the location of the feed tube and by changing the location of the overflow tube.

Also, Lewis, et.al. (8), found that the elutriaton constant was affected by bed diameter. Their results showed that the effect was not evident at large diameters but became significant at small diameters in the range of three inches.

Based on the results of Tanaka and Yoshimura, and Lewis, et.al., it would appear that the correlation of Yagi and Aochi does not completely apply to the six inch roaster used in these experiments. Although the difference is probably due to roaster geometry, it is not certain if the disagreement in calculated residence times and experimental values is due to small bed diameter or feed tube and overlfow locations.

Summary

The work described in this paper dealt with the calculation and prediction of residence times in a laboratory size roaster. The predictions were based on the empirical correlations for elutriation constants of Yagi and Aochi.

Close agreement between calcualted and experimental results was not found. It would appear that in the case of small fluid bed reactors, the affect of geometry on elutriation constant correlations may be significant.

Acknowledgement

The authors wish to express their thanks to ESSO Resources Canada, Limited for their help in generation of the data in this report.

Nomenclature

F_o, F_1, F_2	feed rate of solids, bed overflow rate of solids, and carryover rate of solids by entrainment respectively, gm/sec
$K(R)$	elutriation constant, sec^{-1}
P_o, P_1, P_2, P_b	size distribution function of feed, bed overflow, entrained solids, and solids in bed, cm^{-1}
R	radius of particle, cm
$\bar{t}(R)$	mean residence time of particle size interval, sec
W	bed weight of solids, gm

References

1. F.A. Zenz and D.F. Othmer, Fluidization and Fluid-Particle Systems, p. 203 Reinold Publ. Corp. New York, N.Y., 1960.

2. F.A. Zenz and N.A. Weil, "A Theoretical-Empirical Approach to the Mechanism of Particle Entrainment from Fluidized Beds", A.I.Ch.E. Jol., Dec. (1958), pp. 472-479.

3. D. Kunii and O. Levenspiel, Fluidization Engineering, pp. 326-353, Robert E. Kruger Publ. Co., Huntington, N.Y., pp. 1977.

4. C.Y. Wen and R.F. Hashinger, A.I.Ch.E. Jol., Vol. 6, (1960), p. 220.

5. D. Kunii and O. Levenspiel, Fluidization Engineering, pp. 329-333 Robert E. Kruger Publ. Co., Huntington, N.Y., 1977.

6. S. Yagi and T. Aochi, paper presented at the Society of Chemical Engineers (Japan), Spring Meeting, (1955).

7. Y. Tanaka and K. Yoshimura, "Effect of Equipment Structure on Elutriation from Fluidized Bed", Preprint from meeting of Soc. Chem. Eng. (Japan), (1968).

8. W.K. Lewis, E.R. Gilliand, and P.M. Lang, Chem. Eng. Progr. Symp. Series, Vol. 58, No. 38, (1962), p. 65.

RATE PHENOMENA IN THE OUTOKUMPU FLASH SMELTING REACTION SHAFT

N.J. Themelis[1], J.K. Mäkinen[2] and N.D.H. Munroe[3]

1. Professor of Mineral Engineering, Henry Krumb School of Mines, Columbia University, New York 10027.
2. Senior Research Metallurgist, Harjavalta Works, Outokumpu Oy, Finland.
3. Doctorate Fellow, Henry Krumb School of Mines, Columbia University, New York.

ABSTRACT

In the last two decades, the Outokumpu flash smelting process has become the dominant new technology for copper smelting and has also been applied to the smelting of other metal sulfides. The application of oxygen enrichment and improvements in concentrate burner and furnace design have resulted in ever increasing smelting rates per unit volume of the reaction shaft. This paper examines the rate phenomena taking place in the reaction shaft of the Outokumpu furnace and compares information obtained from a mathematical model of heat and mass transfer rates to results obtained in pilot tests conducted at Pori, Finland, and to actual performance of a number of flash smelters around the world.

INTRODUCTION

The conventional process of extracting copper from its concentrates by reverberatory smelting has the inherent disadvantages of environmental pollution and high energy consumption. Flash smelting, as applied in the Outokumpu flash smelting process, has provided an attractive alternative to reverberatory smelting. Since its inception in 1946 by Petri Bryk and his associates in Finland (1) and commision in 1949 at the Harjavalta smelter, this process has been developed further along a path of increased productivity and energy conservation, longer campaigns between maintenance shutdowns and improved environmental control.

These advances are partly due to the continued research effort by Outokumpu Oy, and partly to improvements made by the many users of this technology. The success of the Outokumpu process is evident in its wide adoption in copper smelters around the world, as illustrated in Table I; it shows, in chronological order, that there have been twenty seven smelters built in both hemispheres and three more are under construction. Most of them are smelting copper concentrates, while four smelt nickel or copper/nickel concentrates. A recent development by Outokumpu is the adaption of the process for lead smelting (2); large scale pilot plant tests have been completed and the process is now ready for industrial implementation.

The Outokumpu flash smelting process consists of dispersing dry concentrates into a stream of oxygen-enriched air or preheated air as it enters in the upper end of a vertical cylindrical chamber, called the <u>reaction shaft</u>. As the gas-solid suspension flows downward through the reactor, the particles undergo controlled partial oxidation, which is accompanied by a large evolution of heat, and melting to fine droplets.

At the bottom of the reaction shaft, the direction of flow of the gas-particles stream is changed by 90° and the gas now flows through a horizontal hearth, or <u>settler</u>. Most of the particles drop out of the gas stream in the settler and separate in two layers, a supernatant slag layer and a bottom matte layer. The process gas then flows upward through the <u>uptake shaft</u> and then to a waste heat boiler and precipitators where the remaining fine dust particles are recovered; the gas then flows to a sulfuric acid plant. An illustration of the flash smelting furnace is shown in Figure 1, which is a schematic drawing of the pilot furnace at the Pori Metallurgical Research Center of Outokumpu.

The smelting reactions in the shaft are highly exothermic and provide most of the heat required for heating the feed materials to the temperature of the reaction shaft (1300-1400°C) and for compensating for the heat losses through the wall of the reaction shaft. The initial concept of the Outokumpu process entailed preheating the air stream to about 450°C, and the injection of fuel oil in the

TABLE I. OUTOKUMPU FLASH SMELTERS AROUND THE WORLD

	Started year	Smelting t/day	% Oxygen in air	Temp. of air, oC
Outokumpu Oy, Harjavalta, Finland	1949	1000	35-80	200
Outokumpu Oy, Harjavalta, Finland*	1959	750	30-80	200
Furukawa Co., Ashio, Japan	1956	550	41	550
Outokumpu Oy, Kokkola, Finland**	1962	1700	21	350
Chimico-Metalurg., Baia Mare, Romania	1966	675	27	520
Dowa Mining Co., Kosaka, Japan	1967	800	21	420
Nippon Mining Co., Saganoseki, Japan	1970	1700	24-27	1000
Sumito Metal, Niihama (Tokyo), Japan	1971	1300	30-40	450
Hindustan Copper, Ghatsila, India	1972	227	21	450
Peko-Wallsend, Mount Morgan, Australia	1972	300	21	450
Mitsui Mining, Hibi (Tamano), Japan	1972	1460	28-30	450
Norddeutsche Affinerie, Hamburg, FRG	1972	1280	30-40	540
Nippon Mining Co., Hitachi, Japan	1972	1200	24	1000
Western Mining, Kalgoorlie, Australia*	1972	1600	25	450
Karadeniz Bakir A.S., Samsun, Turkey	1973	915	21	400
Peko-Wallsend, Tennant Creek, Australia	1973	450	21	350
Nippon Mining Co., Saganoseki, Japan	1973	1900	24	1050
Pikwe, Botswana*	1973	2760	26	290
Hindustan Copper, Khetri, India	1974	430	21	400
Rio Tinto Minera, Huelva, Spain	1975	1200	21	360
Phelps Dodge, Hidalgo, N.M., USA	1976	1900	21	420
Corniczo-Hutniczy, Glogow, Poland	1978	2600	65-80	180
Korea Mining & Smelting, Onsan, Korea	1979	963	40-90	200
Kombinat Norilsk, Norilsk, USSR*	1981	2200	30-40	200
Kombinat Norilsk, Norilsk, USSR	1981	2400	30-40	200
Caraiba Metals, S.A., Camacari, Brazil	1982	1650	40-70	200
Phillipine Smelt. & Refin., Leyte, Phillipines	1983	1500	33	450

*nickel or copper-nickel smelters
**pyrite smelter, not in operation now

291

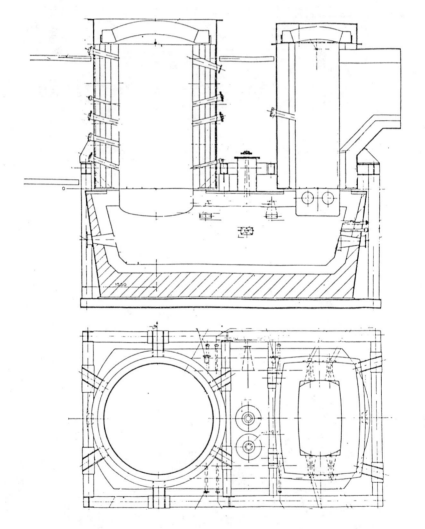

Figure 1. The Outokumpu flash smelting furnace (pilot furnace at Pori, Finland).

reaction shaft, to make up for the deficiency between heat input and output.

Even higher blast temperatures, up to 1040°C were practiced at the Saganoseki smelter of Nippon Mining (3). The higher preheating temperatures resulted in a decrease of the fuel input required and, also, in the volume of process gas from the furnace.

A very significant development came about a few years before the world prices of fuel oil increased and when the comparative cost for producing bulk oxygen went down (4). The use of oxygen to enrich the air stream was found to be more effective and economic than preheating: it served to reduce the amount of fuel requirement in the reaction shaft, produced a smaller amount of process gas, and at the same time it increased the furnace smelting capacity. An illustration of the beneficial effect of oxygen enrichment is given in Figure 2; it shows the comparative volumes of two furnaces and their ancillary gas handling equipment for smelting the same tonnage of concentrates, with and without oxygen enrichment (5).

For a typical chalcopyrite concentrate (28%Cu, 32%S, 30%Fe) and oxygen enrichment of about 40% O_2 in the air stream to the furnace, flash smelting is <u>autogenous</u> for a matte grade of about 60% Cu, which means that no additional fuel is required. Enrichment to a higher level of oxygen would necessitate the addition of large quantities of non-sulfidic, or "inert" materials to the furnace, or somehow increasing the heat losses through the walls of the reaction shaft.

OPERATING PARAMETERS IN REACTION SHAFT

This paper is principally concerned with the phenomena occuring in the reaction shaft of the Outokumpu furnace. On the basis of the above brief history of the development of the Outokumpu process, it is interesting to examine and compare some of the operating conditions in various flash smelting furnaces which use diffferent levels of air preheating and oxygen enrichment. This information was obtained partly from the literature and partly from private communications and was used to calculate certain important operating parameters.

These are: the average gas velocity through the reaction shaft, the corresponding residence time, the fraction of shaft volume occupied by the smelted concentrate particles and the specific smelting capacity in the shaft; this latter term has been introduced by Kellogg (6) for comparing the capacity of different reaction shafts under a given set of process parameters and is expressed as the rate of smelting, in tonnes per hour, per unit volume in the reaction shaft. The capacity of different smelters is normally controlled by other factors, such as converter capacity and volume of the gas handling facilities. The latter factor has been taken into account by the so-called "gas line capacity" and

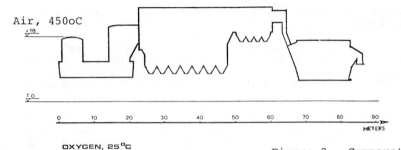

Figure 2. Comparative plant sizes using air and technical oxygen.

Figure 3.

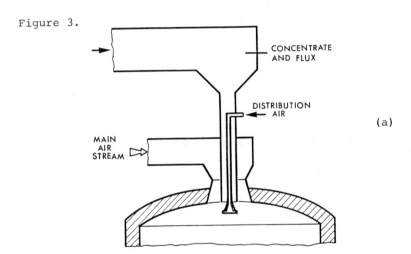

Figure 3. Concentrate burner configuration (a), dispersion cone without central jet distributor (b), and with central jet distributor (c).

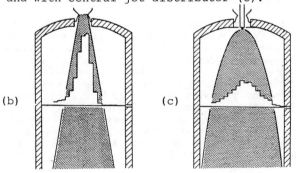

is expressed as tonnes of feed per Nm^3 of air flow. Table II presents a comparison of these operating parameters for seven smelters, four of which have been operated under two different regimes of air preheat and oxygen enrichment.

First, it must be taken into account that these plants smelt different chalcopyrite concentrates and also produce matte grades that range from 50 to 75%; therefore, one would expect some variation in specific smelting capacity even under the same preheat temperature and oxygen enrichment. Nevertheless, this tabulation shows that raising the air preheating temperature results in some increase in smelting capacity (Saganoseki vs. Tamano) while an increase in oxygen enrichment has a drastic effect (Harjavalta, Ashio, Tamano, and Norddeutsche Affinerie vs. Onsan). Highest smelting capacities have been attained by practically pure tonnage oxygen (Pori), but in this case special measures were taken to remove extra heat in the reaction shaft (i.e. boiler tube inserts, etc.).

The average gas velocity through the reaction shaft is calculated to range from a high of 4.7 to a low of 1.4 m/s; the corresponding residence times of the gas flow in the shaft range from 1.4 to 4.8 seconds. However, it should be noted that the residence times shown in Table II have been calculated on the assumption that the gas flow is isothermal at the reaction shaft temperature of 1573K and that it occupies the entire cross section of the shaft along the entire height of this chamber; as will be discussed below, these idealized conditions are not met in reality and therefore, the residence time of the gas in the reaction shaft is considerably shorter than shown in Table II.

The volume of concentrates in the reaction shaft is extremely small and the % loading is calculated to be in the range of 0.004-0.008%, with the exception of Harjavalta, Norddeutsche Affinerie, Ashio and Onsan smelters, which use oxygen enrichments of 42 - 50%. The corresponding figure for the Pori pilot furnace has been as high as 0.020 with pure technical oxygen and 77% Cu matte. Under these circumstances, the amount of concentrate in the shaft at any particular instant is less than 60 kg and it can be calculated that, on the average, adjacent particles are in the order of twenty diameters apart. Therefore, there should be little physical or chemical interaction between adjacent particles and the assumption that the falling velocity of particles can be determined from the Stokes law should be valid.

The calculated specific smelting capacities range form 0.20 to a high of 0.53 tonnes per hour per cubic meter of reaction shaft volume. The high values are for high oxygen enrichment. The corresponding figure for the Pori pilot furnace with pure technical oxygen has been as high as 1.95. As also shown in Table II, the gas line capacities range from 0.82 (air) to 4.31 (94% oxygen enrichment and an air preheat temperature of only 473K).

TABLE II. COMPARISON OF OPERATING PARAMETERS OF VARIOUS SMELTERS

	Harjavalta till 1971	Harjavalta 1983	Pori 1975	Norddeutsche Affinerie 1974(8)	Norddeutsche Affinerie 1984(17)	Onsan 1981 (10)	Ashio (7) till 1978	Ashio (7) 1981	Saganoseki 1981(3)	Tamano 1981(11,12)	Tamano
Conc. feed, t/h	20	44	4.4	52	64	40	18	23	56	47	61
Cu/S ratio	0.75	0.85	0.82	0.84	1.05	0.91		1.00	0.88		0.96
Air flow, Nm³/h	20,300	20,100	1,020	50,000	30,000	16,800	22,000	9,600	43,500	42,900	35,700
Oxygen enrichm., %	21	49	94	21	35	50	21	42	23	21	29
Inlet air temp., K	718	473	273	813	800	473	823	473	1,173	723	723
g Oxygen/g Conc.	0.30	0.32	0.31	0.29	0.23	0.30	0.36	0.25	0.26	0.27	0.24
Matte grade, % Cu	54	67	77	62	62	60	48	60	57	53	58
Sulfur dioxide, %	10	22	62	10	17	22	12	33	12	10	15
Dust as % of conc.	6-9	6-9	10	10	8	4-6	9	5		6-6.5	
Shaft diameter, m	3.9	4.2	1.2	6.0	6.0	4.0	3.1	3.1	5.6	6.0	6.0
Shaft height, m*	7.1	7.1	2.0	7.5	7.5	6.0	7.5	5.7	6.9	7.6	7.6
Ave. gas vel., m/s	2.6	2.4	1.4	2.6	1.55	2.1	4.7	2.0	2.6	2.2	1.9
Ave. res. time, s	2.7	3.0	1.4	2.9	4.8	2.8	1.6	2.8	2.7	3.4	4.1
Shaft loading, %	0.004	0.010	0.020	0.005	0.010	0.010	0.004	0.010	0.006	0.005	0.008
Specific smelting capacity, t/h · m³	0.23	0.45	1.95	0.25	0.30	0.53	0.32	0.53	0.33	0.22	0.28
Gas line capacity, t conc/1000 Nm³	0.99	2.19	4.31	1.04	2.13	2.38	0.82	2.40	1.29	1.10	1.71

*Distance from roof to surface of bath.

TRANSPORT PHENOMENA IN THE REACTION SHAFT

Entry and developed flow regions

In discussing the transport phenomena in the reaction shaft, it is convenient to distinguish between the following two regions of flow: the <u>entry region</u> and the <u>developed flow region</u>. In the entry region, the concentrate particles become dispersed in the air stream and the latter makes the transition from a relatively high- velocity jet, as introduced into the furnace through the <u>concentrate burner</u>, to the "terminal" gas velocity acquired as the gas flow comes to occupy the entire cross-section of the shaft; this is where the developed flow region starts and is assumed to extend until the surface of the melt where the gasf low changes direction to pass through the furnace hearth.

One type of concentrate burner used in the Outokumpu furnace (Figure 3a) consists of an inner nozzle system, which serves to disperse the concentrates in part of the air stream (distribution air) and of a much larger concentric tube, which introduces the bulk of the air flow into the reaction shaft; as illustrated in Figures 3b and 3c, the flared end of the distribution air tube serves to create an "umbrella" of concentrate particles which can be jettisoned as far as the walls of the reaction shaft. Typically, the bulk air flow through such a burner would be about 500 Nm^3/min, while the distribution air would be about 2-5Nm^3/min.

For the above burner configuration, the axial entry velocity of the bulk air stream in the shaft would be in the order of 50m/s. For isothermal air jet flow, this velocity would decay with distance from the entry point, according to the formula (13):

$$u_x = 12.4 u_o r_o / x \qquad (1)$$

where: u_o : entry velocity
r_o : radius of nozzle at entry
u_x : velocity at center of jet at distance
x : axial distance from entry point

According to the above equation, by the time the gas flow reached the end of the reaction shaft (x = 7.1m, Harjavalta), its velocity at the center would still be about 24m/s ; if that were the actual case, the gas would never reach the developed flow region and the residence time of the gas flow in the shaft would be in the order of only 0.3s and not the 3.0 seconds shown in Table II. In fact, the air flow through the shaft is not isothermal but the gas is heated rapidly, due to the oxidation reactions of the entrained concentrates from 200°C (Harjavalta) to a temperature of about 1300°C in a fraction of a second. The effect of this threefold increase in gas volume is to expand the jet cone and decrease its velocity as it traverses the reaction shaft. Also, the heat generated by the chemical reaction creates temperature gradients which in turn produce density gradients. In particular, it would be expected that near the shaft wall there are upward natural convection currents which oppose the downward flow of the main stream.

The above discussion indicates the important effect of the air entry velocity on the residence time of the gas flow through the shaft, especially when the reaction shaft is relatively short and a high specific capacity is required. For instance, in the case of the Ashio smelter, the original burner design provided a hot air blast at an entry velocity of about 100 m/s and a relatively low pressure drop through the burner; however, when the reaction shaft was shortened in 1965, a new burner was adopted, with guide vanes at the inlet and a dispersion cone at the lower end of the concentrate feeding chute.

Particle velocity

The above discussion refers to the velocity of the gas stream. However, from a chemical reaction point of view, the residence time of the particle is more important. Because the flow of solids through the shaft is co-gravity and co-current with the gas flow, the particle velocity is equal to the gas velocity plus the falling velocity of the particle. Chalcopyrite concentrates are ground to a size of about 70 per cent less than 50 um, with particles greater than 100 um accounting for about 5 percent of the feed. The corresponding particle Reynolds numbers in the reaction shaft are less than 0.4 and the particle terminal velocity can be determined from the Stokes equation (13), as noted earlier. On the basis of this equation, the falling velocity of a 50-micron particle in air at 1300°C is only 0.1m/s and that of a 100-micron particle about 0.4m/s. Therefore, when the concentrate is adequately dispersed into the air stream, the residence time of most particles will be close to that of the gas stream.

Chemical reaction, heat and mass transfer

As the concentrate particles enter the shaft, they are first heated by convection from the surrounding air and by radiation from the furnace enclosure, until they reach a temperature at which the oxidation reaction at the particle surface generates more heat than can be dissipated through heat losses to the atmosphere; this temperature is called the ignition temperature and ranges from 400 – 550°C. After this point, the particle temperature rises rapidly until the combined rates of heat transfer by convection and radiation balance the heat generated by chemical reaction.

The heat and mass transfer coefficients between the reacting particle and the gas envelope around it can be expressed by means of the Ranz-Marshall and the Sherwood correlations (13), which relate these coefficients to the dimensionless Reynolds, Prandtl and Schmidt Numbers. Table III shows a tabulation of these numbers and other gas and

TABLE III. MASS AND HEAT TRANSFER PARAMETERS FOR CHALCOPYRITE PARTICLES IN AIR

FILM TEMPERATURE	473K			1173K			1573K		
Gas density, g/cm^3	7.44×10^{-4}			3.00×10^{-4}			2.73×10^{-4}		
Gas viscosity, $g/cm/s$	2.89×10^{-4}			4.64×10^{-4}			5.64×10^{-4}		
Gas thermal conductivity, $cal/cm/s/K$	1.07×10^{-4}			1.77×10^{-4}			2.17×10^{-4}		
Gas diffusivity, cm^2/s	0.35			1.74			2.94		
Particle diameter, cm.	0.0025	0.005	0.001	0.0025	0.005	0.01	0.0025	0.005	0.01
Particle density, g/cm^3	4.2	4.2	4.2	4.2	4.2	4.2	4.2	4.2	4.2
Particle terminal vel., cm/s	4.9	19.8	79.1	3.1	12.3	49.3	2.5	10.1	40.6
Reynolds Number	0.03	0.26	2.04	0.01	0.04	0.32	0.00	0.02	0.16
Prandtl Number	0.76	0.76	0.76	0.73	0.73	0.73	0.73	0.73	0.73
Schmidt Number	1.12	1.12	1.12	0.89	0.89	0.89	0.86	0.86	0.86
Nusselt Number	2.10	2.28	2.78	2.04	2.11	2.31	2.03	2.08	2.22
Sherwood Number	2.11	2.31	2.89	2.04	2.12	2.33	2.03	2.08	2.23

particle properties that are involved in the transport phenomena in the shaft. It can be seen that because of the very low particle Reynolds Number, the corresponding Nusselt (heat transfer) and Sherwood (mass transfer) numbers are close to the value of 2, which is the limiting value representing heat conduction and molecular diffusion through an infinitely thick gaseous boundary layer. Therefore, the heat and mass transfer coefficients between the fine particle and the gas envelop around it can be expressed approximately as follows (14):

$$h = 2k/d_p \quad \text{and} \quad k_d = 2D_v/d_p \quad (2)$$

Although the heat and mass transfer coefficients are enhanced very little by convection, because of the enormous surface areas per unit weight presented by the fine particles (e.g., 290 cm^2/g for a concentrate consisting of 50-micron particles) the resulting heat and mass transfer rates are very high. However, as the particle temperature increases to the furnace temperature, the chemical rate also becomes very high and there is evidence (14,15) that the overall reaction rate is controlled by mass transfer of oxygen through the boundary film around the particle and by diffusion through the reacted layer.

Mathematical model

These phenomena are under further investigation at Columbia University; however, a preliminary mathematical model, based on the differential material and heat balances between the reacting particles and the surrounding gas (14), was used to provide some information as to the expected rates of oxidation of particles of different sizes in the reaction shaft, the effect of oxygen enrichment on the rate of reaction and the effect of other operating parameters. This model assumes that the overall rate of oxidation of the concentrate particles is controlled by mass transfer of oxygen through the gas film and then through the reacted layer to the unreacted core; the model neglects the effects of melting or of breaking up of the particle into fragments, phenomena which have been observed in flash smelting tests by Outokumpu Research.

Figure 4 shows the effect of particle size on the fractional oxidation of chalcopyrite particles in an Outokumpu furnace operated with air preheated to 773 K; the model predicts that 50-micron particles will be 90% oxidized in one second, while the 100-micron particles will be 75% oxidized in the same time interval. Figure 5 shows the effect of oxygen enrichment (air and 34% oxygen) on the oxidation rate of 100-micron particles. As would be expected, oxygen enrichment increases the rate of the reaction, both by accelerating the heating of the gas-particle stream and by increasing the rate of mass transfer of oxygen, due to the higher concentration driving force.

Figure 6 shows the change in oxygen concentration with

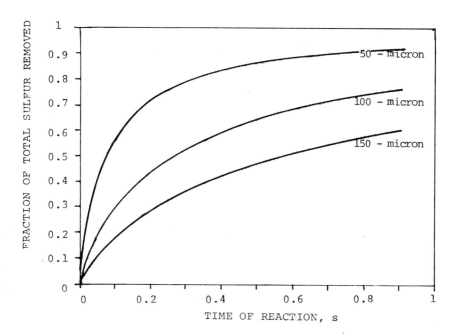

Figure 4   Effect of rate of oxidation of chalcopyrite particles in an air stream.

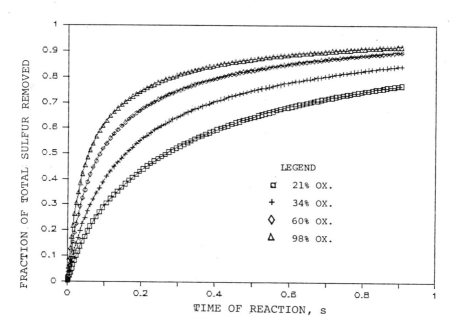

Figure 5   Effect of oxygen enrichment on rate of oxidation.

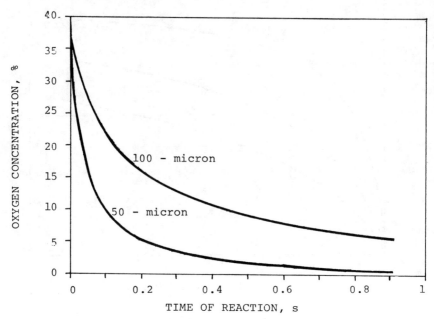

Figure 6 Change in O_2 concentration with time in reaction shaft.

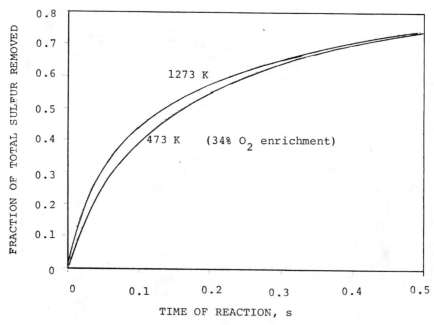

Figure 7 Effect of air preheat on rate of oxidation.

residence time of the gas stream in the reaction shaft (34% oxygen enrichment, 5% excess oxygen over the stoichiometric requirement for complete oxidation); in the case of the finer particles, the oxygen concentration drops very rapidly to 10% of the inlet concentration and this decrease in the concentration driving force is partly responsible for the decrease of reaction rate with time.

Figure 7 explores the effect of air preheat temperature by comparing two operations, one preheating the air to 473 K and the other to 1273 K. It can be seen that the effect of air preheat temperature, at levels above the ignition temperature, is not as significant as that of oxygen enrichment. This is explained further in Figure 8, which shows the predicted temperature profiles for the two cases; both streams reach the reaction shaft temperature within a small fraction of a second. Therefore, all other factors being the same, a higher preheat temperature will not increase significantly the degree of oxidation in a given reaction shaft.

CONCENTRATION AND TEMPERATURE PROFILES IN THE REACTION SHAFT OF THE OUTOKUMPU PILOT FURNACE.

It is interesting to compare the above mathematical projections with actual sampling tests conducted several years ago in the pilot Outokumpu flash smelting furnace at Pori. During these tests, the concentrate feed rate was 590 kg/h. The special probe constructed for sampling the particle stream at different locations within the shaft is shown in Figure 9.

In order to convert the x-axis of Figures 5-8 from residence time to distance traversed in the reaction shaft, it is necessary to know the velocity of gas-particle flow within the shaft. As discussed earlier, there is considerable difficulty in calculating this velocity which may be anywhere from a few seconds to less than a second. It is therefore interesting to compare the above mathematical predictions with actual sampling tests conducted some years ago in the pilot Outokumpu flash smelting furnace at Pori. As shown earlier in Figure 1, the reaction shaft of this furnace was 1.5 m in diameter and there was a distance of 4.0 m between the entry point and the surface of the bath in the settler. However, a still earlier design had an internal shaft diameter of 1.2 m and a height of 6 meters; some of the data presented were obtained in that furnace. Temperature profiles in the gas stream were obtained both by means of thermocouples and using an aspirating thermocouple.

Figure 10 shows the measured concentration profile of oxygen with distance below the roof of the shaft for a smelting rate of 0.5t/h and a calculated average gas velocity of 0.5m/s; the air was introduced at room temperature and the oxygen enrichment was 27%. On the same plot, there is superimposed the concentration profile predicted by the mathematical model. It can be seen that the two sets of data compare well, if the average velocity of

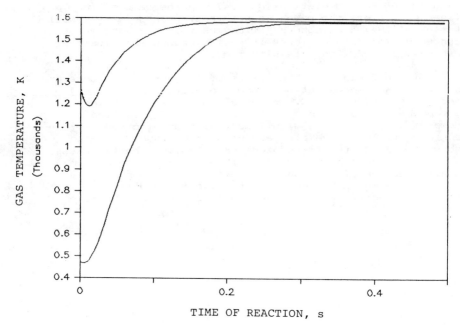

Figure 8 Temperature profile of 34% O_2-enriched air stream entering at 473K and 1273K.

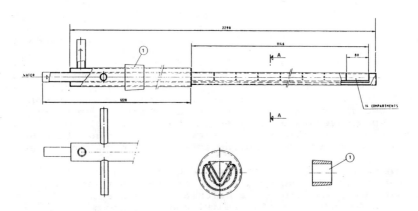

Figure 9 Probe used for sampling solids in Outokumpu pilot furnace.

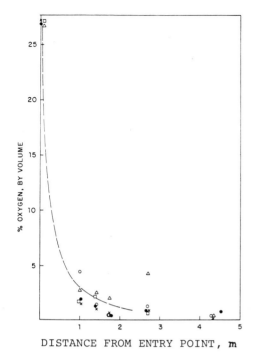

Figure 10 Oxygen concentration profiles in Outokumpu pilot furnace.

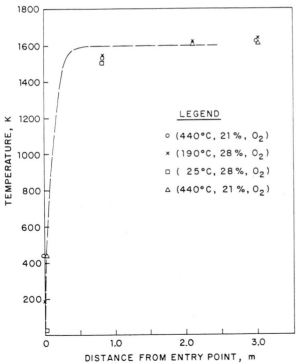

Figure 11 Gas stream temperature profiles in Outukumpu pilot furnace.

LEGEND
○ (440°C, 21%, O_2)
× (190°C, 28%, O_2)
□ (25°C, 28%, O_2)
△ (440°C, 21%, O_2)

gas flow in the Pori pilot furnace is assumed to be 3 m/s.

Figure 11 shows the temperature profiles measured by means of a suction pyrometer inserted to the center of the reaction shaft, at various heights below the roof. The data show that smelting temperatures of 1200°C are apparently attained within the first meter of travel below the roof, independently of the air preheat temperature; again, these data are in agreement with the temperature prediction of Figure 8.

The importance of concentrate burner design can be seen in Figure 12 which shows the mass flux of particles across the diameter of the shaft at a distance of 2.1 meters below the roof of the shaft. With the proper flow of distribution air through the burner(Fig.12), the dispersion of solids across the shaft was fairly uniform.

Figure 13 shows experimental data obtained in the Pori pilot reactor on the progressive oxidation of concentrate particles with distance travelled down the reaction shaft. Here, the y-axis refers to the grade of matte that would have resulted by melting the oxidized particles collected at a given point. At 100% "equivalent matte grade", the fractional oxidation of the particles would be 1, i.e. all of the iron and sulfur content would be oxidized. The different symbols shown in Figure 13 represent samples taken at different times and over a range of experimental conditions.

CONCLUSIONS

A mathematical model was developed to predict the rates of oxidation of concentrate particles in the reaction shaft of an Outokumpu flash smelting furnace under different conditions.This model assumes that the overall rate of oxidation is controlled by mass transfer of oxygen through the gas film and through the reacted layer to the unreacted core. The mathematical predictions were in agreement with data obtained in sampling tests conducted several years ago in the pilot Outokumpu flash smelting furnace at Pori.

The following conclusions were drawn:

1. A residence time of 1 second is sufficient for oxidising a 100-micron concentrate particle to 75% of complete oxidation, which corresponds to a matte grade of about 80% Cu.

2. Even a small amount of oxygen enrichment (34%) accelerates considerably the rate of reaction.

3. The effect of air preheat temperature ,at levels above the ignition temperature, is not as significant as

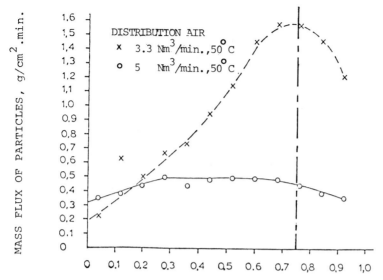

Figure 12 Distribution of local mass flux of particles along the shaft diameter at 2.1m from the point of entry.

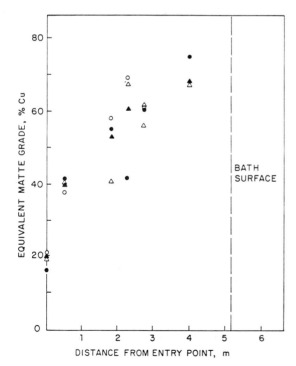

Figure 13 Oxidation of chalcopyrite particles with distance travelled along reaction shaft of Outokumpu pilot furnace. (Four tests at: 0.37-0.48 m/s; 0.49-0.55 t/h; and 27-28% O_2).

that of oxygen enrichment.

In the light of the above theoretical results, some of the operating parameters in various flash smelting furnaces were compared, at different degrees of oxygen enrichment and air preheating. This comparison (Table II) showed that:

1. The average gas velocity through the reaction shaft ranges from 4.7 to 1.4 m/s. On the assumption that this velocity applies throughout the length of the reaction shaft, the corresponding residence time would range from 1.4 to 3.6 seconds. However, because of the high entry velocity of the gas-solids stream into the reactor, it is believed that the actual residence time may be considerably shorter than indicated by the above calculation. Further work is needed to elucidate this matter.

2. The matte grade produced in the various furnaces does not appear to increase with longer residence times in the reaction shaft. This indicates that under the present operating conditions, the residence time is not limiting the smelting capacity. Matte grade is controlled principally by the composition of the feed and the input oxygen to concentrate ratio.

3. The specific smelting capacity and the gas line capacity increase with degree of oxygen enrichment.

4. Increasing the air preheat temperature decreases the fuel consumption when smelting with air of less than about 30% oxygen enrichment. On the other hand, for a given matte grade produced, preheating of the air limits the amount of oxygen enrichment and, therefore, the specific smelting capacity and gas line capacity.

ACKNOWLEDGMENTS

Work at Columbia University on the kinetics of flash smelting reactions is supported by the Minerals and Primary Materials Program of NSF (CPE-81-10526). The authors also gratefully acknowledge the support and contribution to this work of the Outokumpu Metallurgical Research Center.

REFERENCES

1. P. Bryk, J. Ryselin, J. Honkasalo, and R. Malmstrom, "Flash Smelting Copper Concentrates, "Journal of Metals, 10 (1958), pp. 395-400.

2. E. Nermes and T. Talonen, "Flash Smelting of Lead Concentrates," Journal of Metals, 34, No. 11 (1982), pp. 55-59.

3. M. Yasuda, T. Yuki, M. Kato, Y. Kawasaki, "Recent

Flash Smelting Operations at the Saganoseki Smelter," in Copper Smelting - An Update, Edited by D.B. George and J.C. Taylor, The Metallurgical Society of AIME, New York, 1981, pp. 251-263.

4. J. Juusela, S. Harkki, and B. Andersson, "Outokumpu Flash Smelting and Energy Requirement," Paper presented in Symposium, "Efficient Use of Fuels in Metallurgical Industries," Dec. 9-13, 1974, Chicago, USA.

5. J. Honkasalo, T. Tuominen, and J. Juusela, "Outokumpu Flash Smelting - A Modern and Advanced Copper Smelting Process," Paper presented in Symposium, "Copper 83", Metals Society, Nov. 1-2, 1983, London, UK.

6. H.H. Kellogg and J.M. Henderson, "Energy use in Sulfide Smelting of Copper," AIME Symposium, "Extractive Metallurgy of Copper, Pyrometallurgy and Electrolytic Refining," Feb. 1976, Las Vegas, USA.

7. Ibid. ref. 3, O. Fujii, and N. Shima, "Reation Shaft of Ashio Smelter's Flash Furnace Utilizing High Oxygen Enriched Air," pp.165-171.

8. Outokumpu Oy, private communication.

9. L. Lilja, and U. Makitalo, US Pat. L1, 147,535.

10. Ibid. ref. 3, T.J. Kim, "Flash Smelting of Copper in Korea", pp. 33-39.

11. Ibid. ref. 3, M. Hashiuchi, S. Okada and T. Watanabe, "Review of Current Operation at Tamano Smelter," pp. 237-250.

12. M. Mita, T. Watanabe, S. Okada and A. Kora, "Features of Flash Smelting with Furnace Electrodes with special attention to its Energy Requirement", Paper presented at TMS-AIME annual Meeting, Atlanta, GA, March, 1983.

13. J. Szekely and N.J. Themelis, "Rate Phenomena in Process Metallurgy," Wiley-Interscience N.Y (1971).

14. N.J. Themelis and H.H. Kellogg, "Principles of Sulfide Smelting," in "Advances in Sulfide Smelting," Vol. 1. Basic Principles, Edited by H.Y. Sohn, D.B. George and A.D. Zunkel, pp.1-29.

15. F.R.A. Jorgensen, "On the Maximum Temperature attained during the Single Particle Combustion of Pyrite," Trans. Inst. Min. & Met., 90, C10-C16, 1981.

16. Outokumpu Oy, private communication.

17. Norddeutsche Affinerie, private communication.

PROGRESS OF COPPER SULFIDE CONTINUOUS SMELTING

Takeshi Nagano

Mitsubishi Metal Corporation
5-2, Ohtemachi 1-chome,
Chiyoda-ku, Tokyo 100
JAPAN

Summary

Copper pyrometallurgy in Japan has experienced quite fundamental changes in last fifteen years. The severe environmental restriction coupled with increase of energy costs accelerated the development work of the new continuous process in Mitsubishi Metal Corporation. After several years of pilot plant test work the commercial plant was put into operation at Naoshima Smelter in 1974, the design capacity of which was 4,000 tonnes of copper per month.

Many modifications and improvements have been incorporated into the process and 7,200 tonnes per month of copper production was achieved in 1983, proving that continuous process was much more economical than conventional process.

It is now possible to design and operate the continuous process in wide range of capacity. This paper will review the process and describe the progress of the process in these 10 years.

Introduction

It is presumed that the ancient extraction method of copper was the direct smelting of high-grade ores selected by hand-picking, using charcoal as fuel and hand bellows for blowing. Sometime in the 15th century the matte converting process called "Ma-buki" was developed in Japan. By this new technology, the process was divided into two separate steps, matte making and matte converting, thus making it possible to treat low grade ores. The smelting process to make matte from copper ores was similar to the ancient one, in which the molten slag was skimmed and when copper matte appeared in a hearth, it was solidified by water spray. This matte making process, which was called "Su-buki", was repeated until enough amount of solid matte was accumulated for subsequent converting process "Ma-buki". In the "Ma-buki" hearth, the solid matte layers were remelted with charcoal and oxidized with hand bellows blast. There was one small top blow lance installed through the clay cover on the hearth for air blowing. With very mild air blowing to the surface of matte through the lance, iron and sulfur remaining in matte were removed as slag and waste gas, and eventually crude copper was obtained. It is interesting that these basic technologies are employed in Mitsubishi Continuous Process: two step constitution of smelting and converting furnaces and the top blow lancing practice.

In spite of the revolutionary changes in copper industry in the late 19th century by introduction of foreign technologies such as blast furnace and contemporary type converter, the "Ma-buki" process was alive till the 1960's in small smelters. In the 1970's copper pyrometallurgy in Japan experienced quite fundamental changes, facing the increasing copper demand along with the economic growth of the nation, strict environmental control requirement and oil crisis. Most of blast furnaces were replaced by Outokumpu flash furnaces, scrubbers were employed to remove weak sulfur dioxide from the large volume of waste gases such as reverberatory furnace off gas and fugitive gas in converter aisle.

There are many methods in melting and partial oxidation of copper concentrates to form matte and slag i.e., blast furnace, reverberatory furnace, electric furnace, Outokumpu flash furnace, Inco flash furnace, Noranda reactor. However, 97% of copper from sulfide smelting in the world is still produced in traditional converters. To meet recent stringent regulations for both in-plant and ambient pollution control, fugitive gas in converter aisle is most troublesome. Many kinds of effort were made to solve the problem. In Onahama Smelter, whole converter aisle was closed and ventilated with big fan. Converters are covered with secondary hood and gas collecting equipments were installed in matte tapping and slag returning area.

In these circumstances the development work of the new continuous process in Mitsubishi Metal Corporation was accelerated. The pilot plant operation started in 1969 in Onahama Smelter, which continued for 4 years and terminated in 1973 after proving many possible advantages of continuous process for copper production:
a) Capital cost and energy cost for converter gas treatment are reduced with smaller fluctuation of both gas volume and SO_2 concentration in furnace off gas.
b) Temperature changes in converter and heat exchangers in the acid plant are small with stable gas stream, which prolong the catalyst life and reduce the maintenance cost.
c) Spill gases which inevitably accompany the ladle transfer of melt are also much reduced.
d) Continuous operation is less labor intensive by eliminating matte tapping, ladle transfer and charge-discharge procedure of melt.

In November 1972, the directors of the company gave approval for the modernization of its Naoshima Smelter to meet strict environmental regulations. At that time, two reverberatory smelters were operating and all the reverberatory furnace off gases were emitted to atmosphere. One of the smelters had been in operation for more than 50 years and inspite of many improvement works, it had become apparent that it could not continue to be in operation economically under newly announced environmental control requirements. The $40 million modernization included $13 million for replacement of the old reverberatory smelter with the continuous process, $12 million for new acid plant to treat the surviving No. 2 reverberatory off gas mixed with continuous process off gas and $9 million for tankhouse extension.

In March 1974, the plant had been constructed and put into operation. The design capacity of the new continuous smelter was decided to be 4,000 metric tonnes of anode copper per month, which was the same as that of the reverberatory smelter replaced and 3 times scale-up from the pilot plant at Onahama. In 10 years of commercial operation since then there were two major improvements; prolonged campaign life and increased furnace through-put. The furnace campaign life was the item to be determined in the commercial plant operation. Actual campaign life in pilot plant and early commercial operation were three to seven months. With many modifications on the furnace design and operating techniques, more than 2 years campaign life was achieved in February 1983 and stand-by furnaces were demolished. There is a fair prospect of 5-6 years furnace campaign life in the near future judging from recent refractory wear of both smelting and converting furnaces.

In May 1983, after 3 years of experimental work for increasing the through-put into the existing furnace, 7,200 tonnes of anode production, which corresponded to 70% increase from the design figure, was achieved (1). It is now possible to design and operate the continuous process in wide range of capacity, from 20 to 300 thousand tonnes of annual production of copper.

Fundamental idea of continuous process

There are two basic types of chemical, thermal and mass transfer principles in extractive metallurgy of sulfide concentrates. These two types could be distinguished in roasting of zinc concentrate: the flash roaster, where the temperature and oxygen partial pressure change from the top to the bottom and reaction is carried out during concentrate particles are dropping down the roaster; and the fluid bed roaster, where concentrate particles are suspended and react in fluidizing bed with much uniform temperature and oxygen partial pressure.

As magnetite control is most essential for stable operation of copper pyrometallurgical process, to maintain uniform temperature and oxygen potential is most desirable. To put the copper concentrate particles into matte as soon as possible to melt down before they react with oxygen, seems to be effective not to form excess magnetite, because temperature and oxygen potential is maintained rather uniform in melt than in gas phase. To have chemical reaction, thermal and mass transfer in melt was therefore the first fundamental idea at the development of Mitsubishi Continuous Process. Professors Themelis and Kellogg distinguished modern smelting processes into flash-smelting and bath-smelting (2). According to their classification Mitsubishi Process belongs to the bath-smelting.

Multi-furnace system to achieve continuous converting is the second fundamental idea. The constitution to have two steps, matte making and matte converting, has not been changed for 500 years since "Ma-buki" technology was developed. There are some advantages in direct smelting of copper concentrates to blister copper in one reactor such as: a) less heat loss from reactor, b) simple off gas treatment, c) lower capital cost.

But considering the reason why two step system had been predominant for such a long period, it seemed to be rational and rather simple to use two separate reactors to achieve simultaneous production of waste slag which requires an oxygen partial pressure less than 10^{-8} as in the reverberatory furnace, and blister copper which requires 10^{-5} as attained at copper blow of converter operation. There were also some advantages expected in two step system such as a) higher impurity removal, b) lower sulfur content in blister copper.

Then multi-furnace system was selected for the basic idea for the development work of continuous process. Professor Kellogg clearly showed the pyrometallurgical design principle of new sulfide smelting processes of copper (3). ----- He wrote: To aid in understanding the various designs, I have given, in Figure 1, a much simplified representation of the chemistry of progressive oxidation which underlies sulfide smelting of copper. In this figure, the abscissa shows the progressive increase in equilibrium partial-pressure of oxygen as the copper-bearing phase changes from high-iron matte, to low-iron matte, to white metal, to blister copper. The composition of the copper-bearing phase is shown by the curve for % Cu in matte, in the matte and white metal regions, and by the curve for % S in blister in the blister region. Any slag phase present with the copper-bearing phase will contain dissolved copper, the amount of which is shown semi-quantitatively by the curve for % Cu in slag. The slag will also contain mechanically entrained copper or matte, so that the curve for % Cu in slag may be viewed as representing the theoretical lower-limit of copper loss to the slag.

One important conclusion is immediately evident from study of Figure 1: Since a discard slag, with less than 0.5% Cu, requires an oxygen partial-pressure less than 10^{-8} atm., and a blister copper requires an oxygen partial-pressure in excess of $10^{-6.5}$ atm., it follows that no single oxygen partial-pressure (or equilibrium state of the system) can satisfy both requirements. Rather, a successful process for production of discard slag and blister must possess an oxygen partial-pressure which varies at least between 10^{-8} and $10^{-6.5}$, either in position within a continuous reactor, or in time for a batch process. Conventional smelting achieves this by use of the batch converter, where the oxygen partial-pressure varies between the extremes shown on Figure 1 from the start of the slag blow to the end of the finish blow.

To achieve the goal of discard slag plus blister production, a continuous process has two alternatives:
a) Use a single reactor which approximates counter-current plug flow of slag and copper-bearing phase, so that the oxygen partial-pressure varies from one end of the reactor to the other. This is the alternative adopted by the WORCRA process.
b) Use at least two separate reactors of the well-mixed type; one reactor operates at partial pressure of oxygen $<10^{-8}$ to produce a discard slag, the other operates at partial pressure of oxygen $>10^{-6.5}$ to produce blister, the matte from reactor 1 flows to reactor 2, and the slag from reactor 2 is returned to reactor 1. In essence, this is the design adopted by the Mitsubishi Process.

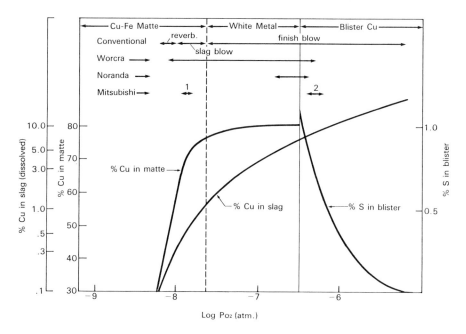

Figure 1   Semi-quantitative representation of copper smelting chemistry

Still another alternative for a continuous process is to abandon the attempt to make a discard slag, and to operate a single well-mixed reactor at partial pressure of oxygen $>10^{-6.5}$ so as to produce blister copper and a slag of high copper content. The slag is then milled to produce a tailing with less than 0.5% Cu, and a copper concentrate which is returned to the smelting reactor. The Noranda Process approximates to this description. -----

To achieve this basic idea to produce discard slag and blister copper continuously from concentrates in two separate steps, it was necessary to develope the new techniques:
a) to have steady stream of matte out of the smelting furnace into the converting furnace.
b) new converting furnace slag with good fluidity to have continuous overflow even with high oxygen partial pressure which is required to produce copper.

These two items were solved with a) the siphon technique which taps matte (blister) from the bottom and the launder connection between the furnaces and b) $CaO-Fe_3O_4-Cu_2O$ ternary slag for the continuous converting furnace operation (4).

Injection smelting of concentrate with coal addition

Smelting furnace, in which raw materials and coal are injected through top blow lances and high-grade matte is produced, is one of the features of Mitsubishi Continuous Process. The history of development work of the injection smelting dates back to 1959 when a laboratory test initiated in the research center of the company. Even at that time there was already the necessity of developing a new smelting process to replace a reverberatory furnace method in the future, which was predominant in copper smelting but had some inherent disadvantages of high fuel consumption and difficulty in economical recovery of SO_2 from the off gas. In order to experiment on this new smelting method a test furnace, 1.0m by 2.3m by 1.3m height, equipped with 4 tuyeres was constructed at Naoshima Smelter in 1964. It was found that if concentrates are injected with air through tuyeres, they can be smelted efficiently, i.e. a) rapid smelting reaction, b) small mechanical dust generation, c) low copper content in slag, d) high oxygen utilization efficiency.

This new method proved to be a potentially high-efficient smelting technology, but there were a number of engineering problems to be solved for commercialization. The major problem was the life of tuyeres, which would determine the length of campaign life of the whole system. The application of lances, which can be supplied during operation without furnace shut-down, was then developed aiming to elimination of disadvantage while keeping advantages of tuyeres.

When lance jet contains certain portion of solid particles like smelting furnace lance practice, the jet penetrates deeper than the gas jet. This phenomenon was found when bottom lining of smelting furnace was damaged just under lances. A series of cold model and hot model tests were conducted to analyze the phenomenon (5) and, at present, this deep penetration of jet is considered as a most important feature for top lance injection smelting. The interaction between the gas jet and depth of cavity in the liquid is correlated to momentum of jet, distance between lance and liquid, etc. (6). With normal lance injection conditions in the smelting furnace (1kg/sec of oxygen enriched blow and 4kg/sec of instantaneous solid charge to each lance), if no solid particle is charged, there would appear a definite cavity and the depth of which is calculated as 35cm. When solid particles are added, the definite cavity may disappear and gas and solid mixture penetrate 110cm in the melt. Figure 2-a) and 2-b) show the definite cavity and deep penetration without cavity which were taken in the cold model test.

a) Definite cavity b) Deep penetration

Figure 2 Behavior of gas-solid jet in liquid

This deep penetration in the actual smelting furnace operation may give a chance of good contact between oxygen and matte to react, liquid and solid to melt down and enough agitation of bath to homogenize the melt in the furnace.

In the original concept, the matte grade was set at the 40% level considering the copper loss in slag. At the 40% matte grade, however, only a small part of the reaction heat of iron and sulfer in concentrates is utilized in the smelting furnace where heat supply is required to melt down the solid. Through pilot plant operations, it was found out that, by proper control of the slag composition and temperature, the slag loss can be maintained in a reasonable level to be discarded if the matte grade is lower than 70%. Then aimed matte grade was determined to be 65% considering some fluctuation in actual operation. By the production of this high-grade matte, the fuel consumption in the smelting furnace was remarkably reduced and, at the same time, the formation of the converting furnace slag was reduced to less than 10% of the raw materials. It was the first plan to return the converter slag in molten state for effective utilization of the heat. But since the quantity of slag gets small, it is now granulated to be mixed with the smelting charge from simple handling point of view.

This 65% matte grade is fairly high, considering that all the converting furnace slag is returned directly and inspite of this, copper loss in the smelting furnace slag is low enough to be discarded which eliminates slag floatation. Although such a low slag loss is mainly attributed to effective cleaning in the slag cleaning furnace, the injection smelting mechanism mentioned above, also seems to play an important role in this phenomenon. As matte and slag produced in the smelting furnace are taken out together by overflow through a common hole, the slag layer in the furnace is thin and apparent residence time of slag in the smelting furnace is short. But with deep penetration of jet into the matte, where uniform temperature and oxygen partial pressure are maintained, the solid converter slag must be melted down and digested rapidly to reduce excess magnetite. With this successful development of top blow lancing, the original plan of injection smelting was carried out.

As regards coal utilization Professor Kellogg stated in his paper entitled "Conservation and Metallurgical Process Design" (7) in 1977 that:
----- The author believes that coal and fuels derived from coal will be the major metallurgical fuels of the future. Coal is no stranger to the metallurgical industry; the steel industry would be helpless without it, zinc and lead smelting still employ it, and copper smelting used it regularly 50 years ago. Though less convenient than oil and gas, it is in some ways a superior fuel for metallurgical purposes. For example, powdered coal burns more rapidly than atomized oil, and this property can have important advantages for some process. The process for fuming zinc from lead blast furnace slags may be cited when the powdered coal-air mixture is injected through the tuyeres, it reacts instantly and completely to reduce zinc oxide in the slag and to supply heat for the process; fuel oil is not an effective substitute for coal in this application unless provisions are made to enhance its combustion rate by special atomization and air preheating. It seems likely that powdered coal would also prove a superior fuel in other injection smelting processes, such as Mitsubishi Continuous Smelting of copper concentrates. -----

Table 1 Typical properties of coal

Proximate analysis	% Natural inherent moisture	2 - 3
	% Volatile	36 - 41
	% Fixed carbon	44 - 52
	% Ash	10 - 12
Lower calorific value	kcal/kg	7,100 - 7,300
Size distribution	+ 35 mesh	10 - 20 %
	+ 70	30 - 45
	+ 100	10 - 15
	+ 200	10 - 25
	+ 400	5 - 15
	- 400	5 - 10

His prophecy has come true in a short time. Early 1979, coal started to be applied for commercial operation after some test operations. Coal, moisture content of which is 20%, is first blended with concentrates in the bedding yard and after dried, injected into the smelting furnace through lances with furnace charge and oxygen enriched air. Coal seems to burn more rapidly than matte oxidation presuming from the fact that even when oxygen supply in lance blast is not enough, combustion of coal completes and matte grade comes down. This combustion feature of coal in injection smelting operation is suitable for simultaneous control of oxidation reaction and melt temperature both in the smelting and converting furnaces. For example when matte grade is high and melt temperature is low, increase of coal feed would consume oxygen in lance blow with priority to matte oxidation, which increases the heat generation and decreases the degree of matte oxidation at the same time maintaining the necessary bath turbulence.

Table 1 shows typical properties of coal utilized in the smelting furnace. The coal is rather coarse as compared with the one which is normally burned through pulverized coal burners (-200 mesh 80-90%). Since coal burns within or on the surface of the melt with good heat transfer and without excess local heating, coal utilization replacing oil brought about several important advantages such as: a) Energy cost was reduced, b) Melt temperature and fluidity were much stabilized resulting increased maximum feeding rate and availability of the system, c) Refractory wear was remarkably reduced.

Behavior of minor element

The distribution coefficients of element x between slag and matte (blister), which are defined as the following equation, are shown in Table 2.

$$L_x^{S/M(B)} = \frac{(\text{Concentration of x in slag : wt \%})}{(\text{Concentration of x in matte (blister): wt \%})}$$

Table 2 Distribution coefficients between molten products

	As	Sb	Pb	Zn
S furnace $L_x^{S/M}$	0.89	1.79	0.31	2.67
C furnace $L_x^{S/B}$	2.17	2.15	1.65	2.00

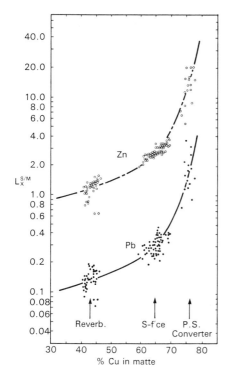

Figure 3 Relation between matte grade and distribution coefficients of Pb and Zn

Table 3 Distribution ratios of elements into gas phase (% to total input)

	As	Sb	Pb	Zn
S furnace	80.5	13.6	55.6	16.7
C furnace	30.0	2.6	44.5	19.8

The relations between these values and the matte grade are also illustrated in Figure 3 for Pb and Zn. The dots around 43% matte grade are the operating practice of the reverberatory furnace, 65% of the smelting furnace and 76% of Peirce-Smith converter at the end of the slagging stage respectively. Rather smooth curves would be due to using the silicate slags in common in all of these three furnaces. Since the distribution coefficients rise remarkably with the increase of matte grade, it is of importance for the removal of impurities to slag, to keep the matte grade as high as possible.

In the converting furnace of Mitsubishi Process the unique ferrite slag is used, which shows rather different distribution behavior as compared with silicate slag. Namely, since Vb elements like As and Sb form very stable calcium compounds with CaO, they show a stronger tendency to be eliminated to slag than silicate slag (8). On the other hand Pb shows smaller distribution coefficient.

Table 3 shows the distributions of these four elements to the gas phase in the smelting and converting furnaces, which were calculated from data collected from a certain period of typical operation. These volatilizations deeply depend on the operating factors like temperature, off gas volume, matte grade and so on. Therefore the distributions of impurities among three phases of gas, slag and matte are estimated from the thermodynamic calculation (9).

In Figure 4 and 5, for example, the distributions of Pb and Zn in the smelting furnace are illustrated, which were calculated using activity coefficients derived from the values in Table 2 and 3. Figure 4 shows how the matte grade affects distributions among the three phases under the fixed charging rate to the smelting furnace of 35t/hr, bath temperature 1,200°C and off gas volume 450 Nm³/min.

Figure 5 presents the effect of off gas volume, when the matte grade is kept at 65% and the other conditions are the same as Figure 4. When the smelting furnace off gas volume changes by varying the oxygen utilization, the distribution must vary according to this figure. For example in the case of Pb the distribution to the gas phase decreases from 56% (point A) to 40% (B) if the off gas volume gets to half from the normal operation (from 450 to 225 Nm³/min.).

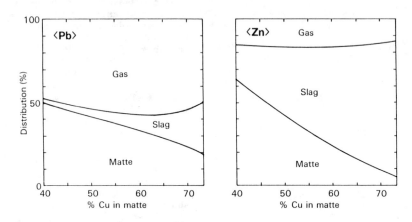

Figure 4 Relation between distribution of Pb and Zn among three phases and matte grade in the smelting furnace

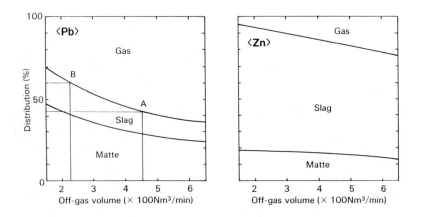

Figure 5 Relation between distribution of Pb and Zn among three phases and off gas volume in the smelting furnace

The distribution of minor elements to the various final products of smelting stage can be calculated using the values of distribution coefficients and volatilization ratios (10). The calculation results of Mitsubishi Process are shown in Table 4 as compared with the conventional process in Table 5. In Mitsubishi Process at Naoshima, the impurity levels are usually not so high that the dusts generated from the furnaces are all returned to the smelting furnace. When the quality of the anode copper is not satisfied, it is very efficient for the elements like Pb, which has high vapor pressure, to bleed off the electrostatic precipitator dusts from the system as shown in Table 4-b). This operating practice is done already in the Kidd Creek Smelter of Canada. In the conventional calcine charge reverberatory furnace process, converter slag is slow cooled and delivered to slag floatation plant for recovery of copper concentrate and dust of converter electrostatic precipitator is bled off and treated in the by-products plant for recovery of valuable metals such as Pb, Zn and Sn. For simplifing the comparison, the distribution was calculated assuming that the converter dust is returned to its original furnace and converter slag is directly returned to the reverberatory furnace. Table 5-a) shows the results.

Table 4 Distribution of minor elements in Mitsubishi Process (%)

a) all dusts are returned to the smelting furnace

Item	As	Sb	Pb	Zn
Discard slag	17.6	84.2	57.4	98.9
Anode	4.6	15.5	39.4	0.3
Acid Plant	77.8	0.3	3.2	0.8

b) E.P. dust is bled off from the system

Item	As	Sb	Pb	Zn
Discard slag	12.6	73.2	18.2	80.5
Anode	3.2	13.2	12.3	0.2
Acid Plant	55.6	0.3	0.9	0.7
Dust	28.6	13.3	68.6	18.6

Table 5 Distribution of minor elements in calcine charge reverberatory process (%)

a) all dusts are returned to its original furnaces and converter slag is returned to reverberatory furnace

Item	As	Sb	Pb	Zn
Discard slag	32.4	74.9	78.2	99.8
Anode	3.8	24.8	19.9	0.0
Acid Plant	63.8	0.3	1.9	0.2

b) converter E.P. dust is bled off from the system and converter slag is treated in the floatation plant

Item	As	Sb	Pb	Zn
Discard slag	28.5	29.8	7.0	39.4
Anode	2.6	19.3	4.9	0.0
Acid Plant	52.5	0.2	0.3	0.1
Floatation Tailing	3.6	45.0	49.3	52.6
Dust	12.8	5.7	38.5	7.9

From these results it is recognized that the distribution of As and Zn to the anode in continuous process is in the same level, Sb is low and Pb is rather high compared with the conventional process.

Intensive operation of Mitsubishi Process

The original designed feed rate of copper concentrates was 25t/hr and the anode production rate had been maintained at around 4,000t/month for several years since the start-up.

In 1980, a new project to increase the concentrate feed rate to the existing Mitsubishi Process up to 40t/hr started and several tests and improvements have been carried out based on a long term plan. The major trials are summarised as follows.
a) The feed rate of 33-36t/hr of concentrates was tested in August 1980 for a week (11).
b) Oxygen enrichment tests were performed in 1979 and 1980 keeping the off gas volume under a certain level in the case of increased feed rate of concentrates (12).
c) A long period commercial operation was carried out at the feed rate of 32t/hr from October 1981 to February 1982 by using tonnage liquid oxygen.
d) An additional oxygen plant with a capacity of 8,000Nm^3/hr, as 80% O_2, was installed and the high through-put test of 40t/hr of concentrates was tried with high oxygen enriched blow in May and July 1982 (13).

With these trials, the following phenomena were confirmed.
1. There was no decrease in the oxygen utilization efficiency with increase of oxygen content of the smelting furnace lance blow up to 60%. The off gas volume and fuel consumption were changed by theoretical amount.
2. The consumption of lance pipe was not affected by higher oxygen enrichment.
3. There was no undesirable observations in the converting furnace such as decrease of the oxygen utilization efficiency.

Eventually it was decided that it would be possible to treat 40t/hr of concentrates at the existing plant without changing the sizes of smelting and converting furnaces.

However, auxiliary facilities such as the dryer and chain conveyors were necessary to be enlarged, and the hearth area of the slag cleaning furnace was increased by 30%. These modifications were completed in March 1983 and high through-put operation started in April.

The capacity of the existing continuous process is now 310 thousand tonnes of concentrate charge and 90 thousand tonnes per year of anode production per year. With these production rate, the running cost per tonne of copper in concentrates charged to the continuous process is about "70%" of the cost of conventional reverberatory system, in which 350 thousand tonnes of concentrates are currently treated, where running cost includes acid plant, oxygen production, maintenance costs and steam generation credit and excludes depreciation, interests, and over head costs.

For a precise comparison with existing calcine charge reverberatory furnace system, operating parameters of the continuous process were calculated for the same production rate as the conventional one, 100 thousand tonnes of annual production of copper, based on actual operation data. Table 6 and 7 show operation parameters and Table 8 gives energy and labor requirement as compared with conventional reverberatory smelter.

Table 6 Chemical composition (%)

	Cu	Fe	S	SiO_2	CaO	Al_2O_3
Concentrates	28.0	26.0	31.0	6.0	-	2.0
Smelting f'ce matte	65.0	11.4	22.1	-	-	-
Smelting f'ce slag	0.6	39.8	-	30.0	4.0	3.0
Converting f'ce slag	15.0	44.2	-	-	17.0	-

Table 7 Operation parameters of continuous process (100,000t/y copper)

Smelting furnace	feed rate	45 t/hr
	lance blast	24,000 Nm^3/hr
	O_2 content	40 %
	off gas	470 Nm^3/min.
	SO_2 content	23 %
	furnace dimension	9 mø
Slag cleaning furnace	slag treated	30 t/hr
	electric capacity	2,100 KVA
Converting furnace	matte fed	20 t/hr
	slag returned to S f'ce	5 t/hr
	blister produced	12.5 t/hr
	lance blast	13,700 Nm^3/hr
	O_2 content	25 %
	off gas	250 Nm^3/min.
	SO_2 content	21 %
	furnace dimension	7 mø

Conclusions

Sulfide smelting of copper has been facing some difficulties in these 15 years i.e., to meet stringent environmental regulations, increased energy cost, labor cost and construction cost. However, as no new effective technologies of copper extraction which takes the place of sulfide smelting was developed in commercial scale of operation, it will continue a role of primary copper production. Several new generation processes for matte making stage from copper concentrates were developed and successfully put into operation resulting in pollution free and less energy requirement. But matte converting is still performed in traditional Peirce-Smith converters in most of the smelters. Mitsubishi Metal Corporation developed a continuous process which consists of intensive injection smelting and continuous matte converting. During 10 years of commercial operation the process was improved in many respects such as long furnace campaign life, high specific capacity, high oxygen enrichment of process air and use of coal rather than oil, etc.

Table 8 Energy and labor requirement

	Continuous (100,000t/y)	Conventional (100,000t/y)
Fuel (G Joul/T-Cu)	6.0	16.8
Power (KWH/T-Cu)		
smelter	515	407
slag floatation	-	87
oxygen plant	309	-
acid plant	211	543
power generated	Δ218	Δ462
Total	817	575
Process fuel equivalent (G Joul/T-Cu)	15.8	23.7
Labor (Man-hour/T-Cu)		
smelter	2.28	3.83
slag floatation	-	0.52
oxygen plant	0.11	-
acid plant	0.37	1.03
Total	2.76	5.38

note: Process fuel equivalent is sum of fuel and power where electric power is evaluated at 12×10^6 Joul/KWH (7).

This continuous process should be one of the solution to continue the economical operation of the sulfide smelting where maximum conservation of energy resources, capital resources, and the environment are required to be achieved.

References

(1) M. Goto and N. Kikumoto, "Intensive operation of the Mitsubishi Process", paper to be presented at Mineral Processing and Extractive Metallurgy, Kunming, People's Republic of China, Oct. 27 - Nov. 3, 1984.

(2) N. J. Themelis and H. H. Kellogg, "Principles of Sulfide Smelting", pp.1-29 in Advances in Sulfide Smelting (vol. 1), H. Y. Sohn, D. B. George and A. D. Zunkel eds., AIME, New York, N.Y., 1983.

(3) H. H. Kellogg, "Prospects for the Pyrometallurgy of Copper", paper presented at First Latin American Congress of Mining and Extractive Metallurgy, Santiago, Chile, Aug. 27 - Sept. 1, 1973.

(4) T. Nagano and T. Suzuki, "Commercial Operation of Mitsubishi Continuous Copper Smelting and Converting Process", pp.439-457 in Extractive Metallurgy of Copper (vol. 1), J. C. Yannopoulos and J. C. Agarwal eds., AIME, New York, N.Y., 1976.

(5) E. Kimura, "Fundamental Research of Lancing Mechanism in Mitsubishi Continuous Smelting Furnace", Transactions of The Iron and Steel Institute of Japan 23 (1983), pp.522-529.

(6) D. H. Wakelin, Ph. D. Thesis, Imperial College, University of London, 1966.

(7) H. H. Kellogg, "Conservation and Metallurgical Process Design", paper presented at Advances in Extractive Metallurgy 1977 Symposium, London, 1977.

(8) K. Itagaki and A. Yazawa, "Thermodynamic Evaluation of Distribution Behaviour of Arsenic, Antimony, and Bismuth in Copper Smelting", pp.119-142 in <u>Advances in Sulfide Smelting</u> (vol. 1), H. Y. Sohn, D. B. George and A. D. Zunkel eds., AIME, New York, N.Y., 1983.

(9) T. Yanagida, S. Kawakita, N. Kikumoto and M. Hayashi, "Lead and Zinc Behavior at S furnace of the Mitsubishi Process and P.S. Converter - Thermodynamic Prediction and Practice", paper presented at MMIJ/Aus IMM Joint Symposium, Sendai, Japan, Oct. 6-8, 1983.

(10) T. Yanagida, S. Kawakita, N. Kikumoto and M. Hayashi, "Material and Heat Balance at the Conventional Calcine Charged Reverberatory Furnace Process", paper presented at The 113th AIME Annual Meeting, "Control '84 Symposium", Los Angeles, Feb. 26 - Mar. 1, 1984.

(11) M. Goto and N. Kikumoto, "Process Analysis of Mitsubishi Continuous Copper Smelting and Converting Process", paper presented at The 110th AIME Annual Meeting, Chicago, Illinois, Feb. 22-26, 1981.

(12) M. Goto, N. Kikumoto and T. Igarashi, "Utilization of Coal and Oxygen in Mitsubishi Continuous Process", paper presented at The 4th Joint Meeting MMIJ - AIME, Tokyo, Nov. 4-8, 1980.

(13) T. Suzuki, T. Yanagida, M. Goto, S. Kawakita, T. Echigoya and N. Kikumoto, "Test Operation for Smelting More Tonnages of Copper Concentrates at The Mitsubishi Continuous Copper Smelting and Converting Process", paper presented at The 112th AIME Annual Meeting, Atlanta, Georgia, Mar. 6-10, 1983.

TOWARD A BASIC UNDERSTANDING OF INJECTION

PHENOMENA IN THE COPPER CONVERTER

J.K. Brimacombe, A.A. Bustos, D. Jorgensen and G.G. Richards

The Centre for Metallurgical Process Engineering
Department of Metallurgical Engineering
The University of British Columbia
Vancouver, B.C.   V6T 1W5
Canada

Summary

Since the Peirce-Smith converter was introduced to the non-ferrous industry in the early part of this century, there has been a paucity of research on process-engineering aspects of the reactor such as gas injection. This paper reviews studies conducted at UBC over the last decade on gas discharge dynamics, high-pressure injection, bath slopping, oxygen/bath reaction kinetics and out-of-stack heat transfer within the converter. The ultimate objectives of this work have been to extend refractory life and to increase converter productivity.

Introduction

Numerous thermodynamic investigations of the chemical reactions in copper converting have been conducted over the years. In his singularly distinguished career, Professor Kellogg, has contributed importantly to this growing body of knowledge with his studies of molten sulphide chemistry (1) and slag chemistry (2); and the metallurgical community is very much in his debt. Thus much is known of the equilibria amongst molten copper, slag and matte phases, eg. distribution of minor elements such as Bi, As and Sb (3) and the influence of oxygen potential on copper losses in slag (4).

In contrast to this state of basic knowledge of equilibria, remarkably little is understood of the rate phenomena in the copper converter, particularly as they are influenced by the design and operation of the reactor. Aspects such as gas discharge dynamics, axial and radial mixing of the phases within the converter, splashing, slopping, heat transfer and the kinetics of reactions amongst gas, liquid and solid phases have only begun to be tackled in a concerted manner. This is unfortunate because operation of the converter has been hampered over the decades by problems such as tuyere blockage due to accretion buildup and the need to clear blocked tuyeres by hand or mechanical punching, severe refractory erosion at the tuyere line and accretion buildup at the converter mouth. The tuyere-line erosion necessitates frequent refractory reline while mouth buildup, coupled with excessive ejection of molten material from the converter, limits the blowing rate and productivity. It is recognized that in many copper smelters converter productivity is not a major concern, relative to other problems like environmental control, but viewed from a general perspective, progress in design of the reactor or improvement in its operation will not be achieved quickly without process engineering knowledge.

Thus a programme of study has been pursued at UBC to investigate the following process engineering aspects of converter operation: gas injection into the bath, accretion growth at the tuyere tip, slopping of the bath, kinetics of bath oxidation and heat losses from the interior of the converter during out-of-stack periods. The study of these subjects has involved several types of laboratory experiments, plant measurements, a physical model and the formulation of mathematical models. The present paper provides an overview of these activities.

Gas Injection into the Bath

The dynamics of gas discharge have been studied both in the laboratory, and on operating converters in five smelters. A reasonably coherent picture of gas behaviour has emerged although it is not as originally supposed.

Laboratory Studies of Injection

In laboratory experiments involving the injection of air from a horizontal tuyere into mercury in a converter-shaped vessel, Oryall (5) measured the distributions of volume fraction of air and the bubble frequency, using an electroresistivity probe. As shown in Figure 1, the probe was connected electrically to the mercury bath such that when the uninsulated probe tip was surrounded by a gas bubble, the circuit was broken. The fraction of time the circuit was open (local gas fraction) was determined using an integrator while the number of circuit openings over a selected time interval (bubble frequency) was measured with a counter. The results thus obtained were surprising in that the gas discharging from the tuyere tip expanded into the bath considerably more rapidly than equivalent water-model experiments had indicated. This can be seen in Figure 2 which shows contours of volume pct. air in the vertical plane passing through the axis of a 3.25 mm I.D. tuyere for air discharging with $N'_{Fr} = 20.5$. This modified Froude number is slightly larger than values normally encountered in operating converters. In Figure 2 the gas is seen to expand nearly as far behind the tuyere tip as in the direction of flow. The expansion angle measured from the contours was 150 to 155° in contrast to that for air injection into water of 21° (6). It was argued that air discharging into the bath of a copper converter would exhibit similar expansion characteristics to air in mercury eventhough the industrial system is nonisothermal and the density of the converter baths is two- to three-fold less than that of mercury. The effect of rapid heating of a cold gas as it discharges into the bath would be to increase expansion beyond that of the isothermal case, whereas the lower bath density would reduce expansion (the influence of surface tension and wetting is also an important consideration and will be taken up in the section on industrial measurements). Although the expansion in the converter baths could not be quantified, it certainly would be larger than indicated from water-model experiments. Therefore in a copper converter, the forward penetration of the discharging air should be small, and because the tuyere exit is flush with the refractory lining, the extensive back penetration should cause the gas to rise in close proximity to the back wall. The latter obviously could contribute to the accelerated refractory wear observed above the tuyere line.

The application of the electroresistivity probe gave new insights into the path of injected gas in the copper converter but did not provide details of the discharge phenomena at the tuyere tip. To study gas discharge

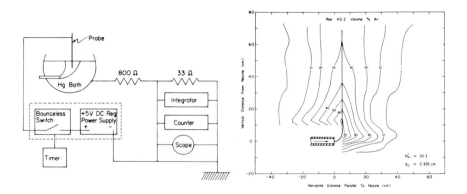

Figure 1. Schematic diagram of apparatus constructed by Oryall (5) to measure the gas-fraction and bubble-frequency distributions in air jets discharging into mercury.

Figure 2. Distribution of volume pct. air in the vertical plane passing through the tuyere axis for N'_{Fr} = 20.3 and d_o = 3.25 mm.

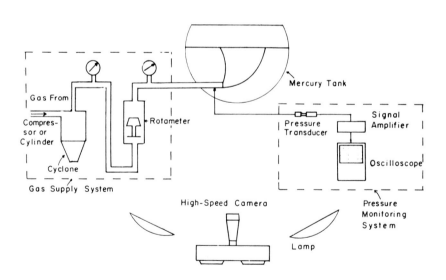

Figure 3. Schematic diagram of apparatus constructed by Hoefele (7) to study the dynamics of gas discharge from horizontal tuyeres into mercury.

dynamics, Hoefele (7) constructed the apparatus shown in Figure 3 which permitted visual observation of events at the tuyere tip and the simultaneous measurement of pressure in the tuyere. Observation of the gas discharge was made possible by injecting the gas through a "half-tuyere" bolted to the Plexiglas sidewall of the mercury tank. High-speed motion pictures were taken of the side wall during injection to record gas discharge events at the tuyere tip. The pressure at the tuyere tip was measured, via a pressure tap through the sidewall, with a piezoelectric transducer the signal from which was amplified and displayed on a storage oscilloscope. The simultaneous observation of gas discharge and pressure measurement was a crucial part of the study because it later assisted in the interpretation of transducer signals obtained from the tuyeres of operating copper converters.

The motion pictures quickly revealed that at low values of N'_{Fr} corresponding to blowing conditions in the copper converter (N'_{Fr} = 10-14, Pressure $\cong$ 60 kPag), the gas discharges in the form of discrete bubbles at a frequency of about 10 s^{-1}. A sequence of photographs from the high-speed films showing the growth of a bubble (shiny area adjacent to half-tuyere) is presented in Figure 4. The bubble can be seen to penetrate only a short distance into the bath and grows back along the tuyere. At 0.054 s, as the bubble is disengaging from the tuyere tip, it is symmetrically placed about the tip. These observations agree closely with the gas fraction contours, Figure 2, measured by Oryall. Following departure of the bubble, mercury flows in against the tuyere tip and, it is important to note, remains there for about 0.04 s until the succeeding bubble begins to grow. Thus if in a copper converter, the gas also is discharging discontinuously from the tuyere, bath must periodically wash against the tuyere tip which obviously could contribute to accretion formation and tuyere blockage. A link therefore was established between gas discharge dynamics and a problem that has plagued the copper converter since its earliest installation.

The temporal variation of pressure at the tuyere exit, recorded under bubbling conditions, as air discharged into mercury, is shown in Figure 5 together with observations made simultaneously of events near the tuyere tip. It is seen that a pressure pulse is associated with the formation of each successive bubble. Each pulse consists of a rapid pressure rise caused by mercury washing against the tuyere mouth which increases the resistance to air discharge. This is followed by a drop in pressure when the bath resistance is overcome and air commences to issue from the tuyere to form a bubble. The pressure continues to decline as the bubble grows until buoyancy and bath motion cause the bubble to rise and become severed from the tuyere. The liquid inrushing against the tuyere mouth again causes the pressure to rise and the cycle is repeated. Thus the bubbling mode of discharge is well characterized by the tuyere pressure signal.

Hoefele raised the injection pressure, and thereby the modified Froude number, to determine the conditions under which the gas discharged continuously into the mercury, ie. without bath periodically washing against the tuyere tip. Thus it was found that continuous discharge or "steady jetting" was achieved only when the pressure was sufficiently high to produce choked flow in the tuyere. Under these conditions the pressure at the tuyere tip was relatively constant with spikes of short duration that appeared randomly. In other experiments involving injection of gases of different density (air, argon, helium) into low-density baths of aqueous $ZnCl_2$ solution (ρ = 1.9 g/cm^3) and water (ρ = 1 g/cm^3), choked flow was not needed to obtain a continuous gas stream issuing from the tuyere. The conditions giving rise to bubbling and steady jetting were delineated in terms of N'_{Fr} and ρ_g/ρ_ℓ on a "jet-behaviour" diagram shown in Figure 6. Also shown are the injection conditions for the copper converter and Q-BOP bottom-injection steelmaking furnace. As stated earlier, the low Froude

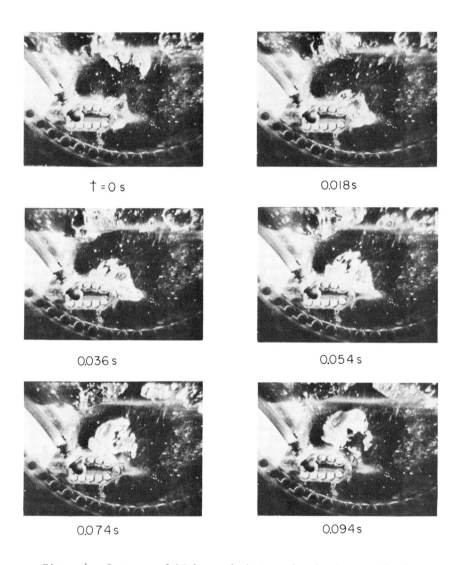

Figure 4. Sequence of high-speed photographs showing growth of an air bubble at the tip of a half-tuyere in mercury (d_o = 4.76 mm, N'_{Fr} = 43, N_{Ma} = 0.3) (7).

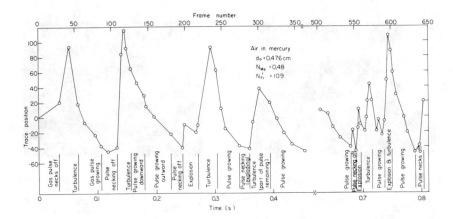

Figure 5. Pressure trace and related observations of air discharging from a 4.76 mm i.d. half-tuyere into mercury (7). $N'_{Fr} = 109$, $N_{Ma} = 0.48$.

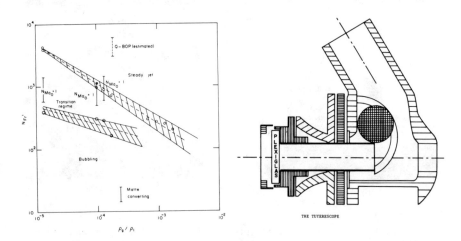

Figure 6. Jet-behavior diagram (7).

Figure 7. Tuyerescope mounted on the tuyere of a Peirce-Smith converter (8).

number of the copper converter, due to the low injection pressure, results in discontinuous gas discharge or bubbling. On the other hand, the Q-BOP (and all other submerged injection steelmaking processes) with much higher injection pressures capable of achieving choked flow in the tuyere, operate in the continuous jetting mode. It is interesting to note that tuyere blockage is not a problem in these steelmaking processes, in contrast to the copper converter. This again suggests that a link exists between the mode of gas discharge and tuyere blockage and moreover that raising the pressure of air injected into the copper converter to achieve steady jetting is one means of preventing tuyere blockage. Obviously steady jetting has this effect because bath is not permitted to wash against the tuyere mouth as in the case of bubbling, and the high gas momentum due to high-pressure injection may dislodge accretions growing over the tuyere mouth. Unfortunately the pressures required to achieve steady jetting are not attainable with the equipment presently employed to supply air to copper converters, which necessitates an expensive retrofit of new compressors and air lines. Nonetheless plant trials have been conducted to determine if high-pressure, steady jetting would prevent tuyere blockage and permit "punchless" converter operation; this work is described in the next section.

Plant Investigations of Injection

Industrial studies were undertaken to check the conclusions reached in the laboratory work and to further our understanding of gas discharge dynamics and accretion growth in an operating converter (8). The investigation utilized a tuyerescope, shown in Figure 7, to allow visual observation of the tuyere tip and to provide a means of inserting a probe into the tuyere to measure pressure fluctuations at the tip. The tuyerescope was constructed of a 130-mm length of 33-mm diameter, stainless-steel pipe one end of which was covered by a removable Plexiglas window. A simple lock-fit device was employed to secure the tuyerescope to the back of the tuyere. To permit sampling of the accretions and insertion of the pressure probe a 9.5 mm hole was drilled in the Plexiglas window.

The tuyerescope provided an excellent view of the bath and the growth of accretions but the dynamics of gas discharge were not easily discernible as compared to the observations of gas injection through the half-tuyere into the mercury bath. Therefore injection behaviour was studied by the measurement and interpretation of pressure fluctuations at the tuyere tip. As was shown in the laboratory work pressure fluctuations in the tuyere can be directly correlated with gas-liquid interaction.

Typical pressure traces generated in an individual tuyere during normal blowing of a copper converter are shown in Figure 8. The pressure measurements were made at two tuyere submergences: 300 mm and 500 mm. At a shallow submergence, Figure 8(a), a relatively constant, low-pressure signal interrupted periodically by pulses of short duration was observed. The pulse frequency was close to 4 s^{-1}. With the deeper tuyere submergence, Figure 8(b), pulses of a greater duration but the same frequency were found. The pressure signal due to bubbling of the type observed by Hoefele (7) was not seen.

By an analysis of pressure traces from the copper converters and a slag fuming furnace (8) three diagnostic aspects of these fluctuations were identified:
(1) the shape of the pressure pulses;
(2) the frequency of the pulses, and
(3) the duration of constant, low-pressure intervals, if any, separating pulses.

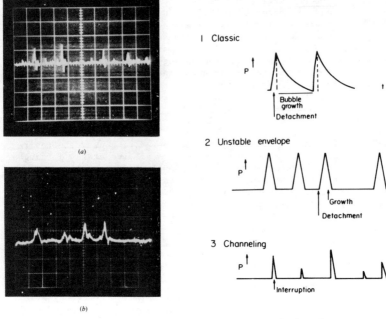

Figure 8. Pressure traces measured at the tip of a tuyere in a copper converter (8). Tuyere submergence: (a) 300 mm and (b) 500 mm. Vertical scale 14 kPa/div; horizontal scale: 100 ms/div.

Figure 9. Idealized tuyere pressure traces for three regimes of gas discharge from horizontal, closely spaced tuyeres (8).

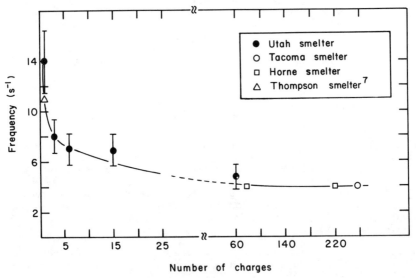

Figure 10. Measured tuyere-tip "bubbling" frequency in a number of different copper converters versus number of charges in the campaign. Zero charge number represents a freshly relined converter.

These helped to distinguish three separate regimes of gas injection in low-pressure industrial systems as shown schematically in Figure 9. Successive pulses having a sudden pressure rise followed by a decline in pressure are characteristics of classic bubbling as described earlier, Figure 5. The rapid pressure rise is due to the increased resistance to gas flow as bath washes against the tuyere tip. Logically, intervals of low pressure are indicative of reduced resistance to gas discharge caused by gas channeling to the surface when the tuyere is shallowly submerged or when more deeply submerged, by the overlap of bubbles growing at adjacent tuyeres. In the latter case an unstable, horizontal gas envelope is established at the tuyere line and each tuyere feeds into it, interrupted only by periodic collapse of the envelope as buoyancy pushes the gas upward. The pulses are generated each time the envelope breaks down near the tuyere and the tuyere pressure drops suddenly once flow into the envelope, or channel to the surface, is re-established. Owing to this tuyere-surface or tuyere-tuyere interaction, successive bubbles as seen in the classic bubbling case cannot easily grow independently, and the slow decay in pressure, Figure 5, characteristic of bubble growth is not found as frequently. Thus the pressure pulses tend to be more symmetrical. In the case of gas channeling the pulses also are of shorter duration presumably because re-establishment of the gas channel with the surface is a more rapid event than coalescence of discharging gas with the horizontal unstable envelope. If this analysis is correct, the mode of gas discharge in the copper converter is either channeling (shallow submergence, Figure 8(a)) or via the unstable envelope at the tuyere line due to strong tuyere interaction (deep submergence, Figure 8(b)). The latter obviously would be enhanced by the presence of a notch at the tuyere line due to refractory erosion. By comparison, in the slag fuming furnace classic bubbling was found eventhough the tuyeres are also closely spaced. Lack of tuyere interaction is likely due in part to the high viscosity of the slag which retards bubble coalescence.

Tuyere pressure measurements commencing with the first charge of a relined converter revealed an interesting change of pulse frequency with increasing number of charges processed, shown in Figure 10. In the first charge, the pulse frequency was about 14 s^{-1} but by the third charge, it dropped to 8 s^{-1} and thereafter there is a slow decline to a steady value of 4-6 s^{-1}. The reason for the reduction of pulse frequency by a factor of 2 in the first few charges has not been determined but may be related to the relining procedure. In the smelter in which the bulk of these measurements were made, the tuyere pipes protrude into the converter 150 mm or more beyond the lining after new refractory has been installed. Coincidentally within about three charges the pipes burn back to become flush with the inside wall. It may be that gas bubbles discharging from the protruding tips of adjacent tuyeres tend to coalesce less readily so that the horizontal gas envelope is not well established and tuyeres operate more independently. Thus the intervals of low pressure are reduced and the pulse frequency is high as bubbles subjected to the influence of both buoyancy and bath motion (9) discharge from the tuyere. After the tuyeres burn back to the wall, the refractory itself may play a role in enhancing tuyere interaction through surface tension effects. If in the early life of the lining, the refractory is not wetted by the bath, bubbles discharging from a given tuyere can easily spread laterally along the wall to link up with adjacent bubbles and form the horizontal envelope. This has the effect of reducing the resistance to gas flow for longer time periods which lengthens the intervals of low pressure and decreases the pulse frequency measured in the tuyere. Later on as the notch at the tuyere line deepens due to refractory erosion, the horizontal envelope is stabilized further. Thus surface tension effects also may strongly influence injection behaviour in the copper converter.

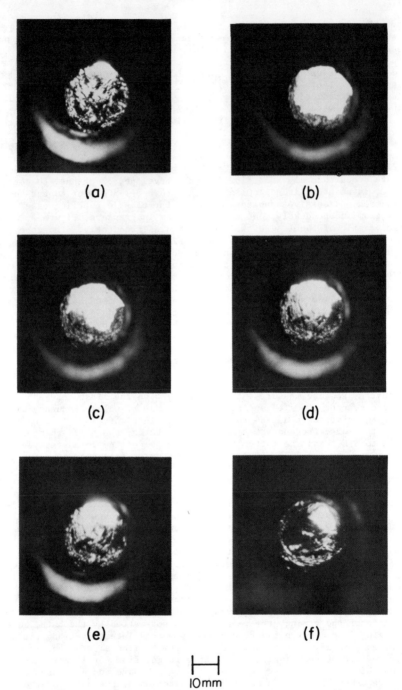

Figure 11. Sequence of photographs taken through the tuyerescope in the copper converter showing the dynamics of accretion formation (8). (a) Tuyere is nearly blocked, (b) accretion has been dislodged spontaneously, and (c)-(f) accretion growth resumes.

Accretion Growth

Accretion growth at the tuyere tip has been studied using photographic techniques and the pressure transducer in conjunction with the tuyerescope (8). A typical sequence of photographs showing accretion formation in a copper converter is shown in Figure 11. Figure 11(a) shows the tuyere nearly covered by an accretion which spontaneously dislodges a few seconds later, Figure 11(b). Figures 11(b) through 11(f) depict the growth of the next accretion over a time span of about 180 seconds. As can be seen the accretion grows upward from the bottom of the horizontal tuyere and exhibits striations roughly concentric with the tuyere. These general features were confirmed by visual observation and cinematography.

Samples of accretions taken during the slag blow were analyzed by X-ray diffraction and the major constituent was found to be $Cu_{1.96}S$, a high-temperature phase metastable at room temperature. Other compounds identified were various copper-iron sulphides. Significantly, no magnetite was found even though special care was taken to detect its presence. This implies that the mechanism of accretion formation at least in the converters studied, is basically the solidification of bath around the tuyere tip.

That growth of the accretion is from the bottom upward can be explained by simple buoyancy effects. Liquid, which in the vicinity of the tuyere is driven upward by the ascending bubbles, presses in on the discharging gas from the bottom of the tuyere and is solidified by the cold gas. At the same time, owing to the low modified Froude number, the upper area of the tuyere is surrounded predominantly by rising gas which also is directed upward by the growing accretion.

If accretions form, in large part by a freezing mechanism, operating variables which affect the local heat balance at the tuyere line should figure importantly in accretion growth and tuyere blockage. The variables would include the level of oxygen enrichment in the gas stream, matte grade, out-of-stack time, cold dope additions and time in the converter cycle. In some converters, particularly those with short out-of-stack times, the refractory at the tuyere line may be hot enough to reduce freezing; and in these cases magnetite formation may play a more significant role than hitherto has been found. Then variables such as fluxing practice and mixing of flux over the length of the converter may affect accretion growth.

The need to dislodge accretions by tuyere punching so that the blast volume flow rate can be maintained, is certainly a contributing factor to refractory erosion at the tuyere line. The accretions adhere to the refractory surrounding the tuyere mouth, and when dislodged by the action of the punch bar, a portion of the adjacent refractory can be broken off. It may be noted additionally that the shape of the puncher head may also influence accretion formation. For example a round puncher head that fills much of the tuyere cross-section can reduce the air passage through the tuyere substantially and draw bath into the tuyere during the punching cycle.

High-Pressure Injection of Air

As mentioned earlier, one possible means of controlling accretion growth and preventing tuyere blockage is to raise the pressure to achieve choked flow in the tuyere, and thereby to change the mode of gas discharge from bubbling to steady jetting. To investigate this possibility, an experimental campaign was organized at the ASARCO Tacoma Smelter in which four thick-walled, 19.1-mm I.D. tuyeres were installed near the end of the No. 4 converter and outfitted with high-pressure air (10). A schematic drawing of a high-pressure tuyere is shown in Figure 12. The tuyeres were connected to an air/oxygen system, Figure 13, capable of delivering air to the tuyeres at 414 kPag giving an overall flow rate of 45.3 Nm^3 per minute. When oxygen enrichment was implemented the O_2 level was maintained normally at 26 pct. With the high-pressure tuyeres in place together with the 48 other low-pressure tuyeres, the converter was operated through a normal campaign. Over a period of three months during which 88 charges were processed, the high-pressure tuyeres remained clear and free blowing without any need for punching. Between charges the tuyeres were inspected and found to be clean and open. Throughout a blow, the air flow rate through the high-pressure tuyeres was constant in contrast to the periodically declining air flow rate through the low-pressure tuyeres caused by tuyere blockage. Thus the concept of punchless converter operation was proved.

The formation of accretions around the tuyere mouth was controlled, when necessary, by adjusting the oxygen level in the air. In one instance solid buildup, likely due to experimentation with lower pressures during magnetite-lining charges, was observed and was removed by raising the oxygen to 28 pct. Although much remains to be learned about the control of accretions during high-pressure injection, it seems clear that conditions can be adjusted, depending on matte grade and other factors mentioned earlier, to build up accretions around the tuyeres to protect the adjacent refractory and thereby to prolong lining life, as is currently practiced in the steel industry. Whether the savings in refractories, coupled with the removal of tuyere punchers and likely increase in converter productivity, are sufficient to offset the increased compression costs for high-pressure costs remains to be evaluated. What is clear, however, is that there is an alternative to tuyere blockage and punching that has been endured for so many decades.

Bath Motion

Gas injected into the bath in a converter imparts motion to the liquid which, depending on conditions, may take the form of slopping. Bath slopping is an oscillatory motion of the liquid between the tuyere line and the breast of the converter (presumably the oscillation also could take place from end to end). This phenomenon is undesirable for several reasons: the ejection of bath from the converter and the buildup of accretions at the mouth are accelerated and refractory erosion may increase due to thermal shock and wear.

It would seem reasonable to suppose that bath slopping is favoured by increasing air injection rates and that the related problems of material ejection from the converter and mouth buildup also become worse. Indeed, in practice these operating problems limit the air injection rate to levels, shown in Figure 14, which are dependent on converter volume (6,11). However, the observed injection rate of about 8.5 Nm^3/min per cubic metre of converter volume has not been linked directly to slopping in previous work.

Thus a study was initiated, using a physical model, to investigate the influence of variables such as air injection rate, tuyere submergence and pct. filling of the converter on slopping (12). The physical model, shown in Figure 15, was effectively a sectional slice of the converter containing five tuyeres and constructed to 1/4th scale from Plexiglas. The tuyeres had a diameter of 16 mm and were of adjustable height and inclination. Slopping and non-slopping of the bath were determined under the various

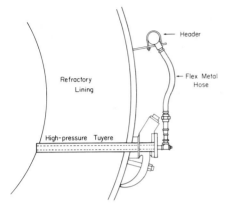 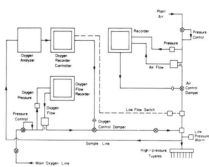

Figure 12. Drawing of high-pressure tuyere constructed from pipe having an ID of 19.1 mm and OD of 44.4 mm (10).

Figure 13. Schematic drawing of air/oxygen gas delivery system constructed for the high-pressure injection campaign (10).

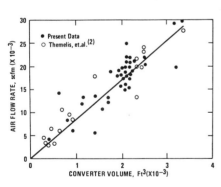 

Figure 14. Air injection rate into a copper converter as a function of converter volume (11).

Figure 15. Photograph of sectional model used to study bath slopping.

 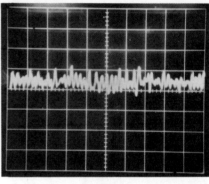

(a) (b)

Figure 16. Photograph of bath under non-slopping conditions
(a). Associated pressure trace measured in the central tuyere
(horizontal scale: 0.5 s/div; vertical scale: 0.18 kPa/div).
(b) Air flow rate: 3.7 ℓ/s; filling: 40 pct.; tuyere submergence: 130 mm.

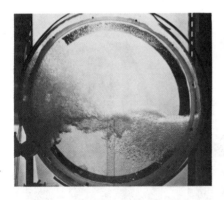 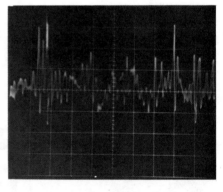

(a) (b)

Figure 17. Photograph of bath under slopping conditions (a).
Associated pressure trace measured in the central tuyere
(horizontal scale: 0.5 s/div; vertical scale: 0.18 kPa/div),
(b) Air flow rate: 20 ℓ/s; filling: 40 pct.; tuyere submergence: 130 mm.

injection conditions by visual observation and by measuring the dynamic pressure in the central tuyere with a piezoelectric transducer as described earlier.

Typical observations of non-slopping and slopping are shown in Figures 16(a) and 17(a) respectively. As expected, when the bath is not slopping, the liquid surface is horizontal with waves moving between the jet spout and the opposite wall. However under slopping conditions, the bath surface moves back and forth with a frequency of about 1 Hz, and some of the injected gas moves to the opposite wall. During slopping, periodic pulses of sound could be heard from the air jet, in concert with the rise and fall of liquid above the tuyere line. The pressure traces recorded on the oscilloscope, Figures 16(b) and 17(b), also clearly reveal the state of bath slopping. Under non-slopping conditions the pressure fluctuates, due to the formation of bubbles, about a constant mean value whereas bath slopping causes a 1-Hz oscillation to be superimposed on the signal. The pressure signals were particularly helpful in delineating the transition from non-slopping to slopping.

Experiments were conducted with varying pct. filling and tuyere submergence to determine the critical air flow rate, above which bath slopping prevailed, for each set of conditions. An attempt then was made to correlate the critical air flow rates in terms of kinetic and buoyancy power input to the bath from the injected air. In a physical-model study conducted earlier by Haida (13), the mixing time in hot-metal ladles and torpedo cars has been correlated, and scaled up to full size, successfully with buoyancy energy* while kinetic energy was found to have a minor influence. Thus the kinetic power and buoyancy power per unit mass of bath were calculated from the following respective relationships (14)

$$\varepsilon_k = \frac{1/2 \, \rho_g u_o^2 Q}{m_b} \quad (1)$$

$$\varepsilon_b = \frac{2QP_a \ln\{\frac{P_a + \rho_\ell g h_b}{P_a}\}}{m_b} \quad (2)$$

for the critical air flow rates determined in the experiments. It was found that the critical flows could not be correlated with the kinetic power, but a good correlation was obtained in terms of buoyancy power input. Indeed the critical-flow data from all the experiments can be represented on a plot of buoyancy power per unit mass of bath against tuyere submergence as shown in Figure 18. The least-squares fit line passing through the data delineates regions of slopping and non-slopping so that Figure 18 is effectively a "slopping-behaviour" diagram. It is seen that more buoyancy energy, eg. greater air flow rate, can be imparted to the bath without slopping when the tuyere submergence is large.

To relate these physical-model results to converter operation, data on operating converters have been taken from the world-wide survey conducted

---

*Although Haida (13) used the term "energy", it strictly is "power" since it also is expressed in units of kW/tonne bath.

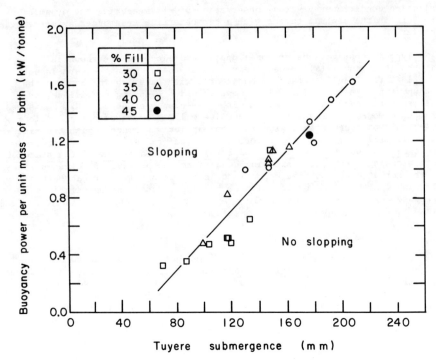

Figure 18. Representation of critical-flow data from physical model experiments on slopping-behaviour diagram.

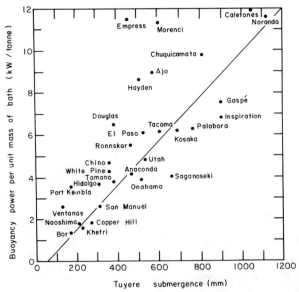

Figure 19. Data from the survey of Johnston et al (11) plotted in terms of buoyancy power against tuyere submergence for different smelters. Solid line is extrapolation from Figure 18 obtained with physical model.

by Johnson et al. (11); the buoyancy power (ambient air temperature assumed) and tuyere submergence for the different smelters have been determined and plotted on the slopping-behaviour diagram shown in Figure 19. The critical slopping line, shown in Figure 18, has been extrapolated to converter tuyere submergences and also is presented in Figure 19. It is seen that most of the converter operations are reasonably close to the critical slopping line. This may indicate, although further study is required, that most converters are being operated at the maximum air flow rate close to the point of bath slopping. It also is evident that a shallow tuyere submergence has the disadvantage of leading to bath slopping at lower levels of buoyancy power (gas flow rates) as well as the disadvantage of causing gas channelling described earlier.

Kinetics of Oxidation Reactions

The oxygen utilization efficiency in the converting process is typically high, usually greater than 80 pct. To explain this industrial observation a simple kinetics analysis was performed by Ashman (9).

During the slag-making blow, FeS is oxidized to FeO according to the reaction

$$FeS + 3/2\ O_2 = FeO + SO_2 \qquad (3)$$

and especially toward the end of the blow, magnetite also is formed as follows

$$6FeS + 10\ O_2 = 2Fe_3O_4 + 6SO_2 \qquad (4)$$

Assuming a converter slag composition of 23 pct. Fe_3O_4 and 30 pct. FeO, the overall slag-making reaction is

$$FeS + 1.57\ O_2 = FeO_{1.14} + SO_2 \qquad (5)$$

At converting temperatures of about 1500 K it may be assumed that the intrinsic reaction kinetics are extremely rapid; therefore the rate of the reaction is governed by mass transport and local equilibrium prevails at the matte-gas interface. Moreover, thermodynamic considerations indicate that Reactions (3) and (4) have large equilibrium constants. Under these conditions the reaction rate must be limited by mass transport of one or both of the reactants: FeS in the bath or O_2 in the gas phase. Criteria for transport control in the different phases are easily derived from the mass-transfer equations and stoichiometric considerations.

[i] FeS transport to the gas-matte interface is rate limiting

$$c_{FeS}^b < \frac{1}{1.57} \frac{k_{O_2}}{k_{FeS}} \frac{p_{O_2}^b}{RT} \qquad (6)$$

[ii] O_2 transport to the gas-matte interface is rate determining

$$c^b_{FeS} > \frac{1}{1.57} \frac{k_{O_2}}{k_{FeS}} \frac{p^b_{O_2}}{RT} \qquad (7)$$

[iii] Both transport steps are limiting

$$c^b_{FeS} = \frac{1}{1.57} \frac{k_{O_2}}{k_{FeS}} \frac{p^b_{O_2}}{RT} \qquad (8)$$

If one assumes a gas temperature of 573 K, oxygen partial pressure of 0.21, matte density of 4.1 g/cm^3, diffusivities of 0.64 cm^2/s for O$_2$ in air and 10^{-5} cm^2/s for FeS in matte and one applies the Higbie surface renewal model with equal renewal times for matte and air to calculate k_{O_2}/k_{FeS}, a value of 1.5 wt.% is found for the RHS of Equations (6) to (8). This is considerably smaller than the value of C^b_{FeS} that normally obtains during the slag-making blow; and hence O$_2$ transport in the gas phase is rate limiting. The oxygen transport can be described mathematically as follows

$$\dot{n}_{O_2} = k_{O_2} A \frac{p^b_{O_2}}{RT} \qquad (9)$$

Toward the end of the slag-making blow when c^b_{FeS} is relatively small, Cu$_2$S oxidation (copper-making) commences

$$Cu_2S + O_2 = 2Cu + O_2 \qquad (10)$$

to produce copper. By similar mass-transfer considerations, it can be shown that this reaction is also governed by O$_2$ transport to the bath-gas interface. Thus through the entire converting cycle, the reactions are limited by mass transfer in the gas phase which is highly turbulent. The resulting mass transport rates are very high leading to large oxygen utilizations in the converter. Most of the oxygen is likely consumed as the gas passes through the bath.

Heat Losses from the Converter

Because accretion formation involves solidification of bath at the tuyere mouth, the temperature of the refractories near the tuyere line, particularly just as the converter is rolled into the stack, can influence tuyere blockage and punching frequency. If, while the converter is out of the stack, sufficient heat is lost through the mouth to cool the inside wall below the solidus temperature of the bath, material will freeze against the refractories when the converter is rolled back into the stack to commence blowing. To quantify the influence of out-of-stack time and other converter variables on the temperature distribution in the refractory wall, especially at the tuyere line, a mathematical model has been formulated (15). Factors such as diameter of the converter, size and position of the converter mouth and the use of a mouth cover have been studied with the model.

For purposes of modelling, the inside wall of the converter has been subdivided into about 200 surface elements; the number was a compromise between computing cost and accuracy (error < 4%). The model calculates the radiative heat exchange between a given element and the rest of the elements in the mantle and the end walls of the converter as well as the converter mouth. The convective and radiative heat losses from the external surfaces of the reactor and conduction through the refractory wall also are included in the model. The model is flexible in that it allows variation in the size and position of the converter mouth. Therefore it has been possible to simulate heat-transfer conditions for both Peirce-Smith and Hoboken converters.

The temperature profile of the inside wall along the tuyere line, predicted by the model for different out-of-stack times, is shown in Figure 20 and is observed to be relatively flat. These predictions were made for a Peirce-Smith converter (central mouth) and similar results were obtained for a Hoboken converter (end mouth). Figure 21 shows the temperature of the tuyere line at two locations - near the endwall most remote from the mouth and facing the centre of the mouth - as a function of out-of-stack time for the two types of converters. If these temperatures are considered relative to the melting point of copper mattes, about 1050 to 1100°C, even short out-of-stack times allow sufficient cooling of the tuyereline to cause the bath to freeze against the wall when the converter is rolled back into the stack. At 10 minutes, freezing could take place at the tuyereline over the entire length of the converter with the central mouth (Peirce-Smith) converter while in the end-mouth (Hoboken) converter, freezing would be confined more to the region closest to the mouth. The reason for the difference is, of course, the larger radiation path between the mouth and the remote endwall in the latter case. Examination of the temperature profile through the refractory wall revealed that temperature changes associated with the converter being out of the stack were localized to within 60 to 80 mm from the inside surface. These local temperature changes obviously generate cyclical thermal stresses that may contribute to refractory erosion.

The effect of mouth area on the tuyereline temperature opposite the mouth during out-of-stack periods is shown in Figure 22. As expected, enlarging the mouth accelerates the drop in tuyereline temperature and increases the total heat loss. Although not shown in Figure 22, the shape of the temperature profile along the tuyereline is not influenced much by the mouth area.

Figure 23 shows the influence of placing a cover over the mouth of the converter while it is out of the stack. Both the tuyereline temperature drop at the mouth and total heat losses are dramatically reduced by covering the converter mouth. After half an hour out of stack with the mouth covered, the tuyereline temperature remains well above 1050°C and the total heat losses have been decreased by a factor of three. Coverage of the mouth also reduces thermal gradients through the refractory at the inside wall as shown in Figures 24 and 25 for the Peirce-Smith and Hoboken converters respectively. In both cases, with the converter mouth uncovered, the temperature gradient central to the mouth swings from negative to positive by 40-50°C/cm during an out-of-stack period of ten minutes or more. On the other hand, coverage of the mouth halves the change in temperature gradient which remains negative. The large change in thermal gradient generates stresses in the refractory, as mentioned earlier, which could contribute to tuyereline refractory erosion. It may be noted that in the Hoboken converter (end mouth), the through-thickness temperature gradient at the tuyereline remote from the mouth, Figure 25, is considerably less than that in the Peirce-Smith converter, Figure 24. If thermal cycling is a significant factor in tuyereline erosion, refractory wear would be greater in the mouth region of a Hoboken converter than near the opposite end when the mouth is uncovered out-of-stack. In the Peirce-Smith converter the wear would be more uniform along the tuyereline. Mouth covers

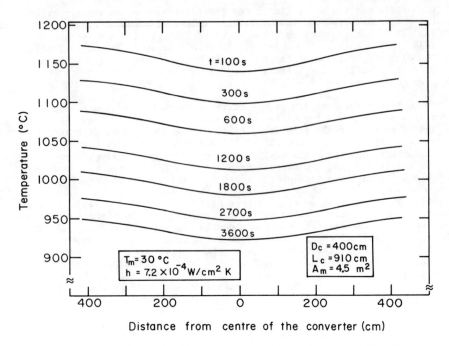

Figure 20. Predicted wall temperature along the tuyere line for different values of out-of-stack time. Converter dimensions: 9.10 m x 4 m dia.; mouth area: 4.5 m^2.

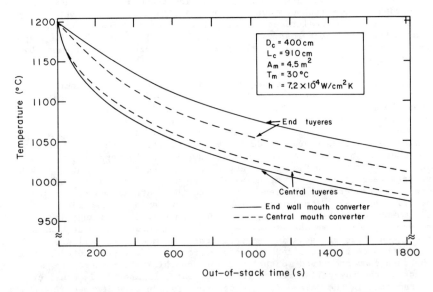

Figure 21. Predicted tuyereline temperature at the endwall remote from the mouth and at a point central to the mouth plotted against out-of-stack time for converters having central and end mouths.

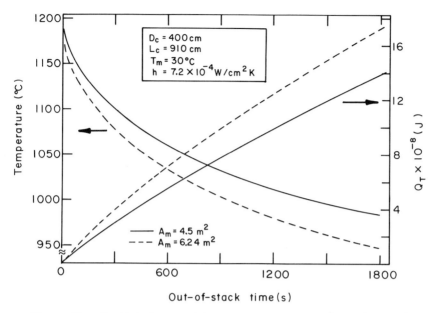

Figure 22. Predicted tuyereline temperature central to the mouth, and total heat losses, plotted against out-of-stack time for two mouth sizes. Central-mouth converter, 9.1 m x 4.0 m dia.

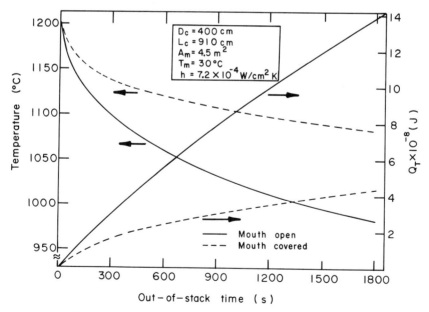

Figure 23. Predicted tuyereline temperature central to the mouth, and total heat losses, plotted against out-of-stack time with the mouth open and covered. Central-mouth converter, 9.1 m x 4.0 m dia.

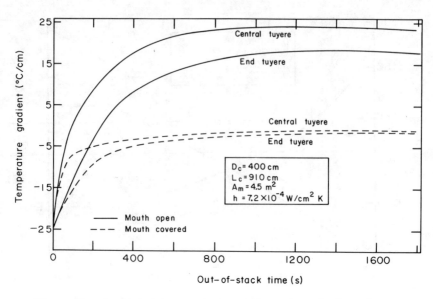

Figure 24. Predicted temperature gradient through the refractory at the inside wall in a central-mouth converter plotted against out-of-stack time.

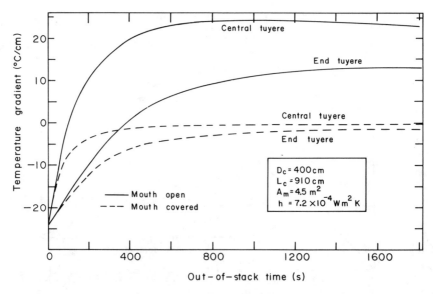

Figure 25. Predicted temperature gradient through the refractory at the inside wall in an end-mouth converter plotted against out-of-stack time.

are not extensively employed in industrial converter operations, but these calculations indicate potential benefits that should be explored relative to the present use of burners for out-of-stack periods.

Summary and Conclusions

Several process-engineering studies of rate phenomena in the copper converter have been reviewed. The studies have involved physical and mathematical models coupled with laboratory and plant trials to examine gas discharge dynamics, bath slopping, oxidation kinetics and heat transfer within the converter. These have shown that in the copper converter, air discharges discontinuously from the tuyeres into the bath. There is significant interaction amongst adjacent tuyeres such that an unstable envelope of gas exists at the tuyereline. Each tuyere feeds into the envelope which breaks down periodically at given locations to release gas bubbles. Owing to the low modified Froude number, the bubbles rise in close proximity to the wall. In cases where the tuyeres are shallowly submerged, the discharging air channels to the surface. The discontinuous gas discharge, characteristic of converter operation, can be made continuous by raising the injection pressure to achieve choked flow in the tuyeres. Under conditions of continuous discharge, or steady jetting, the tuyeres do not become blocked by accretions and can be operated without punching.

The excessive splashing, mouth buildup and ejection of molten material from the converter that limit the air injection rate likely are related to slopping of the bath. The slopping is dependent on tuyere submergence and the buoyancy power input to the bath from the rising gas bubbles. Shallow tuyere submergence gives rise to slopping at lower levels of buoyancy power and therefore is undesirable. Slopping conditions can be delineated on a "slopping-behaviour" diagram and it is useful to examine industrial converter operations in this light.

The rate of oxidation of FeS and Cu_2S during the slag- and copper-making blows respectively is limited by the transport of oxygen from the bulk of the gas phase to the gas-bath interface where the oxidation reaction takes place. This assumes that the reaction itself is effectively instantaneous at converting temperatures. Owing to the high turbulence levels in the discharging air, transport rates are rapid in the gas phase giving rise to a high oxygen utilization.

When the converter is out of the stack heat losses through the mouth cause the inside wall to cool rapidly which may lead to freezing at the tuyereline and tuyere blockage when blowing is resumed. The temperature gradient into the refractory at the inside wall can change by up to 50°C/cm within the first 10 minutes of the converter being out of the stack. The rapid temperature change is localized to within 60-80 mm of the inside wall and may contribute to the refractory wear at the tuyereline. Covering the converter mouth during out-of-stack periods markedly reduces the change in through-thickness temperature gradient at the inside wall as well as heat losses from the converter.

Nomenclature

A	Bath-gas interfacial area, cm^2.
A_m	Area of converter mouth, m^2.
c^b_{FeS}	Bulk concentration of FeS in matte, $moles/cm^3$.
d_o	Tuyere diameter, m.

D_c	Internal diameter of converter, m.
g	Gravitational acceleration (= 9.81 m/s^2).
h	Heat-transfer coefficient at outside surface of converter, W/cm^{2}K.
h_b	Height of bath, m.
k_{FeS}	Mass-transfer coefficient for FeS in the matte, cm/s.
k_{O_2}	Mass-transfer coefficient for O_2 in the gas.
L_c	Length of converter, m.
m_b	Mass of bath, kg.
n_{O_2}	Mass-transfer rate of O_2, moles/s.
N'_{Fr}	Modified Froude number $\left(= \dfrac{\rho_g u_o^2}{(\rho_\ell - \rho_g) g d_o}\right)$.
N_{Ma}	Mach number (= u_o/u_c).
P_a	Atmospheric pressure (= 101.3 kPa).
P_{O_2}	Bulk partial pressure of O_2 in the gas, atm.
Q	Volumetric gas flow rate, m^3/s.
R	Gas constant (= 82.06 cm^3 atm/mole K).
T	Temperature, K or °C.
T_m	Temperature of converter mouth, °C.
u_c	Velocity of sound, m/s.
u_o	Gas velocity in the tuyere.
ε_b	Buoyancy power per unit mass of bath, W/kg.
ε_k	Kinetic power per unit mass of bath.
ρ	Density, kg/m^3 or g/cm^3.
ρ_ℓ	Density of liquid.
ρ_g	Density of gas.

Acknowledgements

The studies reviewed here span over a decade and would not have been possible without the generous support of the Natural Sciences and Engineering Research Council of Canada, Noranda, Inco, Cominco, Asarco, Kennecott, Inspiration Copper, Union Carbide and Canadian Liquid Air. The contributions of individuals in the companies, whether assisting in plant trials or freely participating in discussions, have inspired us, and we are grateful to all.

References

(1) H. H. Kellogg: in Physical Chemistry in Metallurgy, The Darken Conference, R.M. Fisher, R.A. Oriani and E.T. Turkdogan, eds., U.S. Steel, 1976, pp. 49-68.
(2) R.P. Goel, H.H. Kellogg and J. Larrain: Metall. Trans. B, 1980, Vol. 11B, pp. 107-117.
(3) M. Nagamori and P.J. Mackey: Metall. Trans. B, 1978, Vol. 9B, pp. 567-579.
(4) P.J. Mackey: Can. Met. Quart., 1982, Vol. 21, pp. 221-260.
(5) G.N. Oryall and J.K. Brimacombe: Metall. Trans. B, 1976, Vol. 7B, pp. 391-403.
(6) N.J. Themelis, P. Tarassoff and J. Szekely: Trans. TMS-AIME, 1969, Vol. 245, pp. 2425-2433.
(7) E.O. Hoefele and J.K. Brimacombe: Metall. Trans. B, 1979, Vol. 10B, pp. 631-648.
(8) A.A. Bustos, G.G. Richards, N.B. Gray and J.K. Brimacombe: Metall. Trans. B, 1984, Vol. 15B, pp. 77-89.
(9) D.W. Ashman, J.W. McKelliget and J.K. Brimacombe: Can. Met. Quart., 1981, Vol. 20, pp. 387-95.

(10) J.K. Brimacombe, S.E. Meredith and R.G.H. Lee: Metall. Trans. B, 1984, Vol. 15B, pp. 243-250.
(11) R.E. Johnson, N.J. Themelis and G.A. Eltringham: in <u>Copper and Nickel Converters</u>, ed. R.E. Johnson, TMS-AIME, Warrendale, PA., 1979, pp. 1-32.
(12) D. Jorgensen, A.A. Bustos, J.K. Brimacombe and G.G. Richards: Unpublished work, The University of British Columbia.
(13) O. Haida and J.K. Brimacombe: in Proc. Scaninject-III, MEFOS/ Jernkontoret, Lulea, 1984, pp. 5:1-5:17.
(14) T. Robertson and A.K. Sabharwal: in <u>Gas Injection into Liquid Metals</u>, ed. A.E. Wraith, University of Newcastle-upon-Tyne, 1979, pp. I1-I29.
(15) A.A. Bustos, J.K. Brimacombe and G.G. Richards: Unpublished research, The University of British Columbia.

THE BEHAVIOUR OF ARSENIC, ANTIMONY AND BISMUTH IN THE SOLIDIFICATION AND ELECTROLYSIS OF NICKEL-OXYGEN-BEARING COPPER ANODES

O. Forsén, E. Hettula and K. Lilius

Helsinki University of Technology
Institution of Process Metallurgy
SF-02150 Espoo 15
Finland

Summary

In the first part of this paper is studied the behaviour of small amounts of impurity elements in the solidification of copper anodes. In the second part the dissolution of these impure anodes in sulphate based electrolysis is discussed.

The first part is introduced with practical examples, the model of solidification of impurities during the solidification of anode copper. According to this model the solidification structures of anode copper are strongly influenced by both thermodynamic and kinetic factors. Thermodynamics determine the microstructure of anode copper, but kinetics can alter the composition of melt to such that there is formed unequilibrium structures during the solidification, especially when the residual melt is solidified.

The electrolysis experiments have clarified the behaviour of impurities in the electrochemical dissolution in sulphate based copper electrolysis. In this study it is widely investigated to what extent any impurity elements or oxides will dissolve in copper electrorefining. With the aid of the electrochemical experiments it was possible to fulfil quantitatively the qualitative results accomplished by microstructure studies.

Introduction

In copper ores mained today in the world are copper contents decreased and impurity element contents increased. This tendency will be stronger in future. In order to maintain the quality of cathode copper, must the control of impurity elements in copper refining be improved. Through pyrometallurgical processes it is not possible to keep the impurity contents in the anode copper at the present level. Therefore, the copper anodes casted nowadays for electrorefining contain increased amounts of impurities like nickel, antimony, bismuth and arsenic.

During the solidification of anode copper, some impurities are enriched to the solid phase (e.g. Ni) and others (e.g. As, Sb, Bi) to the liquid phase. In the solid copper anode impurities are distributed unhomogeniously being either solute atoms in the metallic copper phase or as inclusions with varied composition in the copper matrix (1-4).

In the electrolysis of anode copper the impurities either dissolve in the electrolyte or form an anode slime. Some impurities react in the electrolyte and form complexes, which can lower the quality of copper cathode produced. Antimony, bismuth and arsenic are a special group of such harmful impurities in the electrorefining process (4-8).

The copper anodes used in Scandinavia are rich in nickel. Metallic nickel is less noble than copper and will dissolve electrochemically, but nickel oxide remains undissolved and reports to the anode slime in sulphate based copper electrolysis (9-10). A very harmful compound is formed in high temperatures in presence of antimony, nickel, oxygen and copper due to the solidification mechanism (8, 10-13).

In practice, successful drift of electrolysis depends on several parameters, such as current densities, temperature and composition of the electrolyte. Anode composition and impurity distribution determine the electrolyte composition, in which it is possible to affect also by the purification of the electrolyte (14-16). Therefore it is essential for the electrorefining process of copper to understand the influence of the solidification microstructure and dissolution behaviour of anode copper.

Experimental

The anodes were prepared by melting cathode copper, Cu_2O, antimony, arsenic, bismuth and cupronickel in a Pythagoras crucible at a temperature of $1200^{\circ}C$ under a nitrogen atmosphere. Impurities that have a low melting point, antimony, arsenic and bismuth were placed on a piece of copper at the bottom of the crucible to prevent their vaporization. The melt was homogenized at $1200^{\circ}C$ and after this quickly removed from the furnace. After four minutes the specimen lost its red hot and was quenched to the water.

A small piece was cut from each specimen for optical microscopy and chemical analysis. The rest was used as anode (Ø 31 mm x 40 mm) for the electrochemical dissolution experiment. A piece of plastic-covered copper wire was soldered at the back of the anode. The anode was then moulded in Araldite and water polished with a paper of 800 Mesh.

The electrochemical dissolution experiments were carried out at a constant current density of 200 A/m^2 at $60^{\circ}C$. Ion-exchanged water with 190 g/l H_2SO_4 (pro analysi) and 2 g/l Cu^{2+} was used as electrolyte. Cu^{2+}-ions were added as sulphate $CuSO_4 \cdot 5H_2O$ (pro analysi). The very low copper content of the electrolyte used in experimentals was chosen, because by doing so it was wished that the secondary reactions of the dissolved ions would not disturb and no passivation of the copper anodes would happen. During the electrolysis copper content of the electrolyte increased to the final content of 3-7 g/l Cu. A platinum electrode with a surface of 25 cm^2 was used as a cathode. The copper that was plated on the cathode was stripped off in a nitric acid solution (50 %) after every fourth day. The amount of copper was then determined by atomic absorption spectrophotometer (AAS). The electrolysis vessel was kept tightly covered, except for when the hydrogen gas formed at the beginning of the process was released.

The microstructure of the anodes were studied from a piece of anode, which was polished by aluminium oxide or diamond paste. Optical microscope and polarization filter was used to identify metallic copper, cuprous oxide and nickel oxide. Also other inclusions were lokalized by the optical microscope. The polished specimens were analyzed by the energy dispersive analysator (EDA) and wave dispersive analysator (WDA) joined to the scanning electron microscope (SEM).

The anode surface morphologies after the electrolysis were studied by the scanning electron microscope. The anode slime formed during electrolysis was centrifugated from the electrolyte. The slime was washed three times with hot water and once with alcohol. After drying the slime was studied by an x-ray diffractometer and a scanning electron microscope to establish its crystallographic composition and morphology. The composition of the electrolyte was analyzed by AAS.

Ternary Cu-Ni-O system

The basis for these experiments is the copper corner of the calculated phase diagram in the ternary Cu-Ni-O system (2), see fig. 1. P. Taskinen has in this thesis (17) determined the equilibrium ternary eutectic point to have the composition of 0.47 wt-% O and 0.32 wt-% Ni in $1065^{o}C$ temperature. According to the equilibrium diagram, the solid copper, which solidifies in the ternary eutectic point consists of 0.32 wt-% Ni and less than 0.004 wt-% O (17).

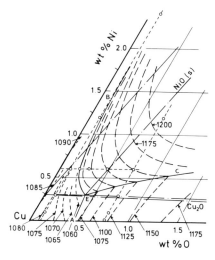

Fig. 1. The phasediagram of Cu-Ni-O system (2).

In the following is shortly reviewed a detailed description of the solidification of copper presented by M. Kytö and P. Taskinen (2). On the basis of primarily solidified phase it is possible to divide the phase diagram into three areas, which are bounded by reactions:

"Cu"(l) = "Cu"(s) + Cu_2O(s)   "AE"

"Cu"(l) = "Cu"(s) + NiO(s)   "BE"

"Cu"(l) = Cu_2O(s) + NiO(s)   "CE"

In this paper "Cu"(l) refers to liquid copper alloy with nominal content of impurity metals and oxygen; "Cu"(s) means solid copper alloy including unoxidized impurity metals and very small amounts of oxygen.

The three eutectic valleys meet in the ternary eutectic point "E", in which occurs the ternary eutectic reaction.

"Cu"(l) = "Cu"(s) + NiO(s) + Cu_2O(s)   "E"

To understand the solidification phenomena of copper anodes it is essential to know the oxidation behaviour of nickel and copper. Therefore it is urgent to be avare about the areas, where nickel oxide and cuprous oxide cannot be crystallized in case of equilibrium solidification. Nickel oxide is not formed in anodes at nominal composition under the tie line, which begins at the point 0.33 wt-% Ni and 0.0 wt-% O and continues towards cuprous oxide (Cu_2O). Cuprous oxide is not formed in anodes at nominal composition over the tie line, which begins at the point 0.3 wt-% Ni and 0.0 wt-% O and continues towards nickel oxide (NiO).

Metallic copper is solidified primarily in the area, which is closest to the copper corner and bounded by reactions "AE" and "BE". The solidification proceeds as growth of copper cells or dendrites depending on the rate of solidification and the distribution of solute atoms. The impurity elements are distributed between solid and liquid phases as their equilibrium distribution coefficient predicts.

Depending on the nickel content of the residual melt, the composition of liquid reach either the eutectic valley "AE" or "BE", in which is formed besides metallic copper either cuprous or nickel oxide. If the ternary eutectic point "E" is ever attained, is there in any case just a small part of the structure as liquid. The residual melt solidifies according to the ternary eutectic reaction "E".

The second area lies at high nickel and oxygen contents, where nickel oxide solidifies primarily. The tie line, which starts from the ternary eutectic point towards nickel oxide, divides this area of primary nickel oxide into two parts.

In the first half with low oxygen contents, that is to the left from the tie line, the ternary eutectic composition is obtained via the eutectic reaction "BE". In the eutectic valley "BE" is formed metallic copper in addition to nickel oxide. Now the anode is almost totally solidified, when the ternary eutectic composition is reached. In the final microstructure of the anode nickel oxide crystals are as inclusions in the metallic copper matrix. The ternary eutectic phase is formed at the grain boundaries of copper dendrites.

In the second half of the area the nominal contents of anodes are to the right from the tie line. The ternary eutectic point is attained via reaction "CE" so that almost the whole structure is liquid, when the ternary eutectic point is reached. The solidification structure is now formed mainly from the ternary eutectic phase, but also cuprous oxide dendrites and nickel oxide crystals are distinguished.

The area, where cuprous oxide forms primary crystals, is located at high oxygen contents bounded by reactions "AE" and "CE". Besides cuprous oxide in the reaction "AE" is generated metallic copper and in the reaction "CE" nickel oxide. While the ternary eutectic point is always reached are there in the solidification structure also finely disperged cuprous oxide and nickel oxide crystals in the metallic copper matrix. The primary and secondary cuprous oxide crystals are dendritic, but the cuprous oxide formed in the ternary eutectic reaction is spherical and small in its shape.

Microstructure of the anodes

In previous chapter was explained how copper solidifies in equilibrium conditions. There has been done practical experiments which verify the principal distribution of elements and solidification of copper to be as required by the phase diagram. The practical solidification of anode copper is presented in detail in previous papers (2, 4, 9-13). Therefore in this paper it has been justified just to refer shortly some structures of anode copper. The aim of this chapter is to introduce generally the solidification mechanism of anode copper at the compositions of copper corner in Cu-Ni-O system.

The specimen under examination here, have constant nickel content of 0.6 wt-% Ni, but varied anode oxygen content from 0.05 to 0.75 wt-% O. The compositions of anodes are far enough from the ternary eutectic point so that solidification mechanism of different types can be explained. The influence of two different oxygen contents on the microstructure of anodes containing 0.6 wt-% Ni is shown in fig. 2.

The solidification of anodes with oxygen content lower than 0.22 wt-% O begins with the crystallization of metallic copper. During the solidification oxygen is enriched to the liquid phase and nickel to the solid phase. The composition of the residual melt reaches the eutectic valley "BE", according to which the residual melt begins to decompose eutectically forming nickel oxide and metallic copper. The final solidification of the residual melt happens through the ternary eutectic reaction "E". At very low oxygen contents (lower than 0.05 wt-% O) all oxygen is consumed to the formation of nickel oxide and the composition of the residual melt does not reach the composition of the ternary eutectic point.

In fig. 2 a) is shown the final microstructure of anode copper containing 0.2 wt-% O and 0.6 wt-% Ni. With oxygen contents higher than 0.22 wt-% O in copper the solidification of anodes starts as crystallization of nickel oxide. If nominal oxygen content in anodes is lower than 0.47 wt-% O, the residual melt decomposes according to the eutectic valley "BE" into metallic copper and nickel oxide. With these oxygen and nickel contents the solidification of anodes ends always with the ternary eutectic reaction "E".

The solidification starts also with the crystallization of nickel oxide when the oxygen content is higher than 0.47 wt-% O. The composition of the liquid reaches now an other eutectic valley "CE" when the temperature is decreased. The residual melt is spilt through the eutectic reaction "CE" into cuprous and nickel oxide. The composition of the residual melt follows

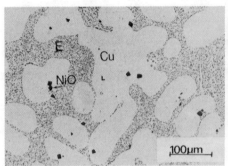 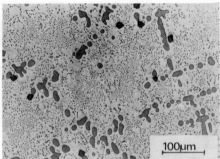

Fig. 2. The microstructure of two anodes both containing 0.6 wt-% Ni.
a) The anode containing 0.2 wt-% O consists of metallic copper "Cu", ternary eutectic phase "E" and nickel oxide "NiO",
b) The anode containing 0.75 wt-% O consists of nickel oxide "NiO", cuprous oxide dendrites "Cu$_2$O" and ternary eutectic phase "E".

the eutectic valley "CE" towards the ternary eutectic point "E". As a result of this reaction the oxygen and the nickel content decrease in liquid phase and the final solidification happens with ternary eutectic reaction. The microstructure of solid copper is shown in fig. 2 b) at nominal content of 0.75 wt-% O and 0.6 wt-% Ni.

Results and discussion

Solidification

The microstructure of anode copper is formed during the solidification of molten copper. According to the thermodynamics molten copper will solidify when the free entalphy of the solid phase is less than the free entalphy of the liquid phase. The equilibrium or phase diagrams are composed as a basis of thermodynamics. However, there are also kinetic factors affecting in the solidification process. Diffusion both in the solid and liquid as well as convection in the liquid will strongly influence to what extension the solidification of anode copper follows the thermodynamics, that is the equilibrium diagram or the phase diagram. The actual microstructure of solid anode copper is a compromise between the thermodynamic and the kinetic factors (18-21).

The solidification of anode copper will happen when the temperature is decreased in casting, or in our case, when taking the crucible out of the furnace to the atmosphere. The first crystals can nucleate either homogeniously or heterogeniously. In the case of homogenious nucleation there has been supposed the first nucleus to be spherical in shape and explained to have a critical radius, which is dependent on the amount of supercooling of the melt and the surface energy at the solid-liquid interface. However, metals usually, anode copper included, do not supercool more than a few degrees under their freezing temperature before they begin to crystallize heterogeniously. In this heterogenious nucleation the first crystals will nucleate on impurity particles, i.e. nucleating agent or mold walls, in our case crucible walls. By heterogenious nucleation it is possible to avoid the thermodynamic barrier to homogenious nucleation and get the crystals to nucleate when the melt is less supercooled than needed for homogenious nucleation. This is due to the smaller volume needed for heterogenious nucleus to be stable than for homogenious nucleus to be stable, the volume of heterogenious nucleus is only a segment of a sphere compared to the volume of a whole sphere of homogenious nucleus, both nuclei having the same critical radius.

In the fig. 3 is presented a schematic description of the solidification mechanism of copper. In fig. 3 a) there are crystals, which have nucleated on the crucible wall. In crystal growth planar solid-liquid interface is soon broken down. At this stage about 1 % of the melt is solidified.

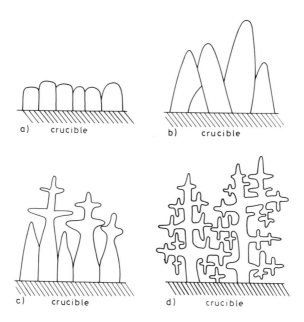

Fig. 3. Schematic description of the solidification mechanism of copper anode.
a) crystals soon after nucleation
b) cellular growth
c) dendritic growth, some secondary arms
d) dendrites at a later stage of solidification

Constitutional supercooling makes the solid-liquid interface unstable, when the solidification proceeds. Any protuberance formed on the interface, is surrounded by supercooled liquid and will be stable and grow, a cellular structure will be formed, fig. 3 b). There will be noticed an unequilibrium distribution of solute elements, which is discussed later on. From the starting point's melt about 5 % has solidified. The growth of cells happens perpendicular to the solid-liquid interface and does not regard any favoured crystal orientations.

Copper is a fcc-structure and its (100) direction will grow faster than other directions. The reason is the difference in accomodation factor. In melting the difference in accomodation factor is small, but in solidification at the close-packed orientations {111} the accomodation of atoms to the solid phase is more difficult than at other orientations. The cellular structure will be broken down and there will be formed dendritic structure. The direction of dendrites is for anode copper (100), as mentioned. Some dendrites have got secondary arms in addition to primary arms, fig. 3 c).

In fig. 3 d) there is shown how the dendrites have grown, when 50 % of the original melt has solidified. There are in addition to primary and secondary dendrite arms also some ternary arms.

The solidification mechanism described above, that is dendritic growth of solid copper, is valid, when the oxygen content of the anode copper is less than 0.47 wt-% O (depends a little from the impurity content of the anode). In this case almost the whole copper melt will be solidified, until the composition of ternary eutectic point is reached. Of course here one must remember, that it is a question of a simplified model, the solidification of the Cu-Ni-O system. When the solidification of the anode copper has proceeded to the eutectic composition there will be major part of the original melt solidified (that is something like 90 % of the original melt). Thus, with low oxygen contents the microstructure of the anode copper is very much influenced by the kinetics, because of the unequilibrium solidification and very low diffusion rate of solute elements in the metallic copper phase.

There will be high levels of nickel in the metallic copper phase and on the contrary low levels of As, Sb, Bi etc. in the metallic copper phase, the latter elements appearing mostly as inclusions at the grain and phase boundaries.

If the oxygen content of the anode is high (that is over 0.47 wt-% O), the primary crystals which grow from the copper melt, are either nickel oxide or cuprous oxide which one, the former or the latter, depends on the nickel content of the anode. These oxides form in certain conditions chainlike structures, dendrites. The solidification proceeds according to the eutectic reaction in which are formed secondary copper oxide and nickel oxide crystals. The solubility of other elements to nickel and cuprous oxide is very low, negligible. When the ternary eutectic composition of the Cu-Ni-O system (simplified model) is reached only a small portion of the original melt is solidified (something like 10 %). The nickel oxide crystals are faceted, with sharp corners and cuprous oxide dendrites are nonfaceted. In this case the microstructure of the anode is quite different compared to the microstructures with low oxygen content, see fig. 2. The matrix is now formed from metallic copper and spherical cuprous oxide, which are the components of ternary eutectic reaction. In the matrix of anode copper are found as inclusions nickel oxide crystals and the primary or secondary cuprous oxide dendrites.

One must notice that the kinetics is now much less determining than when the matrix was out of metallic copper alone. While the melt is mostly molten when the ternary eutectic reaction begins, it is much easier for the impurity and solute elements to act according to thermodynamics. Thus, kinetics has now a minor effect on the elements in solidification.

Distribution of elements in solidification

The equilibrium diagram of the solidifying system is exactly valid only when the solidification takes place in the equilibrium. This is achieved with an extremely slow cooling rate and very finely disperged liquid phase so that there is complete diffusion both in the liquid and in the solid phase. This equilibrium must be maintained throughout the solidification. In practice the equilibrium conditions are not maintained during the solidification of a copper anode, because of the kinetics; the diffusion of the solute elements is not complete neither in the liquid phase nor in the solid phase.

The solidification and therefore the microstructure of the anode copper is profoundly influenced by even small amounts of impurities in the copper melt. Normally these solute atoms tend to remain in the copper melt rather than solidify with the copper matrix. One reason for this is that the melt can accomodate atoms of a different size much easier than the solid copper matrix.

The inhomogenous distribution of impurity components in the experimental solidification of copper anodes can be described with the distribution coefficient k. This coefficient is defined as the ratio of the solid and liquid compositions.

$k = C_s/C_l$

C_s is the concentration of solute in solid phase

C_l is the concentration of solute in liquid phase

The composition of the liquid phase follows the liquidus and the composition of the solid phase follows the solidus of the solidifying system, which has nominal solute element content C_o. The value of k describes the distribution of solute atoms between solid and liquid phases during the solidification. In theoretical considerations in this paper k is assumed to be less than unity; that is solidus and liquidus of the sloping solidifying system downward. However, the description is valid also for alloys in which k is greater than unity.

There are defined different cases for distribution coefficients which depend on the solidification behaviour of the anodes. Only two solidification processes are discussed in the following text. The first one is the equilibrium solidification and the second is a simplified version of a real solidification of anode copper.

An ideal solidification process is a process, where the solidification happens according to the equilibrium, in which case the distribution of the impurity elements during the solidification process can be described with the equilibrium distribution coefficient k_o. Under these conditions the solute concentration of the solid phase C_s is k_o times the concentration of the solute in the liquid phase C_l, and thus strictly obeys the equilibrium phase diagram.

The solute components are enriched in the solid phase if the equilibrium distribution coefficient k_o is greater than unity. The value of k_o for nickel in copper is two (22). The melting point of the alloy raises relatively to the pure copper melt.

On the other hand solute components are enriched in the liquid phase and the melting point of copper alloy is lowered, if the equilibrium distribution coefficient k_o is less than unity. In real solidification process this results in an unhomogenous distribution of the impurity components. The value of k_o for As, Sb, Se, S, O, Bi and Pb is less than unity in the solidification of copper (22).

In a real solidification process, the distribution coefficient is strongly influenced by the kinetic factors, e.g. diffusion both in solid and in liquid phase as well as in convection in liquid phase. In fig. 4 is presented solute redistribution in solidification with limited liquid diffusion and no convection. The solidification begins with the initial solid forming of composition kC_o. Concentration barriers of the alloying components are formed as well in the liquid phase as in the solid phase, when the solidification proceeds. In the melt a layer of high solute concentration, a diffusion layer, is build up adjacent to the advancing liquid-solid interphase, fig. 4 a). If the steady-state is reached the distribution of the components is best described by the effective distribution coefficient k_e, which is the ratio of C_s, the solute concentration in solid at the interphase, to C_1, the mean solute concentration in the liquid outside the diffusion layer. Its value is a function of the equilibrium distribution coefficient k_o and the nature of the solidification. The maximum solute composition of liquid in steady state is C_o/k.

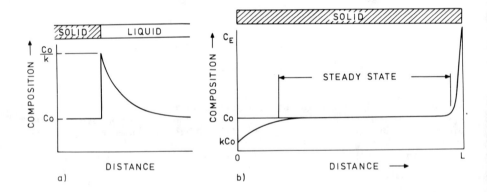

Fig. 4. Solute redistribution in the normal solidification with limited liquid diffusion and no convection, when the solidification is assumed to be one-dimensional.
    a) composition profile during steady-state solidification
    b) composition profile after solidification

The solidification of the residual melt takes place when the temperature is decreased to such a level that solid phase is stable. The enrichment of alloying components in the residual melt is stronger the more the distribution coefficient diverse from unity. Some liquid will remain until an invariant temperature (e.g. eutectic) is reached and the remainder of the liquid then solidifies at eutectic composition C_E.

With the distribution coefficient k it is easy to describe the unequilibrium solidification of the Cu-Ni-O system. In real solidification process with low oxygen contents, a concentration gradient of nickel is formed in the solid copper phase and the amount of nickel is decreased in the residual melt under the level which was predicted by the equilibrium diagram. Oxygen, on the contrary, is enriched in the residual melt where it forms oxides with other elements. The solubility of oxygen in the solid copper phase is

0.03 at-% (22). As a result of this unequilibrium solidification more cuprous oxide and less nickel oxide is formed than expected from the equilibrium diagram.

Practical solidification of anode copper

In the following text is discussed about some examples, where solidification structures, described above, are found in anode copper.

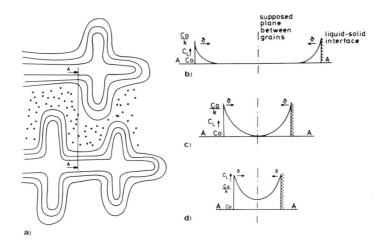

Fig. 5. Schematic description of the enrichment of the solute atoms during the solidification of anode copper.
   a) Two-dimensional microstructure of copper dendrites and eutectic phase at the copper grain boundaries. The isoconcentration surfaces are also indicated as well as the section A-A. Concentration barriers of the section A-A at stages.
   b) Steady-state solidification of copper dendrites
   c) Concentration barriers have dashed against each other
   d) Eutectic reaction

Two solid copper dendrites have adjacently grown in the liquid phase near each other in a schematic fig. 5 a). In this two-dimensional microstructure are also indicated the isoconcentration surfaces of copper dendrites. As previously explained O, As, Sb, Bi, Pb etc. are due to the unequilibrium solidification enriched to the liquid phase and nickel to the solid copper phase. When two copper dendrites grow at the velocity v they become thicker (isoconcentration lines), and there will be formed a concentration barrier into the liquid phase ahead the growing dendrite arms. In fig. 5 b-d) are drawn the concentration barriers in the liquid phase ahead copper dendrites during different stages of solidification. The stage of the innermost isoconcentration line is presented in fig. 5 b) corresponding to the more or less stationary growth of copper dendrites. The concentration barriers will meet each other after a time interval, during which the solidification of copper dendrites has proceeded stationary, corresponding to the middlemost isoconcentration line, fig. 5 c).

Since two barriers have met, is there caused a drastic increase in the concentration of solute elements in the liquid phase, see fig. 5 d). Hence, the composition of residual melt has reached such a high level, that there can happen for example an eutectic reaction, in which metallic copper and cuprous oxide is formed.

As a practical example is introduced the microstructure of an anode containing 0.1 wt-% O and 1.2 wt-% Ni, fig. 6. According to the equilibrium the residual melt should solidify, before the ternary eutectic reaction begins and the structure should contain only metallic copper and nickel oxide. However, some cuprous oxide is formed as a result of the ternary eutectic reaction.

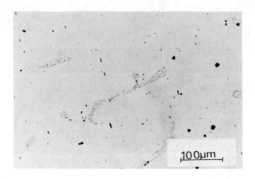

Fig. 6. Microstructure of anode copper containing 0.1 wt-% O and 1.2 wt-% Ni. In copper matrix are some faceted nickel oxide crystals and unequilibrium ternary eutectic phase.

Nickel oxide crystals can be found as inclusions in the solid metallic "Cu"$_s$ matrix. This means that the NiO crystals are formed at the beginning of the solidification process. Fast growing copper "Cu"$_s$ dendrites have surrounded nickel oxide crystals immediately after their crystallization and therefore their crystal size is small. The cuprous oxide formed in the ternary eutectic reaction, is as explained above, in disagreement with the thermodynamical expections that Cu_2O could not be formed at this anode composition.

The behaviour of the antimony in the solidification and electrolysis of anode copper has been described in the previous papers (11-13). In this paper it will be discussed about the mechanism with which micaceous copper, $3Cu_2O \cdot 4NiO \cdot Sb_2O_5$, is formed. This oxide is very harmful, because it forms floating slime in the electrolysis of anode copper.

In fig. 7 is schematically introduced the enrichment of antimony and the formation of micaceous copper. The presentation is similar to the one done with the aid of fig. 5 in case of Cu-Ni-O system. When two copper dendrites have grown, fig. 7 a) there is formed a concentration barrier of antimony and oxygen in the liquid phase adjacent the moving solid-liquid interphase, fig. 7 b). After the concentration barriers have met, the antimony and oxygen contents are increased extremely much, fig. 7 c). While the residual melt contains also copper and nickel, it is evident, that micaceous copper is formed instead of the ternary eutectic structures.

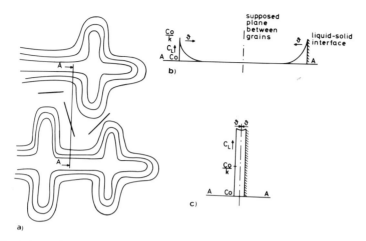

Fig. 7. Schematic description of the enrichment of solute atoms during the solidification of anode copper. a) Two-dimensional microstructure of copper dendrites and micaceous copper at the copper grain boundaries. Concentration barriers of the section A-A at stages, b) solidification of copper dendrites, c) solidification of micaceous copper.

In fig. 8 is the formation appearence of micaceous copper illustrated on the phase diagram of Cu-Ni-O system with constant antimony content 0.1 wt -% Sb. When nickel content lies under 0.3 wt-% Ni no micaceous copper is formed, because nickel is not oxidized. However, with higher nickel contents, there is possible to distinguish several different areas in regard to the existence of micaceous copper. This is because of the solidification mechanism described with fig. 7. With low oxygen content there can be seen an area where the appearance of micaceous copper is very high. With high oxygen contents the amount of micaceous copper is low, because enrichment of some solute elements into the liquid phase, which is obvious reason for the formation of micaceous copper, is not fulfilled to great extent.

In the solidification of Cu-Bi system, which is rich in copper, copper solidifies primarily. Bismuth enriches into the liquid phase and hardly dissolves into the solid copper phase. Therefore, at the grain boundaries of copper is formed a thin bismuth film, when the eutectic Cu-Bi melt is solidified, fig. 9. Actual eutectic structure is not found due to the very good wetting. Copper is crystallized from the eutectic melt to the surface of primary copper dendrites, whereas bismuth solidifies at the grain boundaries of copper dendrites as thin bismuth films. The formation mechanism of the microstructure resembles much that explained in the case of micaceous copper in fig. 7, where antimony was enriched to the residual melt and formed a new phase together with copper, nickel and oxygen.

As another example of the surface activity of bismuth is presented, how bismuth is oxidized in the ternary Cu-Bi-O system. Bismuth oxide shells are found at the surface of cuprous oxide crystals in an anode containing 2.0 wt-% Bi and 0.75 wt-% O, fig. 10 a). The solidification begins as formation of cuprous oxide and proceeds via reaction, where cuprous oxide and metallic copper is formed. Bismuth enriches to the liquid phase. The residual melt, where bismuth content is high, has very low surface energy, compared to copper and for that reason wetting is very good. Bismuth containing shells can lower the overall surface energy of the eutectic structure. The shells were analysed by WDA to contain bismuth, oxygen and possibly some copper (copper can also become from the copper matrix), fig. 10 b-d). Thus,

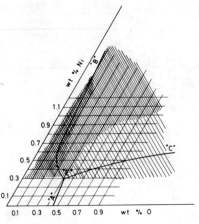

Fig. 8. The schematic appearance of micaceous copper is illustrated at antimony content 0.1 wt-% Sb in the Cu-Ni-O system. The more micaceous copper is found from the anode the darker the area is.

it is question about bismuth oxide shells on cuprous oxide crystals. Similar structures of bismuth and cuprous oxide are also formed in anodes with lower bismuth and oxygen contents.

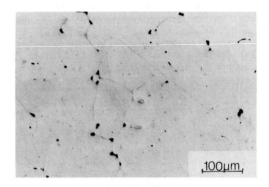

Fig. 9. The microstructure of anode copper containing 2.0 wt-% Bi. Bismuth films can be seen at the grain boundaries of copper dendrites.

Arsenic has the same tendency to enrich to the residual melt as antimony and bismuth. In fig. 11 is presented an etched microstructure of anode copper containing 0.7 wt-% As. The etching solution was ferrichlorid. At the grain boundaries can be distinguished a phase, which was analyzed with EDA to be Cu_3As. Umetzu and Suzuki (24) have found Cu_3As containing 3.18 wt-% As.

In oxygen containing copper anodes arsenic is found to form similar shells as bismuth at the surface of cuprous oxide crystals. In these anodes Cu_3As was not found, because arsenic is oxidized.

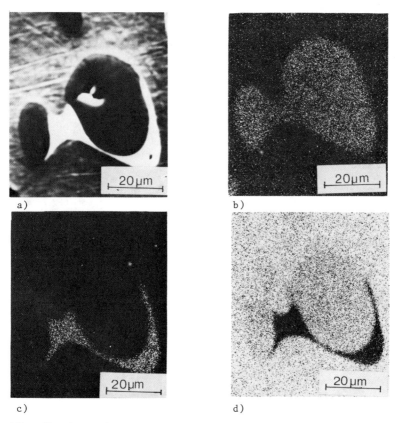

Fig. 10. a) SEM image of anode copper containing 2.0 wt-% Bi and 0.75 wt-% O. Wave concentration maps done by wave dispersive analysis; b) oxygen; c) bismuth and d) copper.

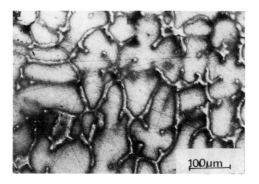

Fig. 11. Etched microstructure of anode copper containing 0.7 wt-% As. Cu_3As appears at the grain boundaries of copper.

Electrolysis

The idea in the electrolysis experiments was to clarify the behaviour of impurities in the electrochemical dissolution in sulphate based copper electrolysis. In the study is considered if the impurity elements or oxides will dissolve and to what extent in copper electrorefining. With the aid of the electrochemical experiments it was possible to fulfill quantitatively the qualitative results accomplished by microstructure studies.

In electrochemical experiments was determined the current efficiency for anode components such as Cu, Ni, Sb, Bi and As. The current efficiency reports the distribution of an element between dissolved phase and undissolved phase. To obtain the current efficiency for each element in the anode, the amount of element dissolved electrochemically from the anode was divided by the theoretical value calculated from Faraday's law and the anode composition. In the calculations impurities are assumed to be totally in metallic form and for that reason to dissolve electrochemically in electrolysis.

The amount of nickel oxide can be calculated as the difference between the amount theoretically dissolved and the amount actually found in solution, because NiO is undissoluble in the electrolyte. The behaviour of such impurities as Bi and As is more complicated while they have oxides which dissolve chemically in electrolysis and after that possibly precipitate through secondary reactions.

In the following text is shortly reviewed some of the electrochemical experiments. A more detailed description is presented in several papers (4, 10, 25).

The influence of As, Bi and Sb on the dissolution of nickel at different oxygen contents is shown in fig. 12.

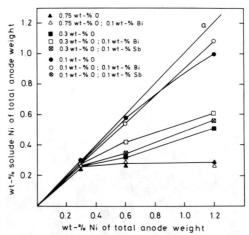

Fig. 12. The influence of bismuth and antimony on the electrochemical dissolution of nickel as a function of nickel content. Line a refers to theoretical conditions, where all nickel is dissoluble.

The received current efficiency for nickel was close to 100 % when the electrolysis was carried out with the nominal contents of 0.3 wt-% Ni and

0.1 wt-% O. The current efficiency of nickel decreased to 90 % with increased anode oxygen content of 0.75 wt-% O. According to this current efficiency 0.27 wt-% Ni will be metallic in the anode and for that reason electrochemically dissoluble in the anode composition of 0.75 wt-% O and nickel content 0.3 wt-% Ni. Arsenic, antimony and bismuth had at this nickel content no real effect on the electrochemical behaviour of nickel . Nickel was clearly observed to behave as was predicted on the basis of the equilibrium diagram of Cu-Ni-O system (17).

There are obtained unevident differences in the amount of metallic nickel, when the nominal nickel content of the anode copper is increased from 0.3 wt-% Ni. Now oxygen content of the anode copper has a decided influence on the portion of metallic nickel. As described earlier in this paper the kinetic factors are less determining in the solidification of the ternary Cu-Ni-O system, when the oxygen content of the anode copper is high, that is 0.75 wt-% O. With this high oxygen content the metallic and thus electrochemically dissoluble nickel content is constant 0.27 wt-% Ni for all nickel contents of the anode copper over 0.3 wt-% Ni. This experimental metallic nickel content (0.27 wt-% Ni) derivates from the equilibrium value of 0.33 wt-% Ni by P. Taskinen (17) due to the supercooling of the melt which leads to greater oxidation of nickel in lower temperatures than the ternary eutectic temperature.

The results with moderate oxygen contents, that is 0.3 wt-% O, fig. 12, give an idea that the equilibrium conditions of the ternary Cu-Ni-O, are not always reached. Nickel is remained in metallic copper phase, because of its slow diffusion rate in metallic copper. This tendency is especially clear with high nickel contents of 1.2 wt-% Ni. Other impurities such as bismut, arsenic and antimony, increase the amount of this dissoluble nickel content of anode copper. Slow diffusion rate of nickel in anode copper shows conclusive effect on the oxidation behaviour of nickel, with low oxygen contents, that is 0.1 wt-% O. Only about 10 % of nickel is oxidized to nickel oxide although there should be 50 % of the nickel oxidized according to the equilibrium diagram, with nominal composition 0.6 wt-% Ni and 0.1 wt-% O.

The anode slimes fallen from anodes with low and moderate oxygen contents, 0.1 wt-% O and 0.3 wt-% O respectively, are shown in fig. 13.

The anode slime from the anode containing 0.3 wt-% O and 0.6 wt-% Ni is presented in fig. 13 a). Cubic nickel oxide crystals are distinguished in the slime as faceted octahedrals. Amorphous copper is formed from the anode containing 0.1 wt-% O and 0.3 wt-% Ni shown in fig. 13 b). Amorphous copper does not come out in x-ray analysis.

In fig. 14 is presented that antimony, arsenic and bismuth increase current efficiency of nickel with constant nickel contents of 0.6 wt-% Ni and varied oxygen contents. The increasing effect of the impurity elements to current efficiency of nickel is not dependent of the oxygen content of anode copper, although the solidification mechanism is different with different contents of oxygen. When the current efficiency of nickel is very high, e.g. with low oxygen content, the influence of impurities to the current efficiency of nickel is diminutive.

In fig. 15 there is shown two typical anode morphologies after the electrochemical dissolution of anodes. The anodes are dissolved unevenly, because of different structures due to the different solidification mechanisms. In fig. 15 a), where the composition of anode is 0.3 wt-% O and 0.6 wt-% Ni, can be seen the original ternary eutectic phase of the solidification structure to have changed to sponge-like structure, which is formed

Fig. 13. SEM image of anode slime
   a) nominal anode composition 0.3 wt-% O and 0.6 wt-% Ni, faceted NiO crystals
   b) nominal anode composition 0.1 wt-% O and 0.3 wt-% Ni, amorphous copper powder.

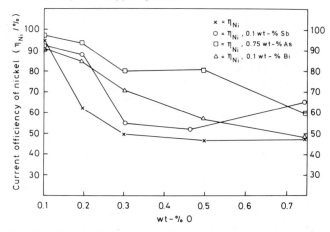

Fig. 14. The effect of arsenic, bismuth and antimony on the current efficiency of nickel as a function of oxygen content with constant nickel content of 0.6 wt-% Ni.

when the cuprous oxide is dissolved chemically from the grain boundaries of metallic copper dendrites. Some nickel oxide crystals are also seen. In the fig. 15 b) the oxygen content is increased to 0.75 wt-% O, while the nickel content is the same 0.6 wt-% Ni. In the solidification structure was found some primarily solidified cuprous oxide dendrites, while the matrix of the anode copper is still metallic. The products of the ternary eutectic reaction, that is metallic copper and cuprous oxide as well as nickel oxide are quite homogeniously distributed all over the structure. While the ternary eutectic phase is the main phase there are no such phase or grain boundaries which would have dissolved faster than the rest of the structure.

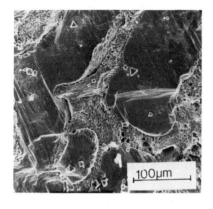 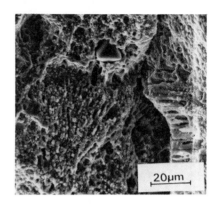

a) b)

Fig. 15. SEM image of anode morphologies after electrolysis
   a) nominal anode composition 0.3 wt-% O and 0.6 wt-% Ni;
      metallic copper matrix, sponge-like eutectic phase and
      NiO crystals
   b) nominal anode composition 0.75 wt-% O and 0.6 wt-% Ni;
      sponge-like eutectic phase having metallic copper as
      matrix, some NiO crystals.

The current efficiency of antimony (25) shows clearly the appearance of micaceous copper, fig. 8, to decrease with increasing anode oxygen content. With anode composition 0.1 wt-% O, 0.6 wt-% Ni and 0.1 wt-% Sb the current efficiency of antimony is only 40 %, but when increasing oxygen content to 0.75 wt-% O the current efficiency of antimony is close to 100 %.

The morphology of anode, which consists of 0.1 wt-% O, 0.6 wt-% Ni and 0.1 wt-% Sb is shown in fig. 16 a). At the grain boundaries of copper dendrites is found a micaceous copper leaf. The micaceous copper leaves are seen in the anode slimes when the anode, mentioned above, was dissolved electrochemically, fig. 16 b). In the electrolysis the appearance of micaceous copper is undesired, because it forms floating slime. This is due to the low specific weight. Further these inclusions containing antimony, nickel, copper and oxygen, are found from the cathode making its impurity content high.

The current efficiency of bismuth in copper electrolysis was measured to be independent on variations in bismuth and oxygen content of the anode copper (4).

In anode copper containing 2.0 wt-% bismuth grain boundaries of copper grains were dissolved in the electrlysis more than metallic copper dendrites, fig. 17 a). That means that while bismuth, which is at the grain boundaries, dissolves easier than metallic copper dendrites, it is less noble than copper dendrites are.

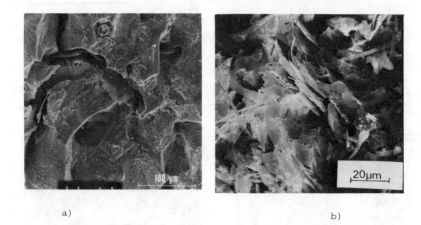

Fig. 16. SEM images with nominal anodecomposition 0.1 wt-% O, 0.6 wt-% Ni and 0.1 wt-% Sb.
   a) anode morphology, micaceous copper at the grain boundaries of copper grains
   b) anode slimes, micaceous copper.

Fig. 17. SEM images
   a) anode morphology with nominal anode composition 0.2 wt-% Bi. Bismuth films at the grain boundaries are dissolved more than copper matrix.
   b) anode slime with nominal anode composition of 0.1 wt-% Bi, 0.1 wt-% O and 0.6 wt-% Ni. NiO crystals and secondary bismuth precipitate.

In the slimes of oxygen, bismuth and nickel containing anodes there were not found bismuth oxide inclusions, which were formed during the solidification. The slime consists mainly of nickel oxide crystals, when the anode composition was 0.1 wt-% O, 0.6 wt-% Ni and 0.1 wt-% Bi, fig. 17 b). Some bismuth was also in the slime, but the shape of bismuth slime was different to that of bismuth inclusions seen in the microstructures of anode copper. Bismuth is therefore expected to dissolve either electrochemically in case of metallic bismuth or chemically in case of oxidized bismuth. First after that bismuth precipitates to slime through secondary reactions in electrolyte.

Bismuth had great influence on the shape of nickel oxide crystals with anode composition 0.75 wt-% O, 0.1 wt-% Bi and 0.6 wt-% Ni, fig. 18 a), changing them to long, faceted chains which are formed because of the dendritic growth, described in the solidification section. The hopper-like nickel oxide crystals are formed, when there is instead of corners faster growing crystal edges. The anode morphology formed was very special, fig. 18 b). Nickel oxide chains came extremely well out of the structure, because they did not dissolve, but formed chain-structures on the anode surface.

a)   b)

Fig. 18. SEM images with nominal anode composition 0.75 wt-% O, 0.1 wt-% Bi and 0.6 wt-% Ni.
a) anode slimes of NiO chains and hopper crystals
b) anode morphology, NiO chains on copper matrix.

No difference was seen in the current efficiency of arsenic in the electrochemical dissolution of anode copper, which contained varied amounts of oxygen, nickel and arsenic. Arsenic dissolved either chemically or electrochemically in the electrolysis (8).

The effect of arsenic on copper anode morphology after the electrolysis was significant. In fig. 19 a) is shown how grain boundaries are dissolved in an oxygen free anode, which contains 0.7 wt-% As. The grain boundaries are rich in arsenic and have dissolved less than other structure, which consists mainly of metallic copper phase. Thus, the metallic copper is less noble than arsenic rich grain boundary areas. One can easily imagine the solute enrichment, that is the unequilibrium distribution of arsenic, as described in this paper, to come out in this morphology structure. In an arsenic containing anode copper with nominal composition 0.2 wt-% As, 0.1 wt-% O, 0.6 wt-% Ni, copper dendrites adjacent to the phase boundary dissolve less. These copper grains, which have the distribution gradient of arsenic state that the more arsenic is enriched into metallic copper phase the more noble it behaves.

With nominal anode composition 0.75 wt-% O and 0.6 wt-% Ni arsenic had the same effect as bismut to the shape of nickel oxide, that is to say there formed NiO chains.

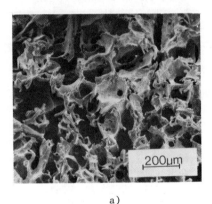 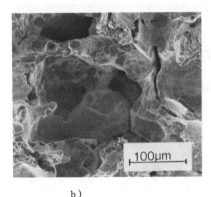

a) b)

Fig. 19. SEM image of anode morphologies.
a) nominal anode composition 0.7 wt-% As. Cu_3As at the grain boundaries remains when copper grains dissolve
b) nominal anode composition 0.1 wt-% O, 0.2 wt-% As and 0.6 wt-% Ni. Near grain boundaries copper dendrites are rich in arsenic and behave as noble metal.

In unequilibrium solidification of anode copper, arsenic is enriched to liquid phase and solidified at the later stage of the crystallization of copper dendrites, resulting in an encreased arsenic content near grain boundaries of metallic copper dendrites.

Conclusions

In this study is introduced with practical examples the model of solidification of impurities during the solidification of anode copper. According to this model the solidification structures of anode copper are strongly influenced by both thermodynamic and kinetic factors. Thermodynamics determine the microstructure of anode copper, but kinetics can alter the composition of melt to such that there is formed unequilibrium structures during the solidification, especially when the residual melt is solidified.

With the aid of this solidification model it is possible to explain the discrepancy between thermodynamically expected and practically solidified structures. Nickel is enriched in solidification and remained to metallic copper phase due to the low diffusion rate in solid copper phase with low oxygen contents of 0.1 wt-% O in Cu-Ni-O system. For example in anodes with nominal anode composition 0.1 wt-% O and 0.6 wt-% Ni, the amount of unoxidized nickel should be 0.33 wt-% O, in electrolysis it was gained to be 0.55 wt-% Ni.

Nickel is oxidized more than the phase diagram predicts with high oxygen contents of 0.75 wt-% O in anode copper. In anodes containing 0.75 wt-% O and 0.6 wt-% Ni, is according to the electrolysis oxidized 0.27 wt-% Ni although the equilibrium value is 0.33 wt-% Ni. This is due to the supercooling of the melt during the solidification.

Arsenic, antimony and bismuth enrich to the liquid phase, when copper melt is solidified selectively. In the residual melt is formed different unequilibrium complexes depending on the anode composition. These impurities, arsenic, antimony and bismuth increase the amount of metallic nickel in anode copper.

In the quaternary Cu-Ni-O-Sb system is formed micaceous copper ($3Cu_2O \cdot 4NiO \cdot Sb_2O_5$). In this paper is explained the formation mechanism of micaceous copper, which behaves as a harmful "floating slime" in electrolysis.

Bismuth and arsenic segregate at the grain boundaries during the solidification in oxygen free copper. In electrolysis the behaviour of these grain boundary phases differ. Compared to metallic copper phase bismuth films were less noble and dissolved faster than copper matrix. In arsenic containing anode the grain boundary phase did not dissolve as much as the copper matrix. In oxygen containing copper anodes arsenic and bismuth are oxidized during the solidification of residual melt. These oxides wet well cuprous and nickel oxide crystals, on which they have formed shells. In the electrolysis these arsenic and bismuth containing oxides dissolved chemically.

ACKNOWLEDGEMENT

The authors wish to thank Jernkontoret, Sweden and the Ministry of Trade and Industry in Finland for their financial support.

We are indebted to professor emeritus M.H. Tikkanen and Mr. R. Lindström (Boliden Ab) as well as other members of Jernkontoret's Committee 6115/80 for fruithful discussions.

REFERENCES:

1. Buhrig, E., Hein, K. & Baum, H., Verteilung von Fremdelementen bei der Kristallization von Kupfer. Metall, 33(1976)6 pp. 592-596.

2. Kytö, M. & Taskinen, P., Thermodynamic Behaviour of Dilute, Ternary System Cu-Ni-O. Report for Jk project 606/77 Jernkontorets Forskning Serie D 227(1977) (in Swedish).

3. Häkkinen, A., Oxidation of Nickel in Oxygen-bearing Copper Metals (in Finnish). M.Sc. Theses, Helsinki University of Technology, Institution of Process Metallurgy, 1980 p. 66.

4. Hettula, E., The Influence of Impurities on the Solidification Structure of Copper Anodes and their Dissolution in Electrolysis (in Finnish). M.Sc. Theses, Helsinki University of Technology, Institution of Process Metallurgy, 1983 p. 105.

5. Braun, I.B., Rawling, J.R. & Richards, K.J., Factors affecting the quality of electrorefining cathode copper. Proc. Int. Symp. Copper Extraction and Refining, AIME, Las Vegas, USA, 1976 pp. 511-524.

6. Claessens, P.L. & Baltazar, V., Behaviour of Minor Elements during Copper Electrorefining, p.F2 in Raffinationsverhaften in der Metallurgie, Hamburg, W. Germany, 1983.

7. Lange, H.-J., Hein, K. & Schab, D., Anodenprozesse bei der Raffinationselektrolyse. Leipzig 1977, Bergakademie Freiberg, Freiberger Forschungshefte, B 195 pp. 29-49.

8. Hämäläinen, T., Dissolution and Passivation of Impure Copper Anodes in Electrolysis (in Finnish). M.Sc. Theses, Helsinki University of Technology, Institution of Process Metallurgy,1983 p. 74.

9. Forsén, O. & Tikkanen, M.H., The Presence and Consequence of Nickel in Copper Anode Alloys containing Oxygen. Part of the report for research project 606/77 Jernkontorets Forskning Serie D 319(1977) (in Swedish).

10. Forsén, O. & Tikkanen, M.H., On the Dissolution of Copper Anodes in Electrolytic Refining, Part I: The behaviour of nickel in oxygen-bearing copper anodes. Scandinavian Journal of Metallurgy 10(1981) pp. 109-114.

11. Forsén, O., The Behaviour of Antimony in Nickel- and Oxygen containing Anodes in Copper Refining. Part of the report for research project Jk 6115/80 Jernkontorets Forskning Serie D 446(1983) (in Swedish).

12. Forsén, O. & Tikkanen, M.H., On the Dissolution of Copper Anodes in Electrolytic Refining, Part II: The behaviour of antimony in nickel-oxygen-bearing copper anodes. Scandinavian Journal of Metallurgy 11 (1982) pp. 72-78.

13. Forsén, O., The Dissolution Behaviour of Nickel in Copper Anodes in Electrolytic Refining. Hydrometallurgy Research, Development and Plant Practice, Edited by K. Osseo-Asare and J.D. Miller, The Metallurgical Society of AIME 1983 pp. 721-736.

14. Abe, S., Burrows, B.W. & Ettel, V.A., Anode Passivation in Copper Refining. Canadian Metallurgical Quarterly 19(1980)3, pp. 59-67.

15. Backx, A., Feneau, C. & Tougarinoff, B., Anodenprobleme bei der elektrolytischen Kupferraffination. Erzmetall 22(1969) Beiheft pp. 120-127.

16. Schab, D. & Hein, K., Untersuchungen zum Gleichgewicht $2Cu^+ = Cu + Cu^{2+}$ unter den Bedingungen der elektrolytischen Kupferraffination. Neue Hütte 15(1970)8 pp. 461-476.

17. Taskinen, P., Liquids Equilibria and Solution Thermodynamics in Copper-Rich Copper-Nickel-Oxygen Alloys. Acta Polytech. Scand. Ch 145(1981), Ph.D. Theses.

18. Flemings, M.C., Solidification Processing. McGraw-Hill, New York 1974 p. 364.

19. Cahn, R.W., Physical Metallurgy. American Elsevier Publishing Company, Inc. New York 1970 p. 1333.

20. Haasen, P., Physical Metallurgy. Cambridge University Press, Cambridge 1978 p. 54-76.

21. Reed-Hill, R.E., Physical Metallurgy Principles. D. Van Nostrand Company, Inc. New Jersey 1967 p. 629.

22. Fromm, E. & Gebhard, E., Gase und Kohlenstoff in Metallen. Springer-Verlag, Berlin 1976 p. 747.

23. Hansen, M. & Anderko, K., Constitution of Binary Alloys, McGraw-Hill, New York 1958 p. 1257.

24. Umetzu, Y. & Suzuki, S., Effect of Impurities on the Copper Anode Potential - Effect of Arsenic, Antimony, Bismuth and Tin. Journal of Mining and Metallurgical Institution of Japan, Nikon Kogyo Kaiski, 77(1961)882 pp. 1087-1093.

25. Forsén, O., to be published: "The Behaviour of Nickel and Antimony in Oxygen-bearing Copper Anode".

SECONDARY COPPER SMELTING

W. R. Opie*, H. P. Rajcevic**, W. D. Jones**

*AMAX Base Metals R&D, Inc.
  400 Middlesex Avenue, Carteret, NJ 07008
**U.S. Metals Refining Co.
  400 Middlesex Avenue, Carteret, NJ 07008
                    USA

Fundamentals of blast furnace smelting of low grade copper and precious metal scrap and residues are reviewed with emphasis on emission control, by-products produced, slag cleaning for high recovery of metal values and the evolution of an off-gas system to eliminate hydrocarbons that result from decomposition of plastics in the blast furnace charge. Problems with tin, antimony and nickel in the smelter-refining cycle are discussed.

Secondary Copper Smelting

Recovery of metal values from low grade secondary copper and precious metal sources such as electrical and electronic scrap, leach residues and foundry residues presents a number of technical problems not encountered in primary copper smelting. In addition to producing copper and precious metals, recovery of by-product metals such as tin, nickel, zinc and lead adds to the economics. Impurities such as iron, antimony, arsenic, chlorides and fluorides must be controlled and organics must be destroyed to the extent that strict furnace emission standards are met.

The Carteret, N.J. smelter of the U.S. Metals Refining Company (a subsidiary of AMAX Inc.) has a "state of the art" flowsheet which was developed over the years to treat the available mix of low grade copper secondaries. Figure 1 shows the units which make up the USMR smelter.

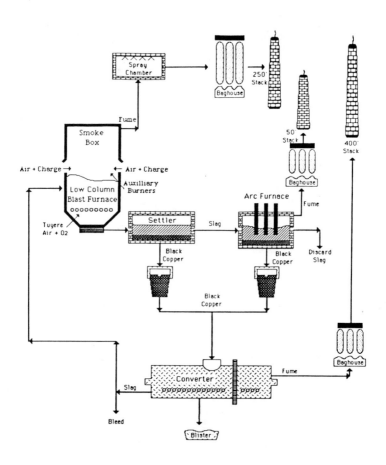

Figure 1 - Schematic of U.S. Metals Refining Co. Smelter

As scrap and residue feeds for the blast furnace are received they are bedded along with flux. Fines are briquetted prior to bedding. Because of the non uniform nature of the charge, control of the chemistry and feed rate is much more difficult than in a primary smelter where the particle size and composition of the flotation concentrate is uniform and the feed rate can be controlled very closely from minute to minute. The practice of horizontal bedding and then using front end loaders to take vertical cuts from the bed and load the furnace feed-skips gives considerable mixing. However the particle size of the charge varies from 2 feet to 0.01 inch and the copper content can vary by ±10% from an approximate 30% average. In contrast to the cold top blast furnaces used for primary smelting of copper concentrates, lead concentrates or iron ore, the secondary furnace operates with a hot top for the following reasons:

1. To volatilize metals such as zinc, lead and tin and oxidize them at the top of the charge column.

2. To decompose organics which include many types of plastics and rubbers and burn the volatile hydrocarbons to completion.

The molten metal and slag from the blast furnace flow into a settler which allows enough residence time for slag metal separation. The metal contains 65 to 70% copper with the balance being iron, antimony, nickel, precious metals and a portion of the tin, lead and zinc which were not volatilized. The slag flows from the settler through an electric arc furnace operating at a higher temperature than the blast furnace (an increase in temperature from about 1300 to 1400°C). By controlling coke addition and residence time 60 to 80% of the metals in the incoming slag are recovered in the furnace fume or the metal product which collects in the furnace bottom.

The metal from the settler and the slag cleaning furnace are blended and treated in a Peirce-Smith converter to produce blister copper suitable for electrorefining. The slag from the arc furnace is granulated and sold for various applications such as road fill, cement manufacture, or grit for metal cleaning.

The oxide particulates from the blast furnace, converter and slag cleaning furnace are collected in separate baghouses and subsequently sold or treated for recovery of their contained metals.

Control of Hydrocarbons

One of the major problems in smelting electrical and electronic scrap is in dealing with the organic materials. Many techniques have been tried to mechanically separate plastic and rubber from copper conductors. Some of these such as chemical treatment, or wire chopping followed by some form of gravity separation are effective for simple conductors but are not applicable to complex units such as panel boards, complex cables, chassis, circuit boards, motors, etc. Generally the mechanical separation techniques produce a "non-metallic" fraction which contains too much metal value to discard and which in itself is a disposal problem.

The USMR operation destroys the organics by burning them as part of the fuel or decomposing them in the upper part of the charge column and then burning the resultant volatile hydrocarbons in the section of the furnace above the charge doors from where the gases enter the flue system.

To accomplish this effectively there must be an excess of oxygen and the reacting gases must be in turbulent flow for long enough time and at a high enough temperature to destroy the combustible volatiles (VOS, TVOS and carbon monoxide).

To insure a high enough oxygen level the tuyere air is enriched with oxygen and the air supplied through the charge doors is many times that required for theoretical combustion. Table I shows typical furnace operating conditions. The temperature of the gases in the combustion chamber above the charge ports is maintained above that needed for hydrocarbon destruction (1100-1225°F). Usually the temperature at the top of the charge column is high enough for complete combustion. However, because of the non-uniformity of the charge there are short periods when the temperature is "quenched" by the cold feed. To assure complete combustion, burners are installed in the upper shaft to maintain high gas temperatures during any short quench periods.

Table I. Cupola Blast Furnace - Operating Data

Feed Rate	= 3.5-4.5 Tons/ft^2/day
Fuel	= 14-18% coke
Tuyere Air	
Rate	= 120-140 CFM/ft^2
O_2 Content	= 22-25%
Temperature	
Metal	= 1200-1300 C
Slag	= 1250-1400 C

Another problem created by the irregularity of the size of components of the charge is tendency for the bed to channel and cause partial tuyere blockage resulting in irregular distribution of the air blast over the length of the furnace. To solve this problem an automatic control system has been installed to monitor air flow through each group of three tuyeres (16 groups) and the furnace draft at four points within the furnace near the charge doors. Using these data a microprocessor equalizes the air flow through each three tuyere unit even though parts of the charge column exert different back pressures. The draft is also controlled based on the least draft measured from the four points in order to minimize air infiltration through the charge doors. In addition to controlling draft and distribution of tuyere air the microprocessor receives and records data from continuous monitoring instruments for oxygen, VOS, benzene and CO emissions from the stack. The net result is destruction of 99+% of the organics in the furnace charge.

Control of Solid Emissions

The metals which volatilize as a result of the "hot top" operation are oxidized in the upper shaft or the hot part of the flue system. The chlorides and fluorides which are products of decomposition of some plastics react with the metal oxides to form particulate oxychlorides and fluorides. These solids either collect in chambers in the flue system or in the baghouses. As long as there is enough volatile zinc, lead and tin in the charge the chlorides and fluorides are not emitted as gases but collect as solids. Table II gives typical composition ranges for the baghouse collect.

Table II. Cupola Blast Furnace Baghouse Oxides Composition

	%
Cu =	8.0-12.0
Pb =	13.0-15.0
Zn =	20.0-35.0
Sn =	1.5-2.0
Cl =	6.0-10
F =	1.0-5.0
Ni =	0.1-0.5
Sb =	0.3-0.8
SiO_2 =	4.0-7.0

Black Copper Converting

Analyses of the black copper from the blast furnace and slag cleaning furnace are given in Table III. Their conversion to blister is more complex than converting a matte from a primary smelter. The iron, nickel, zinc, tin, lead and antimony are the "fuel" in this operation. The major portion of the zinc, tin and lead are volatilized. Coke additions enhance volatilization. Silica flux is added to slag-off the iron, nickel and antimony. Whereas a converter slag from a conventional matte smelter contains about 5% copper, the slag from a secondary smelter usually contains over 30% copper because the converter has to be "blown hard" to eliminate nickel, tin and antimony from the blister. The converter slag is returned to the blast furnace to recover the copper resulting in a large circulating load, particularly of nickel, tin and antimony which are objectionable in blister which is to be electrorefined. A practice at the USMR smelter is to "bleed off" part of the converter slag and smelt it separately to produce a high nickel blister which is cast into anodes and refined in separate sections of the tank house(1). Table IV shows compositions of the products from the converting operation.

Table III. Cupola Blast Furnace Black Copper

	%
Cu =	65.0-70
Ni =	7.5-12.0
Sb =	0.5-1.5
Sn =	2.0-4.0
Fe =	5.0-10.0
Zn =	2.0-4.0
Pb =	2-4

Table IV. Converter Products

	Blister, %	Slag, %	Oxide, %
Cu	94-96	30-35	2-3
Ni	0.5-1.0	10-15	0.5-1.0
Sb	0.1-0.3	0.5-1.5	0.5-1.5
Sn	0.1-0.2	2-4	10-20
Fe	0.1-0.3	20-25	0.5-1.0
Zn	0.05-0.1	1.0-1.5	25-35
Pb	0.05-1.0	2.5-4.0	20-25

Slag Cleaning

The slag from the blast furnace usually contains enough copper and precious metals to warrant further treatment. In contrast to slags from primary smelters which contain prills of copper matte that can be coalesced and settled out or recovered by slow cooling, grinding and flotation, the copper in secondary slags is primarily present as an oxide in solution(2). The base composition is in the $FeO-SiO_2-CaO-Al_2O_3$ system. Appreciable quantities of zinc, nickel, tin and lead as oxides are also present. By adding about 2% coke and raising the temperature 100 to 200°C in an arc furnace a controlled reduction of the oxides occurs. Zinc, lead and some tin volatilize, are oxidized above the bath or in the flue system and are collected in a baghouse. Copper, precious metals, nickel, antimony and some of the tin collect as a low grade black copper in the furnace bottom. Judicious use of coke is essential to prevent too much of the iron oxide, a major component of the slag, from being reduced and thus degrading the black copper. Any nickel, antimony and tin recovered in the arc furnace metal adds to the recycle load through the smelter and thus increases the bleed stream. The addition of lime is essential to maintain and improve fluidity as some of the metallic oxides are reduced. Table V shows composition ranges of electric arc furnace feed and products.

Table V. Electric Arc Furnace Products

	Feed Slag, %	Black Copper, %	Final Slag, %	Baghouse Oxide, %
Cu	1.5-2.0	55-60	0.2-0.5	1-2
Ni	1.0-1.5	5-10	0.2-0.4	0.2-0.3
Sb	1-2	0.5-1.5	0.1-0.2	0.1-0.2
Sn	1-2	2-4	0.05-0.1	1.5-3.0
Fe	30-35	5-7	30-35	0.5-0.7
Zn	2-4	1.5-2.0	0.5-1.0	45-55
Pb	1.5-3.0	1.0-1.5	0.5-1.0	15-20

Summary

Operators of secondary copper smelters have developed a skill of the art procedure to solve problems not encountered in primary smelters. These include:

Handling a non-uniform charge which varies widely in both chemical composition and physical form.

Destroying organics to control volatile hydrocarbons and carbon monoxide emissions.

Recovering by-product metals such as zinc, tin, lead and precious metals.

Controlling impurities such as nickel, antimony and chlorine.

Cleaning slags which contain copper in the oxide form.

Recycling is essential in preventing loss of our valuable metals and in energy conservation. With the proper plant design and operation such metal recovery can be accomplished under strict environmental regulation.

References

1. U.S. Patent 4,351,705: "Refining Copper-Bearing Material Contaminated with Nickel, Antimony and/or Tin", September 28, 1982.

2. Rajecvic, H. P. and Opie, W. R., "Developments of Electric Furnace Slag Cleaning at a Secondary Copper Smelter", J. of Metals, March 1982.

ELECTROCHEMICAL SENSORS USING $Li_2SO_4-Ag_2SO_4$ ELECTROLYTES FOR THE DETECTION OF SO_2 AND/OR SO_3

Q. G. Liu and W. L. Worrell

Department of Materials Science and Engineering
University of Pennsylvania
Philadelphia, PA 19104

SUMMARY

New solid-state electrochemical sensors using $Li_2SO_4-Ag_2SO_4$ electrolytes have been developed to measure SO_2 and/or SO_3 in gas mixtures. Silver embedded in part of the electrolyte is used as the reference electrode. Although single-phase electrolyte sensors yield accurate results for one to two days, a two-phase electrolyte cell exhibits excellent long-term chemical stability. For example, stable and accurate potentiometric responses have been obtained over a six month period. The advantages of the two-phase electrolyte (23 mol% Ag_2SO_4) cell include excellent mechanical strength on temperature cycling and an insensitivity to compositional variations in the electrolyte and to temperature fluctuations.

Experimental tests have shown excellent results in SO_2-air mixtures with the SO_2 concentration ranging from 3 to 10,000 ppm (vol). However, the SO_2 concentration range was limited by a lack of standard gas mixtures, and our sensor cells can be used to measure SO_2 concentrations significantly less than or larger than our experimental range. Results also indicate that the presence of significant concentrations of CO_2 and H_2O in the gas mixture has no effect on the sensor response. Thus the two-phase electrolyte (23 mol% Ag_2SO_4) sensor has exceptional selectivity, reliability and long-term chemical and mechanical stability.

Introduction

In combustion processes which utilize sulfur-containing fuels, SO_2 and SO_3 are combustion products in the exit or stack gases. A combustion process using a fuel with 1.5 wt% sulfur with an excess of 20% air will generate 600 ppm SO_2 in the exit gas; with 0.5% sulfur fuel, the SO_2 concentration is about 200 ppm. Sulfur dioxide is also a significant product in some extractive metallurgical processes. For example, the exit gas generated by a copper reverberatory furnace contains about 1% SO_2. However, the U.S. New Source Performance Standards (NSPS) would require that the SO_2 concentration in exit gases be reduced to 650 ppm[13]. Because of numerous air-pollution and corrosion problems, there is much interest in developing a reliable and rapid method to measure SO_2/SO_3 concentrations in a variety of gaseous environments.

The application of liquid sulfate electrolytes[1] to measure SO_2/SO_3 concentrations in gases was suggested about twelve years ago. However, corrosion problems between molten sulfate electrolytes and their containers have restricted their application. A solid state electrochemical sensor is an attractive method to resolve such problems. The use of solid sulfate electrolytes was first reported in 1977[2,3]. However, stable and reliable electromotive forces were not obtained[4] when solid reference electrodes such as Ag/Ag_2SO_4 or $MgO/MgSO_4$ were used.

In this paper, we summarize our own development of solid-sulfate sensors using electrolyte compositions based upon the $Li_2SO_4-Ag_2SO_4$ system. This system has several unusual advantages. First, silver can be mixed with part of the electrolyte to form a solid $Ag-Ag_2SO_4$ reference electrode. Secondly, the $Li_2SO_4-Li_2SO_4$ system exhibits several two-phase regions in which the chemical activities of Ag_2SO_4 and Li_2SO_4 are fixed at constant temperature. Because of these features, a unique, two-phase sulfate electrolyte sensor for SO_2 and/or SO_3 has been developed in our laboratory.

Experimental Aspects

The sulfate electrolyte mixtures have been prepared from powdered (-325 mesh), anhydrous 99.999% pure Li_2SO_4 and Ag_2SO_4, which have been weighed and mixed to obtain the desired compositions. Three different compositions of Li_2SO_4-based electrolytes have been chosen for the experiments. They are: (A) $Li_2SO_4-Ag_2SO_4$ (5 mol%); (B) $Li_2SO_4-Ag_2SO_4$ (55 mol%); (C) $Li_2SO_4-Ag_2SO_4$ (23 mol%). As shown in Fig. 1[5] electrolyte (A) is a single-phase alpha-Li_2SO_4 solid solution at its working temperature of 700°C. Electrolyte (B) with 55 mol% Ag_2SO_4 is also a single-phase $(Ag,Li)_2SO_4$ solid solution at temperatures between 430 and 570°C. Electrolyte (C) with a composition of 23 mol% Ag_2SO_4 is in a two-phase region at temperatures between 510 and 560°C.

Silver particles (-325 mesh) of 99.99% purity were mixed with the sulfate electrolytes in a ratio of 2:1 by weight to make the reference electrode. The electrolyte and the reference mixture were isostatically pressed at 20,000 psi into pellets using a rubber mold. These pellets were crushed to about -50

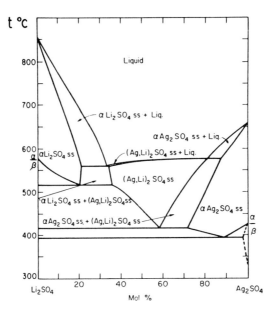

Figure 1. Phase Diagram of the Li_2SO_4-Ag_2SO_4 System[5].

mesh particles, which were then isostatically pressed at 80,000 psi to form the compact sulfate sensor shown at the top of Fig. 2.

At the gas electrode the electroactive species in the gas mixture is sulfur trioxide, not sulfur dioxide[1,2,6,7]. Thus a catalyst must be used to insure that an equilibrium concentration of SO_3 is obtained when the cell is used to determine SO_2 concentrations. The best experimental catalyst to equilibrate reaction (1) is vanadium pentoxide powder, and -50 mesh V_2O_5 was used in cells (B) and (C).

$$SO_2(g) + 1/2\ O_2(g) = SO_3(g) \qquad \log K = 5106/T - 4.845 \qquad (1)$$

However, V_2O_5 melts at 690°C, and platinum wire (0.2 mm diameter) was used as the catalyst in cell (A), which operated at 700°C. As shown at the top of Fig. 2, a platinum mesh was pressed into the electrolyte surface to insure an equilibrium SO_3 concentration at the gas electrode interface. When the unknown gas mixture is equilibrated, the inlet SO_2 pressure (concentration) can be related to the equilibrium pressures of SO_2, SO_3 and O_2 using equation (2).

$$P_{SO_2}(\text{inlet}) = P_{SO_3} + P_{SO_2} = P_{SO_3}(1 + 1/KP_{O_2}^{1/2}) \qquad (2)$$

where P_{SO_2}, P_{SO_3} and P_{O_2} are the equilibrium concentrations at the temperature of the sensor cell and K is the equilibrium constant of reaction (1).

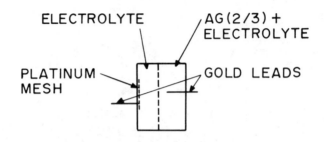
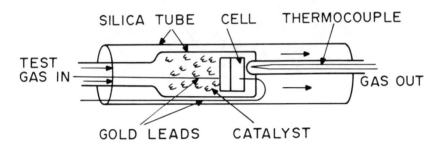

Figure 2. Schematic Diagram of the Sulfate Sensor and the Experimental Apparatus.

If all catalytic materials (V_2O_5 and Pt) are removed from contact with the gas mixture, the sulfate electrolyte sensor will measure the intrinsic SO_3 concentration. In this case, the platinum mesh, shown in Fig. 2, would be replaced with gold mesh.

Experimental gas mixtures of air having SO_2 concentrations of 20, 100, 204, 500, 1000 and 10,000 ppm (vol) were supplied by Airco, Inc. The 20 ppm SO_2 mixture was mixed with air to obtain SO_2 concentrations as low as 3 ppm.

The following three sulfate-electrolyte cells have been investigated in our laboratory.

SO_3, SO_2, O_2 / Li_2SO_4-Ag_2SO_4(5 mol%) / Ag (A)

SO_3, SO_2, O_2 / Li_2SO_4-Ag_2SO_4(55 mol%) / Ag (B)

SO_3, SO_2, O_2 / Li_2SO_4-Ag_2SO_4(23 mol%) / Ag (C)

The Emf (E) for cells (A, B and C) depends on the chemical potentials (μ'_{Li} and μ''_{Li}) of lithium on each side of the electrolyte according to the following equation[8],

$$E = \frac{-1}{F} \int_{\mu''_{Li}}^{\mu'_{Li}} (1-t_e)\, d\mu_{Li} \qquad (3)$$

where F is Faraday's constant, and t_e is the electronic transference number. With negligible electronic conductivity in the electrolyte, equation (3) becomes in terms of lithium activities:

$$E = (RT/F)\, \ln\, (a''_{Li}/a'_{Li}) \qquad (4)$$

The lithium activities are established by the gas electrode and the silver reference electrode according to reaction (5) and (6), respectively

$$2\, Li + SO_3(g) + 1/2\, O_2(g) = (Li_2SO_4) \qquad (5)$$

$$2\, Li + (Ag_2SO_4) = 2Ag(s) + (Li_2SO_4) \qquad (6)$$

where ( ) indicates that the Li_2SO_4 and Ag_2SO_4 are dissolved in the electrolyte. Equation (7) is obtained by using equations (5) and (6) to substitute for a'_{Li} and a''_{Li} in equation (4)

$$E = E^o + RT/2F\, \ln\, (P_{SO_3}\, P_{O_2}^{1/2} / a_{Ag_2SO_4}) \qquad (7)$$

where E^o is calculated using the standard free energies of formation for Ag_2SO_4 and $SO_3(g)$.

Another way to obtain equation (7) is to consider the half-cell reactions which would occur at the gas electrode and the reference electrode if current were to flow through the cell. At the gas electrode, SO_3 is the electroactive species, and the half-cell reaction is

$$SO_3(g) + 1/2\, O_2(g) + 2e^- = SO_4^= \qquad (8)$$

At the silver reference electrode, the half-cell reaction is

$$2\, Ag = 2\, Ag^+ + 2\, e^- \qquad (9)$$

The over-all cell reaction (10) is obtained by combining reaction (8) and (9)

$$SO_3(g) + 1/2\, O_2(g) + 2\, Ag(s) = (Ag_2SO_4) \qquad (10)$$

The Emf of reaction (10) is given by equation (7), where the activity of Ag_2SO_4 in the electrolyte is less than one.

A schematic diagram of the experimental apparatus to measure the concentration of SO_2 is shown at the bottom of Figure 2. The gas flows through a catalyst column to insure an equilibrium concentration of SO_3 according to reaction (1) before reaching the electrolyte surface. Very stable and reproducible Emfs are obtained with the cell design shown in Fig. 2. However, these sensor cells last only several days, because of reactions between the silver in the reference electrode and the SO_2 and and/or SO_3 in the gas mixture. Thus another cell design[9] in which the gas mixture was separated from the reference electrode was used for long-term tests.

The constant temperature zone of the furnace was 6 cm ($+2°C$), and the temperature fluctuation was less than $\pm 1°C$. The catalyst was at the same temperature as the sensor cell.

The Emf was measured by a high impedance 630 Potentiometric Electrometer ($>10^{13}$ ohm) and a 191 Keithley Digital Multimeter.

Results and Discussion

Initial results indicated that the cell Emf was independent of flow-rate when the experimental gas flow rate varied from 5 to 43 cm/min. All subsequent sensor experiments were conducted with a gas flow rate of 20 cm/min.

Although the temperature variation of the Emf for cell (A) has been determined at temperatures between 450 and 750°C, the Emf results at 700°C were more stable and reproducible. It appears that the surface area of the platinum wire catalyst in cell (A) was too low to equilibrate the gas mixtures at lower temperatures. The variation of the cell Emf with SO_2 concentration for cell (A) at 700°C is shown in Fig. 3. The dotted line is calculated using equations (2) and (7), with P_{O_2} = 0.21 atm. in air and the assumption that the activity of Ag_2SO_4 in the 5 mol% Ag_2SO_4 single-phase electrolyte is 0.05. The small discrepancy between the experimental results and the calculated line shown in Fig. 3 is most likely due to the the assumption that the activity of Ag_2SO_4 in the single-phase electrolyte is equal to its concentration (mole fraction).

Results using the Li_2SO_4-55 mol%Ag_2SO_4 single-phase electrolyte (cell B) are shown in Fig. 4. Vanadium pentoxide is used as the catalyst for equilibration of reaction (1) in cells (B) and (C) which are operated at 530°C. For cell (B) the calculated line was obtained using our measured value (0.67) for the activity of Ag_2SO_4 in the 55 mol% Ag_2SO_4 electrolyte[9]. Comparison of the results for the single-phase electrolyte cells (A) and (B) indicate that the results for cell (B) are in much better agreement with the calculated line, presumably because the activity of Ag_2SO_4 is more accurately known at 530°C. However, the Emfs of both cells (A) and (B) decrease after several days, and single-phase electrolytes can not be used in long-term measurements. For example, the Emf decreases about 10 millivolts per day when the SO_2 concentration is 100 ppm in air. Previous results for a cell using a single-phase K_2SO_4 electrolyte and a solid silver reference electrode show considerable drift and instability in the Emf over a 30 day time period[4].

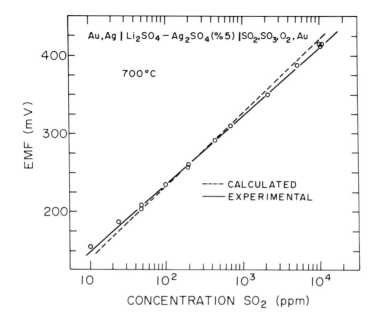

Figure 3. Variation of Emf with the Inlet SO_2 Concentration in Air for Cell (A) at 700°C.

In a single-phase electrolyte cell, the activity of silver sulfate at the reference electrode-electrolyte interface will change with any variation in the concentration of Ag_2SO_4. Any reaction of silver in the reference electrode with the gas mixture to form sulfate or a slight decomposition of Ag_2SO_4 will affect the Ag_2SO_4 activity and the cell Emf. All single-phase sulfate electrolyte cells[4,6,10] have shown potentiometric instabilities in long-term measurements. Furthermore, the solid-state phase transformation which could occur in single-phase sulfate electrolytes can cause microcracks and gas penetration problems during temperature cycling.

The Emf results for the two-phase (23 mol% Ag_2SO_4) electrolyte cell shown in Fig. 4 are in excellent agreement with the line calculated from our measured activity (0.25) for Ag_2SO_4[9] at 530°C. The calculated lines for cells (B) and (C) in Fig. 4 differ because the Ag_2SO_4 activity in the single-phase (55 mol% Ag_2SO_4) electrolyte is about 2.7 times higher than that in the two-phase (23 mol% Ag_2SO_4) electrolyte. The reversibility of cell (C) has been investigated by passing a small current (6 microampere) for 2 to 3 minutes through the sensor cell in both directions. The Emf returned to the original value within two minutes after the current was stopped.

The lack of standard gas mixtures has limited the inlet SO_2 concentration to the 3 to 10,000 ppm (vol) range shown in Fig. 4. Thus the lower and upper SO_2 concentrational limits where the sensor would fail have not been determined. However, qualitative experiments in our laboratory clearly indicate that the lower SO_2 concentration limit of operation is established by the thermodynamic decomposition of Ag_2SO_4, which is 7×10^{-6} ppm SO_2 in air at 530°C. The upper concentration limit of operation is more difficult to estimate, because the lifetime depends upon kinetic factors, which vary with the specific cell design. However, it appears that the sensor lifetime is directly related to the ease or difficulty of the transport of SO_2 and/or SO_3 from the gas mixture to the silver particles embedded in the electrolyte.

There are several reasons why the long-term chemical stability and reliability of the two-phase electrolyte cell (C) should be considerably better than those of the single-phase cells (A) and (B). The activity of Ag_2SO_4 is constant in a two-phase region. Figure 1 indicates that a variation of Ag_2SO_4 concentration from 21 to 35 mol% at 530°C will change the amount of each phase but not their concentration or activity. Thus, the Emf of the two-phase sensor should be very stable and insensitive to changes in the Ag_2SO_4 concentration. Furthermore, the boundaries of the two-phase region between 21 and 35 mol% Ag_2SO_4 are essentially vertical (Fig. 1), which indicates that the concentration of each phase does not change significantly with temperature and that the activity of Ag_2SO_4 is essentially constant with temperature. Finally, a two-phase electrolyte has improved mechanical strength and should not be as susceptible to microcrack formation as a single-phase electrolyte during temperature cycling.

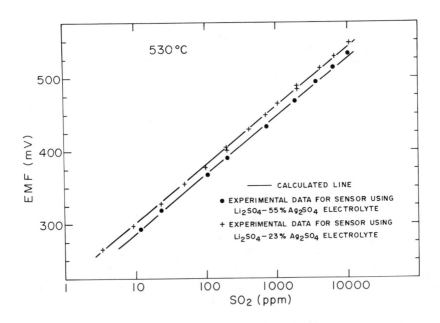

Figure 4. Variation of Emf with the Inlet SO_2 Concentrations in Air for Cells (B) (55 mol% Ag_2SO_4) and (C) (23 mol% Ag_2SO_4) at 530°C.

An improved experimental design using the two-phase electrolyte cell shown at top of Fig. 2 has been used in long-term tests[9] to minimize the chemical interactions between the silver reference electrode and the gas mixture. Results for one of our long-term tests, in which the SO_2 concentration in the gas was changed after every 20-30 days, are shown in Fig. 5. The measured Emfs are within ± 3 mv of the calculated values, which are shown in the parentheses. The measured values are extremely stable even at the end of the 100 day test. Additional tests have shown that stable and accurate Emf values are obtained over a six month period. After such long-term tests, the sensor components show no observable chemical or mechanical degradation which would limit the sensor life. The addition of other gases, such as CO_2 and H_2O, which can attain significant concentrations in some stack gas atmospheres, has no measurable effect in the sensor cell Emf[9,11,12]. Thus the two-phase electrolyte (23 mol% Ag_2SO_4) sensor also has exceptional selectivity and reliability.

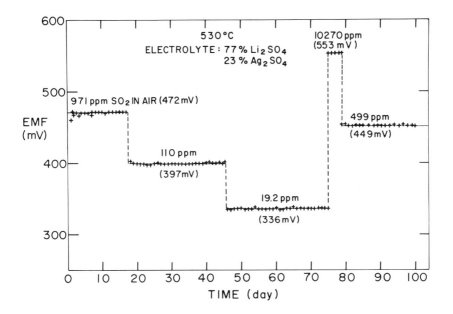

Figure 5. Results of a Long-term Test of the Two Phase Sulfate Electrolyte Sensor

Acknowledgement

Financial support from the NSF Materials Research Laboratory Program (DMR-792367) at the University of Pennsylvania is gratefully acknowledged.

Reference

(1) F. J. Salzano and L. Newman, J. Electrochem. Soc., 119, (1972) pp. 1273-1278.

(2) M. Gauthier, A. Chamberland, J. Electrochem. Soc., 124 (1977) pp. 1579-1583.

(3) M. Gauthier, A. Chamberland, A. Belanger and M. Poirier, J. Electrochem. Soc., 124, (1977) pp. 1584-1587.

(4) M. Gauthier, R. Bellemare and A. Belanger, J. Electrochem. Soc., 128, (1981) pp. 371-378.

(5) Harold A. Oye, Acta Chem. Scand., 18, (1964) pp. 361-376.

(6) K. T. Jacob, D. B. Rao, J. Electrochem. Soc., 126 (1979) pp. 1842-1847.

(7) W. L. Worrell and Q. G. Liu, J. Electroanal. Chem., 168, (1984) pp. 355-362.

(8) C. Wagner, Z. Phys. Chem., B-21, (1933) pp. 25.

(9) Q. C. Liu and W. L. Worrell, to be published.

(10) Nobuhito Imanaka, Gin-ya Adachi, Jiro Shiokawa, Bull. Chem. Soc. Japan, 57, (1984) pp. 687-691.

(11) Q. G. Liu and W. L. Worrell, U. S. Patent Appl. Serial No. 303, 320.

(12) W. L. Worrell and Q. G. Liu, Proc. Int. Meeting Chem. Sensors, Vol. 17, Anal. Chem. Symp. Series, Elsevier, New York, 1983, pp. 332-337.

(13) J. P. Wood, J. J. Spivey, J. of Metals, 35. (1983) pp. 47-54.

PYROMETALLURGICAL REFINING OF

CHLORIDE LEACH/ELECTROWIN COPPER

Ross R. Bhappu

Graduate Student
Colorado School of Mines
Golden, Colorado, 80401
U.S.A.

William G. Davenport

Professor and Head
Department of Metallurgical Eng.
University of Arizona
Tucson, Arizona, 85721
U.S.A.

Summary

A theoretical and experimental investigation into pyrometallurgical refining of chloride leach/electrowin copper is described. The two techniques investigated were (i) vaporization and (ii) gaseous halogenation. Experiments were carried out with a packed bed of electrowon copper crystals and a boat of melted copper crystals. Nitrogen gas and Freon 12/nitrogen mixtures were used as the carrier and reactive gases. Vaporization was found to be effective in removing Bi, Pb and S while halogenation was effective in removing As, Bi, Pb, S and Sb. Silver resisted all attempts at removal. Iron was removed but less effectively than the other impurities.

1. Introduction

The copper which is electrowon from industrial chloride leach solutions is more concentrated in impurities that the copper electrodeposited from industrial sulphate solutions. Improvement of this situation demands either improved chloride solution purification techniques or, as is the subject of this investigation, development of techniques for purifying the electrowon copper product.

This investigation focussed upon pyrometallurgical techniques for refining the electrowon copper. The two basic techniques examined were:

(a) vaporization of impurities (873-1423K);

and

(b) halogenation and vaporization of impurities (873-1423K).

The impurities considered were antimony, arsenic, bismuth, iron, lead, silver, and sulfur. The copper used in the experiments was electrowon from chloride solution in an industrial hydrometallurgical facility. Its analysis as received in the laboratory was:

Impurity	ppm
Ag	300
As	31
Bi	41
Fe	165
Pb	35
S	91
Sb	16

The metal was in the form of electrodeposited crystals 1 to 3 mm in size.

2. Theory

This section discusses the conditions under which removal of the above-mentioned impurities by vaporization and halogenation should be possible. It is based upon published vapor pressures[1], halide free energies[2], and activity coefficients[3].

A. Vaporization Refining

Vaporization purifies copper metal by preferentially evaporating impurity elements from the copper. The general criterion which must be met in order for refining by vaporization to take place[4] is that the ratio of the flux of the impurity element to the flux of Cu leaving the impure copper must be greater than the ratio of the mole fraction of the impure metal to that of Cu, i.e.:

$$\frac{\dot{n}_M}{\dot{n}_{Cu}} > \left[\frac{X_M}{X_{Cu}}\right]_{Copper} \qquad 1$$

(a list of symbols appears at the end of the paper)

In thermodynamic terms, this criterion may be rewritten

$$\left(\frac{n_M}{n_{Cu}}\right)_{gas\ phase} = \frac{P_M}{P_{Cu}} > \left[\frac{X_M}{X_{Cu}}\right]_{Copper} \qquad 2$$

where n_M/n_{Cu} is the ratio of the moles of M and Cu in a gas phase which has come to equilibrium with the copper phase. Furthermore,

$$P_M = \gamma_M^\circ X_M p_M^\circ$$

(γ°, the Raoultian activity coefficient at infinite dilution; p°, the equilibrium partial pressure of pure M)

and:

$$P_{Cu} \simeq X_{Cu} p_{Cu}^\circ$$

so that Inequality 2 may be rewritten:

$$\frac{\gamma_M^\circ p_M^\circ}{p_{Cu}^\circ} > 1 \qquad 3$$

Inequality 3 shows clearly the conditions under which refining by vaporization is thermodynamically possible.

Numerical values of $\gamma_M^\circ p_M^\circ / p_{Cu}^\circ$ for the impurities under investigation are listed in Table I. As can be seen, As and Fe are the only impurities which should be thermodynamically impossible to remove from copper. It can also be noted that vaporization refining generally becomes thermodynamically less effective with increasing temperature.

B. Chlorination Refining

Halogenation purifies copper by preferentially forming halide compounds with the impurities. Furthermore, most of the impurity halides thus-formed are gaseous at the temperatures of this investigation, 873 to 1423K, so that they may potentially be removed from the host copper in gaseous form.

The general refining criterion which must be met in order for refining by halogenation to take place is that the ratio of the flux of the impurity halide to the flux of Cu halide away from the impure copper must be greater than the ratio of the mole fraction of the impurity to that of Cu in the metal phase. For chloride halogenation with an impurity which forms an MCl_3 compound, the criterion is:

$$\frac{\dot{n}_{MCl_3}}{\dot{n}_{(CuCl)_3}} > 3 \left[\frac{X_M}{X_{Cu}}\right]_{Copper} \qquad 4$$

The factor of 3 appears because for every mole of $(CuCl)_3$ vaporized, 3 moles of Cu are removed from the metal phase. The copper chloride trimer $(CuCl)_3$ comprises almost 100% of the gaseous copper chloride species at the temperatures of this investigation.[2]

As in the case of vaporization, (Inequalities 1 and 2 above), the potential for chloride purification may be rewritten in terms of equilibrium partial pressures, in which case Inequality 4 becomes:

$$\frac{P_{MCl_3}}{P_{(CuCl)_3}} > 3 \left[\frac{X_M}{X_{Cu}}\right]_{copper} \qquad 5$$

The conditions necessary for refining may be further elucidated by rewriting this inequality in terms of the equilibrium constants for MCl_3 and $(CuCl)_3$, K_E^M and K_E^{Cu}, giving:

$$\frac{P_{MCl_3}}{P_{(CuCl)_3}} = \frac{K_E^M X_M \gamma_M^\circ (P_{Cl_2})^{3/2}}{K_E^{Cu}(X_{Cu}\gamma_{Cu})^3 (P_{Cl_2})^{3/2}} > 3 \frac{X_M}{X_{Cu}} \qquad 6$$

Also, for copper in which the impurities are at the ppm level, $X_{Cu} \simeq 1$ and $\gamma_{Cu} \simeq 1$ so that Inequality 6 may be written:

$$\frac{K_E^M \gamma_M^\circ}{3 K_E^{Cu}} > 1 \qquad 7$$

A similar development can also be applied to impurities which form MCl_4 gas. In this case the refining criterion is:

$$P_{Cl_2} > \left[\frac{3 K_E^{Cu}}{K_E^M \gamma_M^\circ}\right]^2 \qquad 8$$

which shows that a certain partial pressure of Cl_2 must be maintained in the system for refining to be thermodynamically possible.

Table I Numerical Values of $\gamma_M^\circ P_M^\circ / P_{Cu}^\circ$ for the Impurities of this Investigation.

Values greater than unity indicate that refining is thermodynamically possible. The vapor pressure data are from reference 1 and the activity coefficient data are from reference 3.

Impurity	800K	1000K	1200K	1400K
As	2.5×10^{-1}	8.4×10^{-2}	4.4×10^{-2}	5.4×10^{-4}
Ag	5.7×10^{3}	1.1×10^{3}	3.9×10^{2}	1.7×10^{2}
Bi	4.4×10^{8}	5.5×10^{6}	3.0×10^{5}	3.9×10^{4}
Fe	2.7×10^{-3}	2.8×10^{-2}	1.2×10^{-1}	1.7×10^{-1}
Pb	4.0×10^{8}	4.6×10^{6}	2.4×10^{5}	3.0×10^{4}
Sb	8.4×10^{2}	7.3×10^{1}	1.2×10^{1}	3.1×10^{0}

Numerical values for criteria 7 and 8 are tabulated in Table II. As can be seen:

(a) Fe can theoretically be chlorination refined at all temperatures;
(b) Sb might be removed at 800K;
(c) Pb can be removed at very low partial pressures of chlorine.

The other impurities cannot be removed.

C. Fluorination Refining[5]

Refining inequalities have been similarly developed for fluorination. The principal copper fluoride gas is CuF_2^2 and the criteria for potential refining are:

MF$_2$ Gases

$$\frac{K_E^M \gamma_M^\circ}{K_E^{Cu}} > 1 \qquad\qquad 9$$

MF$_3$ Gases

$$P_{F_2} > \left[\frac{K_E^{Cu}}{K_E^M \gamma_M^\circ}\right]^2 \qquad\qquad 10$$

MF₄ Gases

$$P_{F_2} > \frac{K_E^{Cu}}{K_E^M \gamma_M^\circ}$$
(11)

Table II Numerical Values of $K_E^M \gamma_M^\circ / 3K_E^{Cu}$ for the Impurity Chlorides of this Investigation. Values greater than unity indicate that removal of the impurity is thermodynamically possible. The equilibrium constant data are from reference 2 and the activity coefficient data are from reference 3.

Impurity	800K	1000K	1200K	1400K
AsCl₃	2.2×10^{-5}	8.4×10^{-6}	4.7×10^{-6}	3.5×10^{-6}
SbCl₃	1.5	1.7×10^{-1}	3.9×10^{-2}	1.6×10^{-2}
BiCl₃	1.9×10^{-1}	1.1×10^{-1}	7.8×10^{-2}	8.1×10^{-2}
FeCl₃	6.7	5.1	4.1	3.8
PbCl₄*	4.1×10^{-28}	4.1×10^{-20}	8.0×10^{-15}	3.5×10^{-11}

* PCl_2 values above which refining is thermodynamically possible, Inequality 8.

Numerical values for criteria 10 and 11 are tabulated in Table III. As can be seen, all the impurities listed in the table are potentially removeable from copper at very low partial pressures of fluorine.

Table III Numerical Values of Fluorine Partial Pressures Above Which Impurity Removal from Metallic Copper is Thermodynamically Possible, Criteria 10 and 11. The equilibrium constant and activity coefficient data are from references 2 and 3.

Impurity	800K	1000K	1200K	1400K
AsF_3	6.0×10^{-53}	2.8×10^{-39}	4.9×10^{-30}	1.1×10^{-23}
BiF_3	4.1×10^{-50}	1.1×10^{-38}	4.4×10^{-31}	9.5×10^{-26}
FeF_3	2.8×10^{-69}	4.7×10^{-55}	1.6×10^{-45}	1.0×10^{-38}
SbF_3	1.3×10^{-59}	1.7×10^{-45}	4.2×10^{-36}	1.7×10^{-29}
PbF_4	6.0×10^{-49}	8.6×10^{-38}	2.2×10^{-30}	4.1×10^{-25}

3. Mass Transport Effects

Removal of an impurity from metallic copper is always inhibited by diffusion of the impurity in the metallic copper phase. Cu is always available for vaporization or halogenation at the surface of the copper, but the impurity must diffuse to the surface. Optimum refining conditions are obtained for solid state refining with small copper particles, fortunately the case with the industrial electrowon copper of this investigation. Optimum liquid state refining is obtained with maximum mixing in the liquid metal and a maximum area to volume ratio of the melt.

4. Experimental

The experimental investigation consisted of solid state and liquid state experiments in which the impure copper described in Section 1 was:

(a) heated in nitrogen to examine the effectiveness of vaporization refining;

(b) heated and exposed to Freon 12 (dichlorodifluoromethane)/nitrogen mixtures to examine the effectiveness of halogenation refining.

The experimental temperatures ranged from 873 to 1423 K.

The solid state experiments were carried out in a vertical tube furnace with the copper held in the form of a packed bed. The packed bed reactor consisted of a 28 mm i.d. quartz tube fitted with a quartz frit to support 25 grams of copper. Gas flow for these tests was from the bottom up through the frit, then through the copper packed bed and out of the reactor.

The molten state experiments were carried out in a horizontal tube furnace, again using a 28 mm i.d. quartz tube. In this case, 10 g of copper crystals were melted in a high purity alumina combustion boat. Gas flow for these tests was passed over the surface of the copper metal.

Pure nitrogen (99.999% N_2) was used for vaporization experiments. Nitrogen-Freon 12 mixtures were used for halogenation experiments. Freon 12, which is a trade name for dichlorodifluoromethane was used as the source of both chlorine and fluorine because of its ease of use and its non-toxic form. A flowrate of 500 cm^3/min of total gas was used in all experiments.

Vaporization experiments were run for ½ hour, 1 hour and 2 hours. Halogenation experiments were run for ½ hour and 1 hour. Both types of experiments were run at temperatures of 873 K, 1073 K, 1273 K and 1423 K. Sample mass was measured both before and after each test to determine the magnitude of copper loss during the experiment.

5. Results

The results of the experiments are presented alphabetically in Figures 1 to 4. The behavior of each metal is summarized below.

A. Antimony

Antimony was not removed at all by vaporization. It was removed extensively by halogenation, with 10 volume % Freon being much more effective than 5 % Freon. Solid state and liquid state refining were equally effective at the 10 % Freon level.

B. Arsenic

One third to two thirds of the arsenic in the original electrowon copper was removed by vaporization. The final arsenic concentrations after these experiments were not at an acceptable commercial level. Halogenation removal of arsenic with 10 % Freon/90 % nitrogen* gas was very effective at removing arsenic from both solid and liquid copper.

C. Bismuth

In the solid state experiments, bismuth was removed to below 1 ppm by both vaporization and halogenation. Vaporization and 5 % Freon/95 % nitrogen experiments with liquid copper were not nearly as effective.

D. Iron

Vaporization removal of iron in the solid state was ineffective. Vaporization removal in the liquid state brought the iron down from 165 ppm in the starting material to about 10 ppm in the product. This was unexpected from the thermodynamic predictions in Table I.

Halogenation removal from solid copper was also ineffective. Liquid state halogenation refining was about as effective as liquid state vaporization refining.

E. Lead

Lead was removed to low levels by all forms of refining at 1000°C and above. Halogenation treatments of liquid copper removed lead to negligible levels.

6. Discussion

An important factor in any copper refining process is the amount of copper lost during the refining. This was monitored carefully during the experiments.

In general, copper losses were:

Experiment Type	Copper loss, (% of original mass)
vaporization	< 0.5 %
5 % Freon/95 % N_2	< 1 %
10 % Freon/90 % N_2	< 5 %

It is clear from these data that a minimum concentration of chlorine and/or fluorine, consistent with effective refining, should be used.

Two other impurities were also examined in the experiments, silver and sulfur. Silver was not removed at all in any of the experiments. Sulfur was removed to below 3 ppm in solid state experiments at 1000°C, both by vaporization and halogenation. Liquid state experiments were not quite as effective.

---
* all Freon 12 concentrations are expressed in volume %.

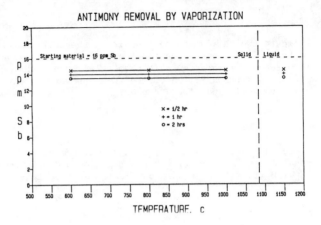

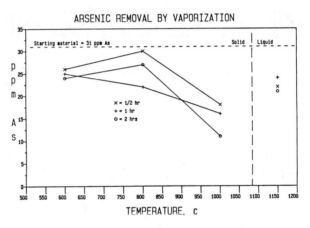

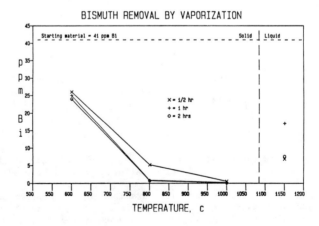

Figure 1. Vaporization Removal of Sb, As and Bi from Copper

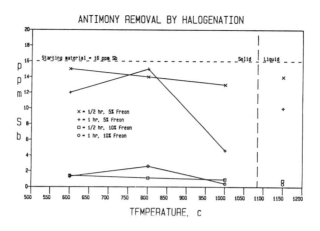

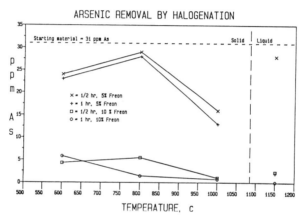

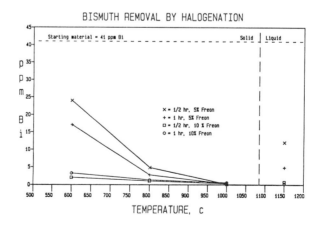

Figure 2. Halogenation Removal of Sb, As and Bi from Copper

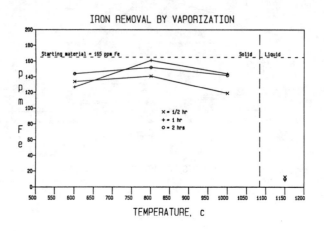

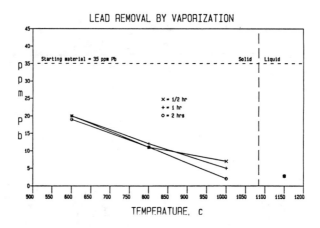

Figure 3. Vaporization Removal of Fe and Pb from Copper

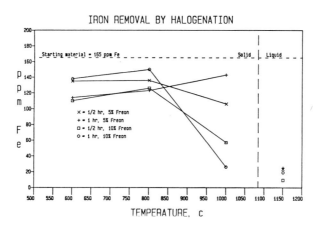

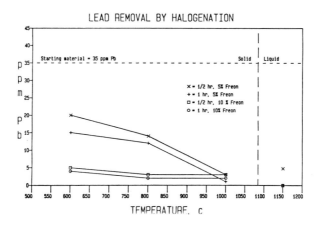

Figure 4. Halogenation Removal of Fe and Pb from Copper

Effectiveness of the Process

A target of the investigation was to ascertain if vaporization and/or halogenation could remove the impurities in chloride leach/electrowon copper down to the levels of the London Metal Exchange specifications for electrolytic copper.

Table IV compares the results of this work with those LME specifications.

Table IV Comparison of the Experimental Best Results of this Investigation with the London Metal Exchange Requirements for Electrolytic Copper.

Impurity	LME Specification[6]*	Vaporization Minimum	Halogenation Minimum
Ag	25 ppm	300 ppm	300 ppm
As	5	11	<1
Bi	2	<1	<1
Fe	10	9	9
Pb	5	2	<1
S	15	1	1
Sb	4	14	<1

* Certain combinations of impurities might require levels lower than these.

As can be seen from the table, only silver completely defies removal, although iron only barely reaches the required level. Vaporization is effective with Bi, Pb and S while As and Sb require halogenation to reach LME levels.

Comparison with Thermodynamic Predictions

Table I predicted that all impurities except As and Fe can be removed from copper by vaporization. The experimental results, however, showed that Ag and Sb were not removed at all. Contrarily, Fe was extensively removed, albeit only in the high temperature liquid phase experiments. Using the magnitude of the $\gamma_M^\circ P_M^\circ / P_{Cu}^\circ$ values as a guide it can be seen that the refining tendency from strongest to weakest is:

Bi
Pb
Ag
Sb
As
Fe

which is essentially the order of the observed experimental effectiveness.

Iron, which is at the bottom of the above list showed quite anomolous behavior. As predicted by thermodynamics, essentially no removal was obtained during the solid state experiments. However, excellent iron removal was obtained from liquid copper. A possible explanation is that the iron removal might have been due to an oxidation phenomenon during melting.

Tables II and III indicated that chlorination or fluorination should be effective in removing:

antimony	(fluorination)
arsenic	"
bismuth	"
iron	(chlorination, fluorination)
lead	" "

at even small partial pressures of halogen gas. Experimentally this has been found to be true except for the case of iron. Iron is resistant to removal from copper, especially from solid copper. This may be because iron is present in the electrowon copper as an oxide or oxy- chloride which resists gasification.

7. Conclusions

1. Vaporization is an effective way of removing bismuth, lead and sulfur from chloride leached - electrowon copper.

2. Halogenation is an effective way of removing antimony, arsenic, bismuth, lead and sulfur from this copper.

3. Silver resists removal by both vaporization and halogenation. Thermodynamics indicates that it should be removeable by vaporization so that enhanced kinetics in a vacuum system might be effective. Thermodynamic data for silver halide gases are not available so that it is not known whether silver removal by halogenation is a thermodynamic impossibility or whether the halogenation is slow kinetically.

4. Iron is removed from copper by halogens as predicted thermodynamically, but with somewhat limited effectiveness, perhaps due to iron oxide formation during storage of the electrowon crystals. Iron is quite effectively removed from melted electrowon crystals by exposure to nitrogen as well as by exposure to halogens. Oxidation during melting may be partially responsible.

8. Future Work

Experiments are continuing with chloro- and fluoro- carbons to isolate the effects of chlorine and fluorine.

List of Symbols

γ_M°	activity coefficient of impurity M at infinite dilution
K_E^{Cu}, K_E^{M}	equilibrium constants for Cu halide and impurity halide formation
M	impurity
n	number of moles in gas phase
$\dot{n}_M$, $\dot{n}_{Cu}$, $\dot{n}_{MCl_3}$	flux of M, Cu, MCl_3 into the gas phase
P_M, P_{Cu}	equilibrium partial pressures of M and Cu over the impure copper
P_M°, P_{Cu}°	equilibrium partial pressures of pure M and Cu
P_{MCl_3}, $P_{(CuCl)_3}$ etc.	equilibrium partial pressures of halides over the impure copper
P_{Cl_2}	partial pressure of chlorine in the system
X_M, X_{Cu}	mole fractions of M and Cu in the impure copper

Acknowledgments

The authors would like to thank the Arizona Mining and Mineral Resources Research Institute, Dr. O. E. Childs, director, for its financial support of this work. They would also like to thank J. M. Toguri, A. W. Fletcher and W. J. Mitchell for their many helpful ideas and technical support.

References

1. L.B. Pankratz, Thermodynamic Properties of Elements and Oxides, Bureau of Mines Bulletin 672, United States Department of the Interior, Washington, D.C., 1982.

2. L.B. Pankratz, Thermodynamic Properties of Halides, Bureau of Mines Bulletin 674, United States Department of the Interior, Washington, D.C., 1984.

3. A. Yazawa, "Thermodynamic Considerations in the Removal of Impurities in Copper Smelting", Bulletin of the Research Institute of Mineral Dressing and Metallurgy, Tohoku University, 23(1) (1967), pp. 67-75.

4. R. Harris and W.G. Davenport, "Vacuum Distillation of Liquid Metals: Part 1. Theory and Experimental Study", Metallurgical Transactions B, 13 (1982), pp. 581-588.

5. H.U. Schutt and J.M. Toguri, "Removal of Impurities in Copper by a Halide Carrier Technique", Metallurgical Transactions, 236 (1965), pp. 230-231.

6. Private communication, S.K. Young, Magma Copper Company, San Manuel, Arizona, January 1984.

Iron and Steel

CARBOTHERMIC REDUCTION OF MINERALS IN A PLASMA ENVIRONMENT

J. J. Moore, M. M. Murawa, K. J. Reid

Mineral Resources Research Center
University of Minnesota
Minneapolis, Minnesota 55455

Summary

The application of plasma technology to process metallurgy is briefly discussed. Experiments involving in-flight reduction of minerals (taconite, chromite, in particular) with a solid-based carbon reductant are described. The resultant metallic and slag product analysis is discussed with respect to thermodynamic and kinetic considerations.

Introduction

The simplest definition of plasma is that it is an assembly of charged particles which exhibit collective action. In process metallurgy, the main attention is confined to partially ionized gaseous plasmas in the temperature range 2,000 to 20,000 K. Since a significant level of ionization occurs in the plasma state, plasmas conduct electricity. Therefore, one method of generating a plasma is to establish an electrical discharge through a gaseous medium and this is the basic method used in a variety of plasma torch configurations.

Although low pressure plasmas, inductively coupled plasmas and microwave plasmas have areas of special application in metallurgical research, their potential for use in large scale commercial process metallurgy is small compared with that of atmospheric pressure plasma arc systems.

For the production of arc plasmas at least two electrodes are required, a cathode (negative potential) and an anode (positive potential). The design and development of plasma torch systems is a technology to itself, a technology upon which the successful application of plasma in process metallurgy is critically dependent. For plasma to be economically acceptable in process metallurgy, plasma devices must be able to run for weeks and have low electrode replacement costs.

There are four basic geometries in current use for electrodes, as illustrated schematically in Figure 1. The tubular and ring configurations generally employ water cooled electrodes with rapid rotation of the arc attachment in order to minimize local overheating and electrode wear. The rod and button geometries are generally made from thoriated tungsten embedded in a water-cooled copper holder. Temperatures at the arc attachment point are much higher for tungsten electrodes than for copper electrodes.

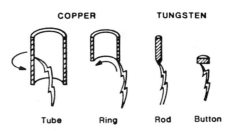

Figure 1. Basic Electrode Geometries

The copper and tungsten electrodes are termed nonconsumable since the rate of wear is extremely small when compared with the consumption rates of graphite electrodes. Consumable graphite electrodes with all four basic geometries have been used in experimental work but the only configuration with substantial commercial application is the rod electrode as used in electric arc furnace practice. There are two main modes of operation, (i) nontransferred mode in which the plasma arc is ejected from the torch and (ii) the transferred mode in which the plasma arc is transferred to an external electrode which may be a molten bath, a solid workpiece or an independent counter electrode. Figure 2 illustrates the two cases.

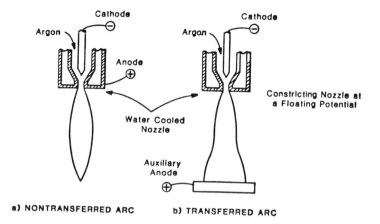

Figure 2. Illustration of Transferred and Nontransferred Modes of Operation.

There have been several reviews on the application of plasma technology to process metallurgy operations[1-6] and many examples of plasma torch designs are illustrated in the earlier reviews.[1,2,3] There are three main methods employed for stabilizing the plasma: vortex stabilization using a high velocity swirling gas stream; magnetic stabilizing, using a magnetic field to produce rapid rotation of the arc; wall stabilization, using an elongated channel to constrict the arc.

In the transferred mode the external electrode (anode) can take the form of a conducting hearth as in most melting/smelting applications or as a ring-type counter electrode to produce an expanded conical plasma.

Three fundamental methods of coupling the energy contained within the plasma to materials have emerged with respect to process metallurgy applications, the basic geometries of which are schematically shown in Figure 3.

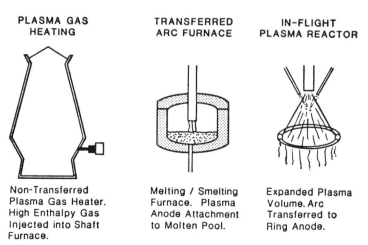

Figure 3. Schematic of Three Fundamental Types of Plasma Application.

419

i) Plasma gas heating, in which a plasma torch is used to heat a process gas to temperatures above those attainable by conventional fossil fuel combustion. The high temperature gas is subsequently used to convey the required reaction enthalpy to the reactants. Examples of this approach are the SKF PLASMARED[7] and PLASMASMELT[8] processes and the CRM PIROGAS[9] process.

ii) Molten pool transferred arc furnaces, in which metal/slag reactions take place in the molten pool with the major energy input provided via the anode attachment zone on the molten surface. Examples of this approach include the molybdenite studies at Union Carbide[10] and McGill[11] and the chromite reduction work at MINTEK.[12]

iii) In-flight plasma reactors, in which solids are introduced directly into the plasma medium and undergo metallurgical reactions while in flight. Examples of this approach are the TAFA/IONARC zircon dissociation process[13] and the PLASMA HOLDINGS SSP technology.[14]

The drive for exploring the application of these plasma systems in the reduction of minerals is influenced by several factors:

i) the improved reaction kinetics available at the elevated temperatures achieved within the plasma medium,

ii) the highly reactive species created within the plasma medium at these high temperatures,

iii) the improved slag/metal chemistry resulting from the wider range of flux compositions allowable in plasma systems relative to conventional submerged arc practice.

iv) the use of a cheap, available reductant such as medium or low grade coals rather than expensive coke.

v) the need for radically new approaches in extractive metallurgy in order to minimize capital costs, maximize energy efficiency, and have minimal adverse environmental impact.

The key factors in determining the application of plasma processes in a given location are the relative cost of electricity compared with the cost of fossil fuels and the potential for improved energy recovery from the effluents of plasma systems relative to conventional pyrometallurgical processes. All three modes of plasma energy transfer apply to a greater or lesser extent in any plasma reactor configuration. However, the in-flight system has the potential to fully exploit the excited and ionic species that exist within the plasma volume itself.

Experimental Program and Results

The plasma metallurgy research facilities at the Mineral Resources Research Center (MRRC) include both in-flight and molten bath plasma smelting reactors.[4] As indicated above, even with the molten bath plasma smelting configuration there may be some in-flight reduction of the mineral as it passes, with the carbon-based reductant, through the plasma arc into the molten bath.

The experimental program was designed to evaluate the extent of in-flight reduction achievable within a plasma medium and to establish the operating reduction mechanisms in order that such in-flight reduction may be optimized. In all experiments conducted a solid-based carbon reductant, e.g., activated carbon or coal, was intimately mixed with the mineral and fed into the plasma environment. Reduction experiments to date have investigated taconite and chromite intimately mixed with either activated carbon, lignite char, graphite or various coals. In-flight reduction experiments have also been conducted with Cu-Ni-Fe sulfide minerals, e.g., Duluth gabbro copper-nickel concentrates[15] and chalcopyrite,[16] mixed with carbon in an oxygen smelting environment and also with a lime-carbon reduction smelting environment. The current paper will emphasize the work conducted on taconite and chromite with some small reference made to the chalcopyrite-lime-carbon work.

Major problems in effecting reduction reactions between a mineral and carbon within a plasma environment include:

    i) the introduction of the mineral and carbon reductant into the plasma,

    ii) achieving sufficient interaction between the reactants before they exit the plasma due to the momentum generated by the plasma arc,

    iii) achieving uniform reduction of minerals within a plasma medium with its inherently high viscosity and temperature gradients.

These problems were addressed in the current experimental program by producing an expanded, diffuse plasma medium as schematically presented in Figure 3c. A full description of the plasma reactor used in these experiments has been given elsewhere[17] and, therefore, only a brief description will be given in this paper.

The plasma generated from a conventional plasma torch is transferred to an annular anode and made to orbit by electromagnetic means. The plasma arc is, therefore, expanded into a conical form and orbited at a rate, in this set of experiments, of between 2,000-10,000 rpm. The reactor configuration is schematically shown in Figure 4 and the operating parameters given in

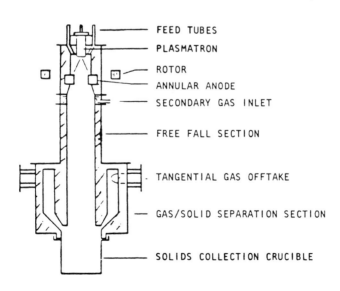

Figure 4. Plasma Reactor Configuration.

Table I. The compositions of the minerals, fluxes and reductants used in these experiments are given in Table II and their corresponding size distributions in Table III. The efficiency of plasma-particle interactions has been found in previous experiments to depend on particle size and feed rate.[18] Thus, these two factors were controlled within limits known to provide acceptable plasma loading of the minerals.

Table I. Experimental and Operating Parameters

Arc length	5"
Anode diameter	5"
Arc voltage (DC)	140-220 V
Current	400-290 A
Power	56-70 kW
RPM	2000-10,000
Plasma gas flow: Ar	7-20 SCFH
Feed rate	72-266 g/min
Carrier gas flow: AR	0-40 SCFH
O_2	36 SCFH
Free fall zone length	36"
Secondary gas flow: Air	0-498 SCFH
O_2	0-300 SCFH

Table II. Composition of Minerals, Reductants and Fluxes Used in Plasma Reduction Experiments

MINERALS:

Taconite Concentrate (wt %)		Chromite Concentrate (wt %)	
Total Fe, Fe_T	66.81	SiO_2	1.05
Fe^{3+}	44.49	Al_2O_3	14.50
Fe^{2+}	22.16	FeO	24.10
FeO	28.5	MgO	11.50
Fe_2O_3	63.6	Cr_2O_3	46.30
Fe_3O_4	92.1	CaO	0.17
Metallic Fe	0.16	Cr:Fe	1.70
SiO_2	6.23		
S	0.004		
C	0.37		

REDUCTANTS (wt %):

	Activated Carbon	Coal
Carbon	63.00	49.46
Sulfur	-	4.50
Al_2O_3	4.80	3.51
SiO_2	8.40	6.50
Fe	2.40	4.00
Ca	4.70	0.15
Mg	1.30	0.11
Na + K	0.13	0.40

FLUXES/SLAGGING CONSTITUENTS (wt %):

	CaO	SiO_2
Heavy Metals	0.11	0.8
Alkali Metals	1.00	
Balance	CaO	SiO_2

Table III. Screen Analysis of Mineral Concentrates and Reductants

Mesh Size	Microns	Taconite and Activated Carbon, Wt %	Chromite and Coal, Wt %
48	300	-	-
-48 +65	-300 +210	-	25.44
-65 +100	-210 +149	-	29.46
-100 +150	-149 +105	0.77	18.54
-150 +200	-105 +74	2.72	9.79
-200 +270	-74 +53	7.88	5.59
-270 +325	-53 +44	11.34	2.74
-325	-44	77.29	8.44
Total		100.00	100.00

The minerals were thoroughly mixed with the carbon-based reductant, i.e., activated carbon for taconite and chalcopyrite or coal for chromite, with or without certain levels of SiO_2 and/or CaO and introduced into the plasma by means of a screw feeder and four feed tubes equispaced around the plasma torch. The plasma-particle interaction time was calculated to be less than 100 milliseconds. The products of reaction, on leaving the plasma medium, were rapidly quenched in the free fall section in order to "freeze-in" the plasma effects as much as possible and, thereby, more clearly determine the extent of plasma-particle interactions. These products were subsequently collected in the collection crucible and subjected to bulk chemical analysis to determine the degree of metallization, i.e., $\frac{wt\% \text{ elemental metal}}{wt\% \text{ total metal}}$ x 100%, and mineralogical and metallographic examination using optical microscopy, scanning electron microscopy (SEM, STEM), and electron probe microanalysis (EPMA) in order to facilitate interpretation of possible reduction mechanisms.

In each of these experiments, the reduced metallic phases were normally distributed as fine particles embedded in partially reduced mineral or a slag phase (Figures 5 and 6). The composition of the slag phase was controlled by the slagging constituents in the mineral, e.g., SiO_2, and the fluxing additions used. The slag constituents in the taconite product were largely associated with the $FeO-SiO_2$ system (Figure 5 and Table IV) and those in the chromite product were largely associated with the low melting point $MgO-SiO_2$ system and partially reduced chromite (Figure 6 and Table V).

However, the morphology of the reduced metallics distributed in the slag or partially reduced mineral tended to be in one of two forms, depending on the feed particle size and the efficiency achieved in plasma-particle interactions. For chromite particles greater than 100 μm fed at a rate of 72 g min^{-1} into the expanded, conical plasma, the reduced metallic particles tended to be distributed along certain channels or planes in the mineral (Figure 6b). A similar form of distribution of reduced Cu particles was also found[16] on feeding predominantly 50 μm particles of a chalcopyrite-lime-carbon mixture into the expanded plasma at 160 g min^{-1} (Figure 6c). On the other hand, feeding taconite particles predominantly less than 44 μm at 72 g min^{-1} resulted in a random distribution of the reduced metallics situated predominantly in FeO in the $FeO-Fe_2SiO_4$ slag (Figure 5).

It was also found that thermodynamic principles applicable under conventional, lower temperature, slag-chemistry conditions and to the formation of metal carbides could also be applied under these in-flight plasma

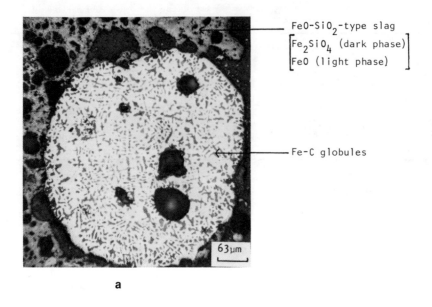

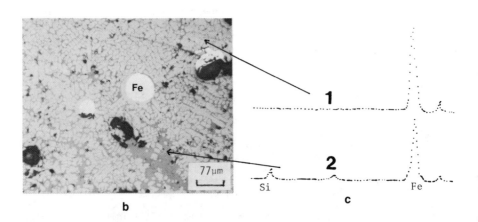

Figure 5. In-Flight Reduced Taconite Products. High Carbon Iron Surrounded by $FeO-Fe_2SiO_4$-Type Slag. (a) Large Fe-C globule in slag; (b) Small Fe-C spherule in slag; 1 = FeO light phase; 2 = Fe_2SiO_4 darker phase.

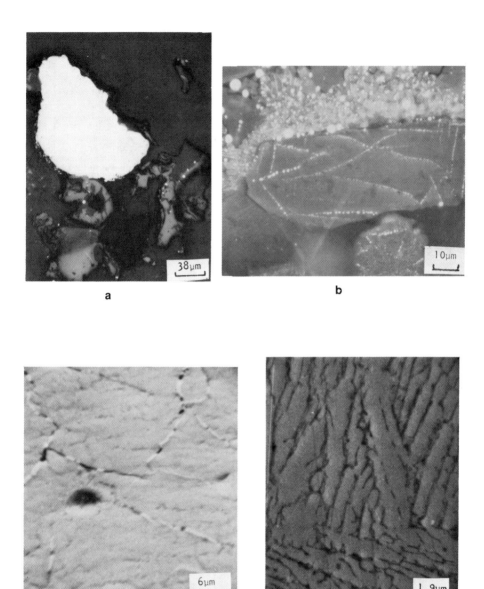

Figure 6. In-Flight Plasma Products. (a) Cr-Fe-C metallic globule embedded in Mg_2SiO_4 + partially reduced chromite; (b) metallic (Cr,Fe,C) spherules nucleated on certain channels or planes in chromite mineral; (c) metallic Cu (+2% Fe) spherules nucleated on certain channels or planes in chalcopyrite from chalcopyrite-lime-carbon experiments; (d) primary carbides, $(CrFe)_7C_3$, in metallic product resulting from in-flight reduction of chromite.

Table IV. Charge Composition and Metallic and Slag Products for In-Flight Plasma Reduction of Taconite

Feed Rate g/min	Taconite (g)	Activated Carbon (g)	Composition of Metallics (wt %)		Typical Composition of Slag (wt %)			Metallization (%)
			Fe	C		FeO	SiO$_2$	
72	500	105	98.2	1.8	Wustite	95-98.5	1.5-1.3	75
					Fayalite	60.5-74	38-21	

Table V. Charge Compositions and Metallic Products for Chromite Reduction within the Plasma Reactor

Feed Rate (g/min)	Experimental Charge Mixtures (g)				Typical Composition of Metallics (wt %)				Metallization (%)	
	Chromite	Coal	SiO$_2$	CaO	Cr	Fe	Si	C	Cr	Fe
200	500	418.7	1.05	–	48.9	39.8	5.5	5.8	20.1	25.7
200	500	418.7	137.5	15.5	49.0	39.4	5.5	6.2	40.1	52.1

Typical Composition of Slag Products (wt %)

	MgO-rich Slag	Partly Reduced Chromite
MgO	58	18
SiO$_2$	19	1
Cr$_2$O$_3$	4	42
Al$_2$O$_3$	2	21
FeO	8	15

conditions.[19,20] The importance of slag chemistry was demonstrated with the chromite reduction experiments. The rate controlling step in conventional submerged arc carbothermic reduction of chromite has been determined to be the dissolution of acidic Cr_2O_3 in the slag as basic CrO according to the reaction,[21]

$$Cr_2O_{3(spinel)} + C = 2CrO_{(slag)} + CO$$

Hence, this rate controlling step is favored by use of an acidic slag. Table V shows the improved metallization, i.e., $\frac{\text{metallic Cr}}{\text{total Cr}} \times 100\%$, achieved with a SiO_2 addition to the chromite concentrate.

Also, the presence of 6.2% SiO_2 in the taconite concentrate resulted in the formation of fayalite, $2FeO \cdot SiO_2$. The FeO "tied up" as fayalite is less readily reduced than the "free" FeO. Stoichiometrically, 11.5% of the FeO should be present as fayalite. If only the "free FeO" is available for reduction this would result in 83% metallization which corresponds well with the 75% metallization actually achieved. Table IV gives the taconite reduction conditions used, analysis of products and degree of metallization achieved.

A recent examination of the thermodynamics of carbothermic reduction of oxides within a plasma environment[19] has indicated that, for oxides more stable than FeO, the metal carbide (reaction 1) is more likely to be produced rather than the elemental metal (reaction 2), e.g.,

$$3MO + 4C = M_3C + 3CO \tag{1}$$

$$MO + C = M + CO \tag{2}$$

The tendency to produce carbides was also found to increase with increase in temperature for these more stable oxides and, therefore, with in-flight plasma reduction.

These theoretical considerations have been confirmed in the current in-flight, plasma reduction experiments with taconite and chromite. The reduced metallic products from the taconite experiments varied between a hypereutectoid steel, e.g., 1 to 1.8% C and a hypoeutectic cast iron, e.g., 3 to 3.8% C (Figure 5a) while chromite reduction produced high carbon white irons with a large volume of primary carbides (Figure 6d). Typical product analyses are given in Table VI.

Discussion

Although the residence time of the reactants within the plasma medium is calculated to be less than 100 ms it is apparent that greater than 70% reduction of taconite to iron and 50% reduction chromite to chromium is readily achievable in-flight with this small laboratory plasma unit. In another set of similar experiments, greater than 60% reduction of chalcopyrite to copper has been achieved in-flight using a chalcopyrite-lime-carbon feed mix into the plasma reactor.[16] These results indicate the considerable kinetic advantage of employing the plasma environment as the reaction medium rather than using plasma simply as a heat source. This research approach needs to be expanded in order to gain a more complete understanding of the fundamental conditions which control the reduction of minerals in-flight within the plasma.

Table VI. SAM Point Analysis of Metallic Particles

Specimen	Phase	Fe (Wt %)	C (Wt %)	Cr (Wt %)
Typical metallics produced from in-flight, plasma reduced taconite				
i) Hypereutectoid Steel*	Metallic	98.2	1.8	
ii) Hypoeutectic Iron**	Metallic	96.2	3.8	
Typical metallics produced from in-flight, plasma-reduced chromite		55	12	33

*Figure 5a
**Figure 6d

A second important conclusion from these results is that, although the plasma medium contains highly reactive species, e.g., molecules, atoms and ions, which are probably not present under lower temperature, conventional smelting conditions, fundamental thermodynamic principles may be applied in order to improve the degree of metallization. In this respect it is feasible that, due to the intimate mixing of the reactants, slag-metal-mineral microcells are established within the plasma which, because of their high surface area and reactive states effect a rapid reduction mechanism. It must be pointed out here, however, that the crucial ingredient is to achieve efficient introduction of the reactants into the plasma and subsequently effect efficient plasma-mineral interactions. It is thought that the expanded conical plasma generated with this particular system aids in these latter requirements. With respect to these points it is possible to speculate possible reduction mechanisms that prevail with this mode of in-flight plasma-mineral reaction. Examination of the plasma-treated products reveals two morphologies in which the reduced metallics are present.

The first is that in which fine metallic particles are initially nucleated along specific channels or planes within the unreduced or partially reduced mineral (see Figure 6b). This morphology tended to predominate when larger particles were introduced into the plasma or when smaller particles were fed into the plasma at higher feed rates, indicating inefficient plasma-particle interaction.

The second form of reduced metallics was as larger islands surrounded by low melting point slag phases and/or the partially reduced mineral (see Figures 5, 6a). This morphology tended to predominate when smaller particles were introduced into the plasma medium. In both morphologies evidence of coalescence of these reduced metallic species was present.

A possible speculation as to the reduction mechanism based on these observations is that the large mineral particles, i.e., greater than 100 μm, develop microcracks or channels on being subjected to the turbulent plasma medium and very rapid heating rates. Outgassing of any contained gaseous species which may be produced at these elevated temperatures then proceeds along these channels which become microchannels of plasma. Such channels are likely to be highly reactive areas along which reduction of the mineral is initiated, producing fine spherules of liquid metallics. Given sufficient time within the plasma and continued efficient plasma-particle interaction, these liquid metallic spherules coalesce in order to reduce their surface energy, eventually forming large globules of reduced metallics. The presence of carbon in various activated states in contact with these liquid metallic (i.e., Fe and Cr) spherules and globules provides rapid saturation of the metallics with carbon. As discussed earlier, the driving force for carbide formation increases with increase in stability of the metal oxide and increasing reaction temperature.

The smaller mineral particles, i.e., less than 50 μm, may tend to melt completely on introduction into the plasma medium, thereby partly or totally inhibiting the formation of microchannels of plasma within the particles as discussed above. In this condition very fine slag-metal microcells are established within the plasma. Such microcells provide favorable reduction kinetics and as such effect the necessary reduction reactions producing metallic products embedded in a slag phase. The slag phase will be controlled by the composition of the ore and by any flux additions, e.g., SiO_2 or CaO, made to the feed material. Since the metallic phase will usually have a higher melting point than the slag phases, the metallic phase will solidify first and will be surrounded by a slag phase. It is also likely that reduction of the mineral, e.g., Fe_3O_4 in taconite, will proceed through one or several intermediate stages during reduction, e.g., $Fe_3O_4 \rightarrow FeO \rightarrow Fe$. Therefore, it may be expected to find the metallic phase situated predominantly in a partially reduced mineral or slag phase, e.g., Fe spherules in FeO and Cr in partially reduced chromite or low melting point magnesium silicates.

Again, due to the high diffusion rate of carbon (which may be present in certain nonequilibrium reactive forms within the plasma) in liquid Fe, a high saturation of Fe and FeCr metallics with carbon is to be expected.

A schematic representation of these two possible reduction mechanisms is given in Figures 7 and 8. The limiting particle size which determines which of these reduction mechanisms predominate is likely to depend on the thermal and physical properties of the mineral, and it is also feasible that both of these mechanisms may operate within a mineral system.

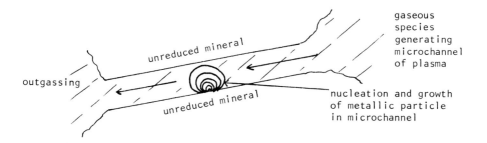

Figure 7. Nucleation and growth of metallic spherule in microchannel of plasma.

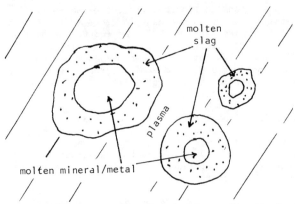

Figure 8. Microcell of molten mineral and molten slag developed in plasma environment.

Conclusions

1. Substantial reduction of taconite (75%) and chromite (40% Cr, 52% Fe) has been achieved in-flight in a plasma medium using a solid carbon-based reductant with a residence time of the order of 100 ms.

2. Slag chemistry thermodynamic principles used in conventional pyrometallurgical processes have been applied to determine the degree of reduction achieved within the plasma environment, i.e., addition of SiO_2 enhances chromite reduction and restricts taconite reduction.

3. Two reduction mechanisms have been postulated based on the production of microchannels of plasma in larger mineral particles, e.g., greater than 100 μm for chromite, and the formation, slag-metal microcells in the finer mineral particles, e.g., less than 50 μm for taconite.

References

1. I.G. Sayce, "Plasma Processes in Extractive Metallurgy," Adv. Ext. Met. Symp. Inst. Min. Met., 1972, p. 241.

2. S.M.L. Hamblyn, "Plasma Technology and Its Application to Extractive Metallurgy," Minerals Sci. Engng., Vol. 9, No. 3, July 1977, p. 151.

3. N.N. Rykalin, "Thermal Plasma in Extractive Metallurgy," Pure & App. Chem., Vol. 52, 1980, pp. 1801-1815.

4. K.J. Reid, "Plasma Metallurgy in the 80's," Plenary Paper, MINTEK Conference, Johannesburg, March 1984.

5. C.W. Chang, J. Szekely, "Plasma Applications in Metal Processing, Journal of Metals, February 1982, p. 57.

6. K.U. Maske, J.J. Moore, "The Application of Plasmas to High Temperature Reduction Metallurgy," High Temperature Technology, $\underline{1}$, (1), 1982, pp. 51-63.

7. S. Santen, "Plasma Technology Gives New Lease of Life to Swedish DR Plant," Iron & Steel International, 53, No. 1, 1980.

8. S. Santen, "Plasma Smelting," Proceedings 6th International Symposium on Plasma Chemistry, ISPC-6, Montreal, 1983.

9. N. Ponghis, R. Vidal, A. Poos, "PIROGAS - A New Process Allowing Diversification of Energy for Blast Furnace," High Temperature Technology, Vol. 1, No. 5, August 1983, p. 275-281.

10. J.H. Harrington, "Reduction and Dissociation of Molybdenum Compounds in a Transferred Plasma Arc," 4th Int. Symp. on Plasma Chemistry, Zurich, August 1979.

11. W.H. Gauvin, G.R. Kubanek, G.A. Irons, "The Plasma Production of Ferromolybdenum - Process Development and Economics," J. Metals, Jan 1981, pp. 42-46.

12. T.R. Carr, N.A. Barcza, K.U. Maske, J.F. Mooney, "The Design and Operation of Transferred-Arc Plasma Systems for Pyrometallurgical Applications," Proc. 6th International Symposium on Plasma Chemistry ISPC-6, Montreal, July 1983.

13. M.L. Thorpe, P.Y. Wilks, "Electric-Arc Furnace Turns Zircon Sand to Zirconia," Chemical Engineering, Nov 15, 1971, pp. 117-119.

14. J.K. Tylko to Plasma Holdings N.V., U.S. Patent 4,361,441.

15. K.J. Reid, M.M. Murawa, N.M. Girgis, "Plasma Smelting of Copper-Nickel Concentrate for Minimal Environmental Impact," IUPAC Conference ISPC-6, Montreal, 1983.

16. A.R. Udupa, J.J. Moore and K.A. Smith, "Initial Studies of Chalcopyrite-Lime-Carbon Reactions within a Plasma Environment," TMS paper selection number A84-41, 1984. Paper presented at Annual AIME Meeting, Los Angeles, 1984.

17. J.J. Moore, K.J. Reid, J.K. Tylko, "Reduction of Lean Chromite Ore Using a New Type of Plasma Reactor," paper presented at AIME Conference, Chicago, 1981, and published in Extractive Metallurgy of Refractory Metals, Ed. H.Y. Sohn, T. Smith, pub. by AIME, 1981.

18. K.J. Reid, J.K. Tylko, J.J. Moore, "The Application of the Sustained Shockwave Plasma (SSP) in Process Metallurgy," Second World Congress of Chemical Engineering, Montreal, Canada, October 1981.

19. K. Upadhya, J.J. Moore, K.J. Reid, "Application of Thermodynamic and Kinetic Principles in the Reduction of Metal Oxides by Carbon in a Plasma Environment," submitted for publication in Metallurgical Transactions B.

20. J.J. Moore, K.J. Reid, M.J. Murawa, K.U. Maske, A.R. Udupa, "Carbothermic Reduction of Minerals in a Plasma Medium: The Effect of Slagging Constituents," International Symposium on Metallurgical Slags and Fluxes, TMS-AIME, 1984.

21. N.A. Barcza, T.R. Curr, W.D. Winship, C.P. Heanley, "The Production of Ferrochromium in a Transferred-Arc Plasma Furnace," 39th Elec Furnace Conf, Houston, Dec 1981.

DIRECT METHOD TO PREPARE LOW CARBON FERROCHROME

S. E. Khalafalla[1] and J. E. Pahlman[2]

Twin Cities Research Center
Bureau of Mines
Minneapolis, MN

A method for preparing ferrochromium alloys containing less than 2 pct carbon has been devised in a single vacuum furnace reactor. This method can conserve chromium and should reduce the capital and energy requirements relative to chromite electrothermal reduction.

Reduced products containing less than 2 and as low as 0.01 wt pct carbon were obtained by reacting pellets of chromite and carbonaceous reductant mixtures at pressures of 0.1 to 1 torr and temperatures of 1,230° to 1,320° C. The extent of reduction increased with increased temperature and decreased pressure; however, operating conditions were limited due to the onset of significant chromium vaporization at higher temperatures and lower pressures. Foundry coke, anthracite, and carbon black were found to be superior to graphite as reductants. The reduction of chromite to metallics was found to proceed via the carbide intermediates Fe_3C or Cr_3C_2. Lime accelerates the rate of solid state reduction of chromite and can shorten the furnacing time at 1,300° C under vacuum from 13 hours to less than three hours. Additional lime and silica are added to the hot reduced chromite after releasing the vacuum and raising the furnace temperature to 1,700° C. Soaking the charge for about 20 minutes gave a button of low carbon ferrochromium alloy beneath a well-defined slag layer.

[1] Research supervisor.

[2] Group supervisor.

Introduction

Stainless steel is important to the economic growth of the United States since it is used for many industrial applications where severe process environments are encountered. Because chromium is required in the manufacture of stainless steel, and the United States is dependent upon foreign resources for ore, chromium is considered a critical material. Domestic chromite resources are small in size, low in grade, and high in iron. Consequently, they are neither competitive with foreign ore nor able to satisfy U.S. chromium demand. To help reduce chromite imports, conservation of chromium in metallurgical processing of the chromite ore is required. As part of its effort to develop technology, which could both minimize the requirements for mineral commodities through conservation and reduce the capital and energy requirements of mineral processing, the Bureau of Mines has investigated a more efficient chromite reduction method. In this method, the reduction of chromite ore with carbon is carried out at reduced pressures. The reduced product is then melted to produce a ferrochromium alloy containing less than 2 percent carbon.

The prevalent method of processing chromite for use in the stainless steel industry involves submerged arc smelting of chromite with carbonaceous reductants to produce high-carbon ferrochrome (HCFeCr) with 4 to 10 percent carbon. To produce low-carbon stainless steels, an iron charge together with HCFeCr and sometimes charge chrome or blocking chrome are processed in the Argon-Oxygen Decarburizer (AOD).[3] Chromium recovery in the production of HCFeCr by submerged arc smelting ranges from 85 to 90 percent, but in the AOD it is greater than 98 percent. Refractory consumption is high in the AOD with a typical lining lasting around 100 hours or for 50 to 60 heats ([3]).[4]

The high refractory cost is the major disadvantage of the AOD. Decarburization of the melt in the AOD also requires large quantities of argon. By using a ferrochrome alloy with less than 2 percent carbon as feed to the AOD, the decarburization can be carried out in a shorter time with a lower argon consumption. Thus, argon is conserved, and the shorter blowing time in the AOD will result in less refractory attack per heat, thereby allowing more heats to be conducted before relining the AOD. Also, by preparing a ferrochrome alloy with less than 2 percent carbon by the vacuum reduction and simple melting process for feed to the AOD, the overall chromium utilization in the production of low-carbon stainless steels is increased from the current levels of 85 to 90 percent to greater than 98 percent.

Before the advent of the AOD, low-carbon ferrochrome (LCFeCr) was used directly in the stainless steel furnace as a major source of chromium. The current trend is to use LCFeCr only for final adjustment of the chromium level in the stainless steel bath after AOD processing. Low-carbon ferrochrome is produced by the Perrin ([7]) Duplex (modified Perrin) ([7]), or Simplex process ([4-5], [9]).

In the Perrin process, chrome ore and lime are melted in a slag furnace to form a chrome-rich slag containing about 30 percent Cr_2O_3. In addition, chrome ore, silica, and coke or another carbonaceous reductant are smelted in an alloy furnace to produce a chrome- and silicon-rich alloy containing about 45 percent

---

[4] Reference to specific equipment does not imply endorsement by the Bureau of Mines.

[5] Underlined numbers in parentheses refer to items in the list of references at the end of this report.

silicon and about 38 percent chromium. In a stepwise process, the chrome- and silicon-rich alloy is first reacted with an intermediate slag (14 percent Cr_2O_3) to produce a final waste slag (<1 percent Cr_2O_3) and an intermediate alloy 25 percent Si. The intermediate slag (14 percent Cr_2O_3) in the first step is produced along with a final alloy (70.7 percent Cr; <0.05 percent C) by reacting the intermediate alloy (25 percent Si) of the first step with the chrome-rich slag.

The duplex process is similar to the Perrin process except that in the Duplex process the two stages of mixing are carried out in the same ladle. The waste slag also has a higher Cr_2O_3 content of 3 percent as compared to <1 percent in the Perrin process.

In the Simplex process (4-5, 9), finely divided HCFeCr is reacted with finely divided oxidants such as Cr_2O_3, iron oxide, chrome ore, SiO_2 or oxidized HCFeCr under 2 torr pressure for 25 hours to produce LCFeCr. Use of chrome ore as the oxidant introduces gangue from the ore into the process; use of iron oxide as the oxidant reduces the chromium content of the alloy; while use of SiO_2 results in a final alloy with about 6.5 percent silicon. Complex programed heating rates are used in the process to avoid blockage of the solid state reaction caused by the formation of the eutectic phase of chromium and carbon at 1,265° C.

Reduction studies at reduced pressures for Cr_2O_3 with carbon (1, 8) have been reported in the literature. Boericke (1) found the reduction of Cr_2O_3 by carbon to involve four distinct, reversible reaction steps:

$$1/3\ Cr_2O_3 + 13/9\ C \quad 2/9\ Cr_3C_2 + CO, \qquad (1)$$

$$1/3\ Cr_2O_3 + 27/15\ Cr_3C_2 \quad 13/15\ Cr_7C_3 + CO, \qquad (2)$$

$$1/3\ Cr_2O_3 + Cr_7C_3 \quad 1/3\ Cr_{23}C_6 + CO, \qquad (3)$$

and $$1/3\ Cr_2O_3 + 1/6\ Cr_{23}C_6 \quad 27/6\ Cr + CO, \qquad (4)$$

with the overall reaction being

$$1/3\ Cr_2O_3 + C \quad 2/3\ Cr + CO. \qquad (5)$$

Boericke's Cr_4C compound has been more recently identified by Downing (2) as $Cr_{23}C_6$.

He measured the equilibria in these reaction steps and derived thermodynamic values for each reaction from equilibrium and requisite auxiliary thermal data.

Reactions 1 and 2 begin just below 1,300° C at atmospheric pressure; reactions 3 and 4 occur at much higher temperatures or require the removal of CO by either evacuation or dilution with inert gas flushing to proceed at 1,300° C.

Samarin and Vertman (8) found the reduction rate of chromia to increase with increasing temperature in the range of 870° to 1,370° C. Soot was found to be a more effective reducing agent than graphite, and the rate of reduction at 1,300° C was found to increase with decreased pressure. Also, they found the rate of reduction to increase with decreasing grain size of the chromia.

The objective of the present investigation was to determine (1) the factors that influence the carbothermal reduction of chromite, (2) whether reduced products containing less than 2 weight-percent carbon could be obtained by this method, (3) what carbides are intermediate products in the reduction process, and (4) whether a ferrochromium alloy could be produced by simple melting of the reduced product.

Energy Considerations for Ferrochrome Alloy Production

Simplex Process

Production of LCFeCr by a vacuum reduction-melting process should have an advantage over production of LCFeCr by a submerged arc smelting-vacuum oxidation process (Simplex process). Private communications from industrial sources indicated that the energy expenditure for production of 1 long ton of high-carbon ferrochrome by three-phase submerged arc smelting in the Simplex process ranges from 4,500 to 6,400 kilowatt-hours, and an additional 4,000 to 6,000 kilowatt-hours is required in the process for the oxidation of a part of the HCFeCr and vacuum decarburization of the remaining HCFeCr with the oxidized HCFeCr to produce 1 long ton of LCFeCr. Thus according to these industrial sources, energy expenditure for production of 1 long ton of LCFeCr by the Simplex process ranges from 8,500 to 12,400 kilowatt-hours.

Although 2 tons of material must be heated in the vacuum furnace for each ton of ferrochrome alloy produced in the vacuum reduction-melting process, the energy expenditure with respect to the vacuum decarburization step in the Simplex process should be about the same; that is 4,000 to 6,000 kilowatt-hours per long ton of LCFeCr produced since the furnacing time for the vacuum reduction is less than 50 percent of the furnacing time for the vacuum decarburization step in the Simplex process. Melting of the vacuum-reduced products of this investigation was estimated by the same industrial sources to require 900 to 1,200 kilowatt-hours per long ton of ferrochrome alloy produced. Thus, the total energy expenditure for production of 1 long ton of LCFeCr by the vacuum reduction-melting process is estimated to range from 4,900 to 7,200 kilowatt-hours. Comparison of these values with those for the Simplex process (8,500 to 12,400 kilowatt-hours) indicates a potential energy saving of 3,600 to 5,200 kilowatt-hours should be realized in the production of 1 long ton of LCFeCr by the vacuum-reduction melting process instead of by the Simplex process.

Althernatively, it appears that an additional energy savings of about 450 to 900 kilowatt-hours could be realized by adding the low-carbon-reduced products with the iron charge in the furnace to make stainless steel. The iron charge would consist of a high scrap iron to pellet ratio in order to maintain a reasonable slag burden in the furnace. In this case, the furnace for melting the reduced products would not be required, and the gangue material in the reduced products would constitute much of the necessary slag burden in the furnace, and replace the part of the slag burden usually caused by the gangue in the pellets of the normal iron charge. The energy saved would be that necessary to heat and melt the gangue material in the reduced product melting furnace.

Perrin Process

Energy consumption for LCFeCr by the Perrin process is about 8,950 kilowatt-hours per long ton of LCFeCr (7). Comparison of this energy consumption with the estimated energy consumption of 4,900 to 7,200 kilowatt-hours per long ton of LCFeCr produced by the vacuum reduction-melting process indicates an energy saving of 1,750 to 4,050 kilowatt-hours should be realized in the production of 1 long ton of LCFeCr by the vacuum reduction-melting process instead of by the Perrin process.

Submerged-Arc Smelting and AOD Processes

Production of low-carbon stainless steels by addition of LCFeCr produced in the vacuum reduction-melting process should have an advantage over a process in which HCFeCr is produced in a submerged arc furnace, decarburized together with the iron charge in the AOD, and then added to the stainless steel bath to produce low-carbon stainless steels.

The estimated energy consumption of 4,900 to 7,200 kilowatt-hours for production of 1 long ton of LCFeCr by the vacuum reduction-melting process is comparable to the 5,400 to 6,300 kilowatt-hours required for production of 1 long ton of HCFeCr. (These estimates were obtained from industrial sources in a private communication.) Thus, the energy savings between these two processes is the energy expenditure for decarburizing the high-carbon ferrochrome-scrap melt in the AOD. Production of low-carbon stainless steels by first producing a ferrochrome alloy with less than 2 percent carbon by vacuum reduction and simple melting, and then processing this alloy with scrap in the AOD should result in an energy savings equal to that required for decarburization of high-carbon ferrochrome to the carbon content of the vacuum reduction-melting process ferrochrome product.

Thermodynamic Analysis of Chromia-Carbon Reactions

Carbothermal reduction of the chromium in chromite $(Fe,Mg)O \cdot (Cr,Al)_2O_3$ is complex, but guidelines for its reduction behavior can be obtained by investigating the carbothermal reduction of the constitutent oxides, FeO, and Cr_2O_3. The free energy for the reduction of FeO by carbon to either iron or Fe_3C is negative above 800° C, while the free energy for the reduction of Cr_2O_3 by carbon to carbide or metal is negative above 1,100° C. This means that the iron in the chromite should be reduced first, and the limiting reduction reactions should be those for the chromium in the chromite. Figure 1 gives the standard free energy change, ΔF^o, for several reduction reactions of Cr_2O_3 as a function of temperature (lines A, B, and C). Superimposed on this diagram is a grid of dashed lines representing the quantity, $-RT \ln P_{CO}$, as a function of temperature for CO partial pressures of 1, 0.1, 0.01, 0.001, and 0.0001 atmospheres. At equilibrium, $\Delta F^o = -RT \ln P_{CO}$, thus the intersection of a solid line with a dashed line gives the equilibrium temperature for that reaction at the pressure represented by the dashed line.

It can be seen in figure 1 why reduction of Cr_2O_3 to Cr_7C_3 (line C), is more likely to occur than the reduction of Cr_2O_3 to chromium metal (line B). At any temperature, the reaction

$$1/3 \ Cr_2O_3 + 9/7 \ C \quad 2/21 \ Cr_7C_3 + CO \quad (6)$$

is favored over the reaction

$$1/3 \ Cr_2O_3 + C \quad 2/3 \ Cr + CO \quad (5)$$

by about 5 kilocalories per mole of CO formed, thus carbide formation is more likely than direct reduction to chromium metal.

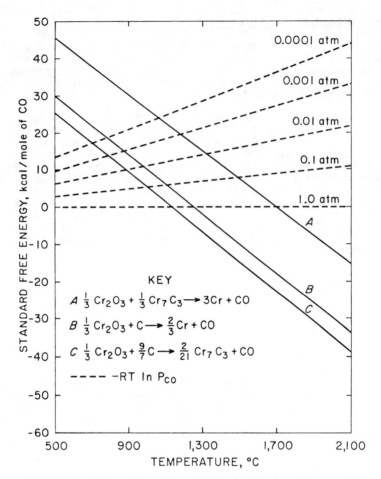

FIGURE 1. Thermodynamic analysis of chromia-carbon reactions.

Line A represents the standard free energy change for the reaction

$$1/3\ Cr_2O_3 + 1/3\ Cr_7C_3 \rightarrow 3Cr + CO, \qquad (7)$$

and shows why the reduction will have to be carried out at reduced CO pressure in order to obtain chromium metal at a temperature below 1,700° C. Reduced CO pressure can be attained by flushing with inert gas or by evacuation as in the present investigation.

Materials

A minus 20-, plus 40-mesh table concentrate containing, in weight-percent, 9.6 Fe, 41.4 Cr, 2.7 SiO_2, 15.6 MgO, 7.3 Al_2O_3, and 1.2 CaO was employed in this investigation. The Cr/Fe ratio for this concentrate is 4.31. The chromium-bearing mineral in the concentrate was chromite $(fe,Mg)O \cdot (Cr,Al)_2O_3$, the principal gangue mineral was serpentine with some carbonate present. Reductants employed in this investigation were graphite, carbon black, foundry coke, and anthracite. The foundry coke and anthracite contained about 92.5 and 77.0

percent fixed carbon, respectively; the graphite and carbon black were essentially pure carbon.

Experimental Method and Apparatus

Carbothermal Reduction Studies

Concentrate and reductant materials were comminuted to the desired mesh sizes. Two-hundred-gram lots of concentrate were blended with 2 grams of chromium trioxide (CrO_3) as a binder and the amount of carbonaceous reductant required for the reduction of Fe^{2+} and Cr^{3+} in the concentrate and Cr^{6+} in the binder to metallics. The blended mixture was moistened to make a paste, pushed through a 14-mesh (Tyler) screen, and dried at 105° C. The spaghetti-like, dried product was broken and screened to give a porous minus 10- plus 20-mesh product. The undersize was repelletized.

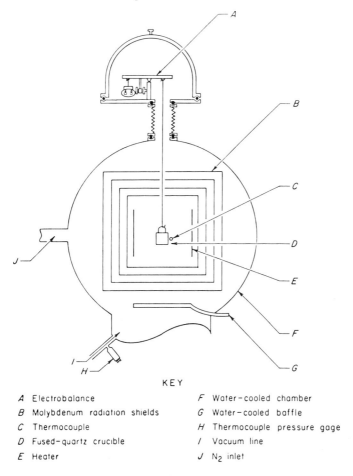

KEY

A Electrobalance
B Molybdenum radiation shields
C Thermocouple
D Fused-quartz crucible
E Heater
F Water-cooled chamber
G Water-cooled baffle
H Thermocouple pressure gage
I Vacuum line
J N_2 inlet

FIGURE 2. Schematic of vacuum furnace and electrobalance.

Figure 2 shows a schematic of the vacuum furnace and electrobalance. Samples of the pellets were put into the quartz crucible suspended from the electrobalance. The chamber was evacuated to less than 0.02 torr and backfilled with high-purity nitrogen to the desired pressure. The nitrogen was passed over hot copper turnings (480° C) to remove oxygen before entering the chamber. The sample was slowly heated with the tungsten heaters to the desired operating temperature, and the weight loss was continuously recorded on a strip-chart recorder. The temperature of the sample was monitored and controlled by a calibrated tungsten-rhenium thermocouple positioned near and behind the quartz crucible. This temperature was recorded on a second strip-chart recorder. Periodic checks of the temperature of the sample made with an optical pyrometer gave good agreement with the temperature determined by the thermocouple. The chamber pressure was monitored with the thermocouple pressure gage and controlled by adjusting a vacuum throttle valve.

The percent total reduction R(t) to iron and chromium metals at time t was calculated from the weight loss data by using the following relationship:

$$R(t) = \frac{\Delta W(t) - \delta}{\Delta W(\infty)} \times 100, \tag{8}$$

where $\Delta W(t)$ is the observed weight loss of sample at time t,

δ is weight loss due to volatile material,

and $\Delta W(\infty)$ is the theoretical weight loss for total reduction of Fe^{2+} and Cr^{3+} in the concentrate and Cr^{6+} in the binder to metallics.

A weight loss δ due to volatile material was obtained by heating the sample to 800° C; this weight loss was subtracted from $\Delta W(t)$. Except for the volatiles the sample did not lose any more weight until the temperature of the sample was between 1,080° C and 1,100° C. Zero time was taken at that time.

Polished sections of the reduced products were examined microscopically and analyzed on a microprobe. X-ray diffraction patterns and chemical analyses were made of reduced products in powedered form.

Melting of Reduced Products

Twenty 25-gram samples of pellets of minus 400-mesh concentrate and minus 400-mesh graphite were heated to 1,300° C under 1 torr pressure. The reduction was allowed to proceed to near completion in each case. The reduced products from these tests were blended and split into four 80-gram samples, which were mixed with additives such as CaO, SiO_2, and/or CaF_2 and put in fused-alumina crucibles (3.5 centimeters ID by 9 centimeters long). The crucibles were placed in a second fused-alumina crucible (5.5 centimeters ID by 10 centimeters long) situated in the center of an induction furnace on a graphite pedestal and surrounded by graphite susceptors. The double-crucible configuration was heated to 1,700° C and soaked for 20 minutes. An optical pyrometer monitored the temperature of the melt. The sample was cooled to room temperature in the crucible, which was then cut in half to show the melted products. The quantity of major constituents in the mixtures were calculated from chemical analysis and weight loss data. These calculated percentages are listed in table 1.

TABLE 1. Calculated compositions of reduced product-additive mixtures

Analysis	Mixture 1 (additive: 14.0 g CaO)	Mixture 2 (additives: 4.5 g CaF$_2$ 14.0 g. CaO)	Mixture 3 (additives: 14.0 g CaO 12.9 g SiO$_2$)	Mixture 4 (additives: 14.0 g CaO 12.9 g SiO$_2$ 4.5 g CaF$_2$)
Constituent, percent:				
Chromium	45.0	43.0	39.6	38.0
Iron	10.4	9.9	9.1	8.8
MgO	17.0	16.2	14.9	14.3
CaO	16.2	15.5	14.2	13.7
Al$_2$O$_3$	8.0	7.6	7.0	6.7
SiO$_2$	2.9	2.8	14.7	14.1
CaF$_2$	0	4.6	0	4.0
Weight of mixture[1] grams	94	98.5	106.9	111.4

[1] 80 grams of reduced products was used in each mixture.

Results and Discussion

Effect of Temperature on the Reduction of Chromite Concentrate

Twenty-five grams of pellets made from minus 400-mesh graphite and minus 400-mesh concentrate were heated to varying temperatures under 1 torr pressure. The carbothermal reduction of chromite at various temperatures in the range 1,230° to 1,320° C are shown in figure 3. Maximum temperature in each test was attained within 20 minutes. Figure 3 shows that the rate of reduction increases with temperature. It should be noted that there is a maximum temperature for reduction under a given pressure before the incipient vaporization of chromium. Table 2 shows vapor pressures of chromium metal at various temperatures. To prevent measurable chromium vaporization at a given temperature, the reduction should be carried out at a pressure that is at least 10 times greater than the vapor pressure of chromium at that temperature.

TABLE 2. Vapor pressure of chromium metal at various temperatures

Pressure, torr	Temperature, ° C
1	1,504
10^{-1}	1,342
10^{-2}	1,205
10^{-3}	1,090
10^{-4}	992
10^{-5}	907

Source: CRC Press, Handbook of Chemistry and Physics, Cleveland, Ohio, 54th ed., 1974, ed. by R.C. Weast, p. D-161.

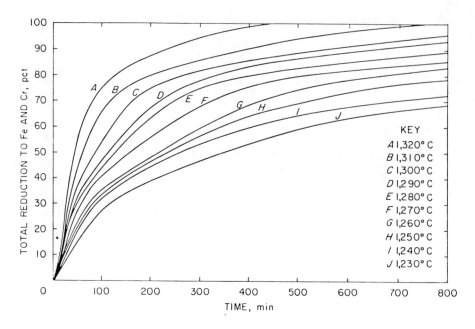

FIGURE 3. Effect of temperature on the reduction of chromite at 1 torr pressure (minus 400-mesh graphite, minus 400-mesh concentrate).

Effect of Pressure on the Reduction of Chromite Concentrate

The carbothermal reduction of chromite, under pressures ranging from 0.1 to 1 torr are shown in figure 4. Twenty-five grams of pellets made from minus 400-mesh graphite and minus 400-mesh concentrate were heated to 1,300° C under pressures of 0.1, 0.4, 0.6, 0.8, and 1 torr. Figure 4 shows that larger degrees of reduction are achieved by decreasing the pressure. For any given temperature of operation there is a minimum pressure below which the metallic chromium begins to vaporize. This is illustrated by curve A surpassing 100-percent reduction because of additional weight loss due to chromium vaporization. Chemical analysis of the reduced product corraborated this observation. A material balance showed a loss of chromium in the reduced product. The different shape of curve A after 93-percent reduction is also probably a result of chromium vaporization.

Effect of Reactivity of Carbonaceous Reductants on the Reduction of Chromite Concentrate

Tests with four carbonaceous materials determined the effect of the reactivity of these reductants on the reduction of chromite. Twenty-five grams of pellets made from minus 400-mesh concentrate and a minus 400-mesh reductant were heated to 1,270° C under 1 torr pressure. Figure 5 shows that in decreasing order of effectiveness, foundry coke, carbon black, and anthracite were all superior to graphite as reductants in the carbothermal reduction of chromite.

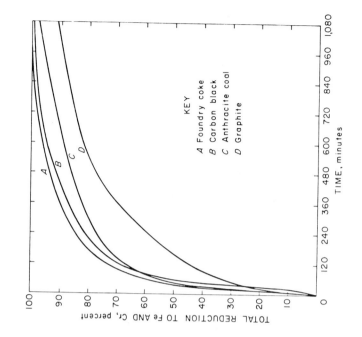

FIGURE 5. Effect of reactivity of carbonaceous reductant on the reduction of chromite at 1,270° C and 1 torr pressure (minus 400-mesh reductant, minus 400-mesh concentrate).

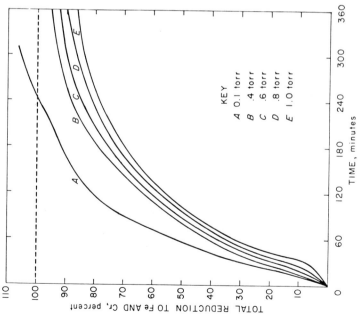

FIGURE 4. Effect of pressure on the reduction of chromite at 1,300° C (minus 400-mesh graphite, minus 400-mesh concentrate).

Effect of Particle Size of Reactants on the Reduction of Chromite Concentrate

A series of 6-hour tests determined the effect of reductant and concentrate particle size on the reduction of chromite. Ten grams of pellets made of minus 100-mesh concentrate with minus 100 plus 200-, minus 200 plus 270-, minus 270 plus 325-, and minus 325 plus 400-mesh graphite, and minus 400-mesh graphite with minus 100-, minus 200-, minus 270-, and minus 400-mesh concentrate were heated to 1,300° C under 1 torr pressure. The results of these tests plotted in figure 6, indicate that for a minus 100-mesh concentrate particle size the rate of reduction increases with decreasing graphite particle size. A further increase in the rate of reduction is obtained by decreasing the concentrate particle size to minus 400 mesh while keeping the graphite particle size at minus 400 mesh. The fastest reaction was obtained with pellets of minus 400-mesh concentrate and minus 400-mesh graphite. The large gap between curves D and E indicates that the rate of chromite reduction is greatly accelerated with very fine graphite particles. In the latter case, only graphite particles between minus 325 and plus 400 mesh were available as reductants; whereas, in the former case, only graphite particles smaller than minus 400 mesh were available as reductants.

Evidence of Carbide Intermediaries in the Reduction of Chromite

Tests were conducted to determine whether the reduction of chromite, like the reduction of chromia (1), proceeds via the carbide intermediates Cr_3C_2, Cr_7C_3 and $Cr_{23}C_6$ (reactions 1 through 4). Ten grams of pellets from minus 400-mesh graphite and minus 400-mesh concentrate were heated to 1,300° C under 1 torr pressure. The reduction was allowed to proceed to various stages of completion.

These partially reduced products were chemically analyzed for chromium, graphitic carbon, and total carbon. Total carbon was determined by a standard combustion technique using an automatic, carbon-combustion analyzer. Determination of graphitic carbon involved reacting the reduced product with a mixture of acids overnight to dissolve everything but the graphitic carbon. The resulting solution was filtered, and the residue analyzed for carbon by the standard combustion technique. The amount of carbidic carbon in the reduced products was determined from the difference between total carbon and graphitic carbon analyses.

Figure 7 shows the percentage of the carbon in the initial (unreduced) sample preesnt as carbide in the partially reduced products versus percent weight loss for these partially reduced products. This curve shows the presence of carbides in all the reduced products, and thus supports the reduction of chromium in chromite via carbide intermediates.

X-ray diffraction patterns were also made for partially reduced products. Peak height data from these X-ray diffraction patterns for Cr_7C_3. $Cr_{23}C_6$ and iron and chromium metallics are graphically shown in curves A, B, and C, respectively, of figure 8 as functions of percent weight loss of the sample. Peak heights for graphite and chromite peaks decreased with increased weight loss. Diffraction peaks for Fe_3C and Cr_3C_2 were not observed in any of the diffraction patterns. Although peak height data are not quantitative, they do support the following reactions occurring consecutively as the chromite is reduced:

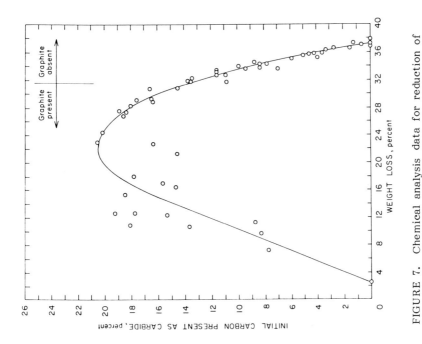

FIGURE 7. Chemical analysis data for reduction of chromite with graphite.

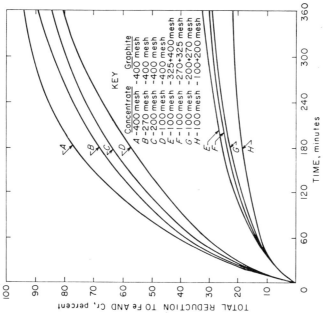

FIGURE 6. Effect of reactant particle size on the reduction of chromite (1 torr pressure, 1,300° C).

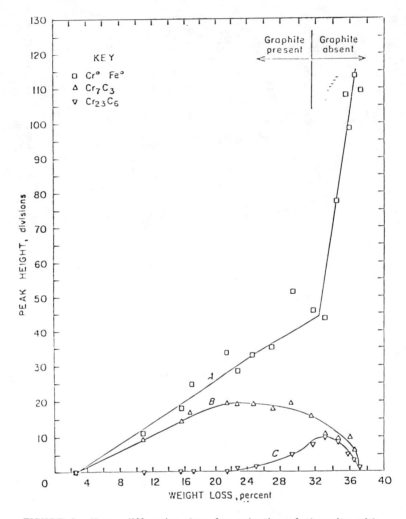

FIGURE 8. X-ray diffraction data for reduction of chromite with graphite.

$$1/3 \ Cr_2O_3 + 9/7 \ C \quad 2/21 \ Cr_7C_3 + + CO, \tag{9}$$

$$1/3 \ Cr_2O_3 + Cr_7C_3 \quad 1/3 \ Cr_{23}C_6 + CO, \tag{3}$$

and
$$1/3 \ Cr_2O_3 + 1/6 \ Cr_{23}C_6 \quad 9/2 \ Cr + C). \tag{5}$$

The reaction shown in equation 9 is supported by the decrease in peak heights for graphite and chromite and the increase in peak height for Cr_7C_3 up to about 23 percent weight loss (curve B). The decrease in Cr_7C_3 peak height (curve B) with the appearance of $CR_{23}C_6$ peaks (curve C) at about 23 percent weight loss and the subsequent increase in $Cr_{23}C_6$ peak height (curve C) and continued decrease in Cr_7C_3 peak heights (curve B) with increased weight loss supports reaction 3. The initial decrease in $Cr_{23}C_6$ peak height (curve C) and drastic change in peak height for iron and chromium metallics (curve A) at about 33 percent weight loss and subsequent increase in the peak height for iron and chromium metallics with continued decrease $Cr_{23}C_6$ peak height supports reaction 4. The absence of graphite after about 31.5 percent weight loss indicates additional weight loss is due to reactions 3 and 4.

The reduciton of irom from chromite would be expected to occur first and proceed simultaneously as follows since the free energies for these reactions are more negative than the chromium reduction reactions and are almost identical in value.

$$FeO + C \quad Fe + CO, \tag{10}$$

and
$$3FeO + 2C \quad Fe_3C + CO. \tag{11}$$

As the temperature is raised above 1,152° C, the reduced iron would most likely be in the liquid state since there is an eutectic composition of iron-carbon alloys containing 4.3 percent carbon, which is liquid above 1,152° C.

Thus, as Urquhart (10) suggests in his review article on production of HCFeCr in a submerged-arc furnace, the reduction of chromium from chromite may proceed by the following reaction:

$$Cr_2O_3(s) + 3C(s) \quad 2[Cr]Fe\text{-}c \ (\ell) + 3CO(g), \tag{12}$$

where $[Cr]Fe\text{-}c \ (\ell)$ represents chromium dissolved in the liquid iron-carbon alloy. Furthermore, since it is known that the presence of chromium reduces the activity of carbon in liquid iron melts, mixed carbides may precipitate from the melt $[Cr]Fe\text{-}c \ (\ell)$ either at the temperatures of the experiments or upon cooling down to room temperature.

Microprobe analysis of a metallic-like phase of a reduced product (19.7 percent weight loss) showed that this phase contained Cr and Fe in a ration of 2.5:1 with about 9 percent of the phase composition not accounted for and presumably carbon. Al, Mg, Si, and Ca were absent from this metallic-like phase. It is concluded that a mixed carbide of the type $(Cr,Fe)_7C_3$ containing about 9 weight-percent carbon is one of the intermediates in the reduction of chromite. Microprobe analysis of the metallic-like phase of another reduced product (36.7 percent weight loss) showed that this phase contained Cr and Fe in a ratio of 2.8:1 with about 5.4 percent of the phase composition not accounted for and presumably carbon. Here again, Al, Mg, Si, and Ca were absent from this metallic-like phase. It is concluded that a mixed carbide of the type $(Cr,Fe)_{23}C_6$ containing about 5 to 6 weight-percent carbon is another of the intermediates in the reduction of the chromite.

The presence of the mixed carbides $(Fe,Cr)_7C_3$ and $(Fe,Cr)_{23}C_6$ and absence of a mixed carbide $(Fe,Cr)_3C_2$ analogous to Cr_3C_2 in chromia reduction may be explained on the basis of miscibility of the carbides. Chromium carbide (Cr_7C_3) can dissolve iron to replace up to 55 percent of the chromium. The cubic carbide $(Cr_{23}C_6)$ can accommodate substitution of up to 25 percent of the chromium by iron, whereas Cr_3C_2 can accommodate very little iron (10). If Cr_3C_2 is formed in the reduction of the Cr_2O_3 in chromite, it would probably immediately dissolve in the liquid iron-carbon alloy. Since the structure accommodates very little iron the subsequent precipitation of $(Fe,Cr)_3C_2$ is not possible.

Figures 9 and 10 each show photomicrographs (X 160) of four partially reduced products. Figure 9A depicts a sample after the initial weight loss of 2.4 percent (below 800° C) but before any reduction has taken place. The irregularly shaped grains are chromite and the cigar-shaped or elongated particles are graphite. No carbide or metallic phases are present. Figure 9B shows a sample after 7.3 percent weight loss. The reduction has begun, as evidenced by the appearance of fine particles of carbide phase (white). As the reduction continues (panels C and D), more carbide phase is present, and carbide particles are coalescing into larger rounded grains. Graphite and chromite grains are still present; however, a metallic-like phase is present in the cracks and holes that have appeared in the chromite grains. As the reduction proceeds further (fig. 10, panels A, B, and C), graphite is disappearing and is finally absent in the reduced product shown in 10C. The chromite grains have become more and more porous owing to loss of Fe and Cr. A metallic-like phase continues to be present in these defects. Carbide phases have coalesced into larger and larger particles. Some carbide particles contain both a high-relief and low-relief phase, which are differentiated because of dissimilar polishing hardness. Explanation for this difference in hardness may be found in (1) differences in orientation of the same carbide phase, (2) difference in the iron and chromium ratio of adjacent grains of the same carbide phase, or (3) intergrowths of two different carbide phases. As the reduction nears completion (fig. 10D), the small carbide grains have coalesced into large particles that are lower in relief than the surrounding residual chromite particles or the carbide phases of reduced products that have undergone less reduction.

The coalescing of carbide particles into larger, well-rounded grains suggests that the carbide phase was at one time in the liquid state. This would tend to support the hypothesis that the reduction Cr_2O_3 in the chromite is done by carbon in the liquid iron-carbon alloy.

The low-relief metallic phase of the almost completely reduced products and several high-relief carbide phases of other partially reduced products were microscopically examined in more detail. Under polarized light, the carbide phases of the partially reduced phases were observed to be anisotropic (not cubic) in structure. Since iron-chromium alloys and $(Fe,Cr)_{23}C_6$ are cubic in structure and are isotropic in reflection under polarized light, it is probable that the carbide phase present in the partially reduced products is the tetragonal structure $(Fe,Cr)_7C_3$. The metallic phase in almost completely reduced products was much softer than the carbide phases of partially reduced products as evidenced by the low relief and fine scratch marks on the metallic phase in figure 10D. Also, this phase was isotropic under polarized light. It is believed that the metallic-like phase in almost completely reduced products is either the cubic carbide $(Fe,Cr)_{23}C_6$ or an iron-chromium alloy.

X-ray diffraction patterns for the partially reduced products showed that the major chromite peak broke into three distinct, closely related peaks for reduced products with about 20 or more percent weight loss. One of these peaks corresponded to the d-value of the main chromite peak, one corresponded to the $MgAl_2O_4$ spinel peak, and the third one could not be positively identified. For

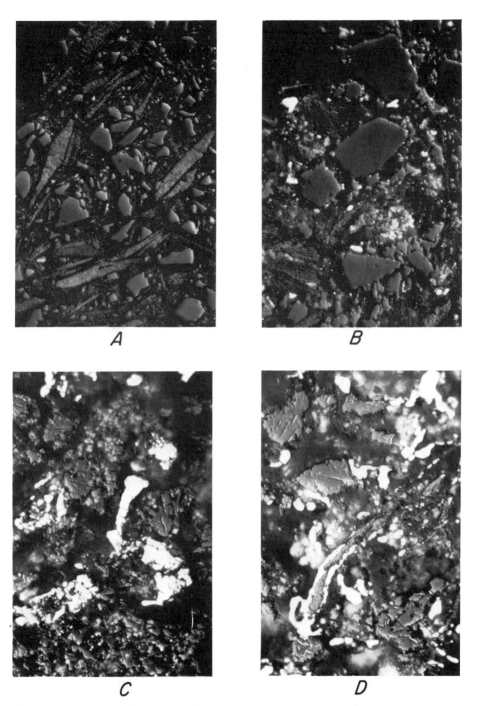

FIGURE 9. Photomicrographs of reduced products at X 160 after various percentages of weight loss. A, 2.4 percent; B, 7.3 percent; C, 17.0 percent; and D, 21.5 percent.

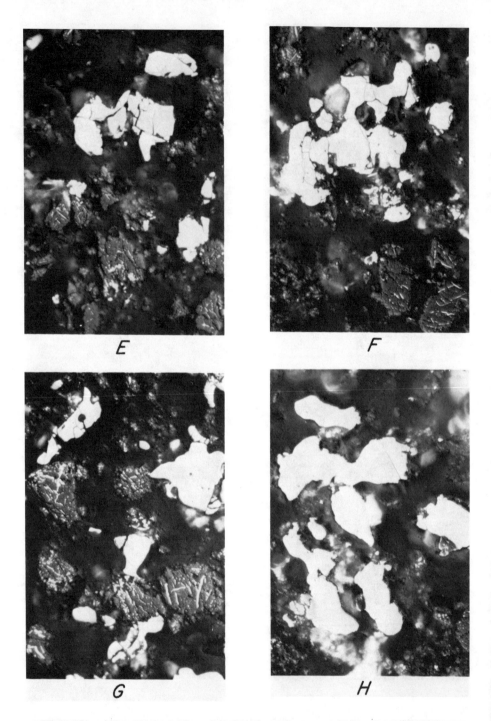

FIGURE 10.- Photomicrographs of reduced products at 160X after E. 26.9 pct wt loss, F. 29.5 pct wt loss, G. 31.5 pct wt loss, and H. 36.7 pct wt loss.

almost completely reduced products, the X-ray patterns showed the presence of MgO and the $MgAl_2O_4$ peak was the highest of the triplet peaks.

Total carbon analyses of graphite-reduced products indicated that a weight loss of about 33.8 percent was necessary to achieve reduced products containing 2 weight-percent carbon. A weight loss of about 35.8 percent was necessary to achieve reduced products containing 1 weight-percent carbon. Weight loss of about 36.8 percent was necessary to achieve reduced products containing less than 0.25 weight-percent carbon. Sixty reduced products containing less than 2 weight-percent carbon were obtained during this investigation; of these, 40 contained less than 1 weight-percent carbon, 20 contained less than 0.25 weight-percent carbon, 14 contained less than 0.1 weight-percent carbon, and 1 contained 0.01 weight-percent carbon.

Promoters for Carbothermal Reduction of Chromites

The effect of alkaline earth oxide additives on the carbothermal reduction of chromite was investigated in some detail. Kinetic curves are expressed as the plots of the percent reduction of chromite versus time. The additives were introduced at the 1:5 atom ratio with respect to Cr^{3+} in the chromite mineral, i.e., one atom of Ca, Sr, or Ba in each of CaO, SrO, and BaO was present in the pellets for five atoms of chromium. The beneficial effect of these additives on the chromite reduction rate is obvious from the kinetic curves in figure 11. Thus, while the time required to achieve 90 percent chromite reduction, $t_{0.9}$, is 760 minutes (more than 12 hours), it is shortened by more than one half, to 370 minutes, in the presence of lime. The addition of Sr and Ba at the same atomic level of Ca further shortens the required $t_{0.9}$ to 100 and 70 minutes, respectively. Thus the furnacing time can be reduced by a factor of about 10 by the addition of barium as BaO. It is interesting to note that the degree of acceleration of chromite reduction correlates with the atomic size of the additive. The ionic radii of Ca, Sr, and Ba are 0.99Å, 1.13Å, and 1.35Å, respectively. The time for 90 percent reduction is shortened from 760 minutes in the blank test to 370, 100, and 70 minutes by the presence of Ca, Sr, and Ba, respectively.

The effect of additive concentration on the kinetics of carbothermal reduction of chromite is illustrated in figure 12. When Ca in the form of lime was added at the atomic ratios of 1:10, 1:5, and 1:3.3 with respect to Cr, the time required for 90 percent reduction was shortened from 760 minutes in absence of Ca to 490, 370, and 170 minutes, respectively, with increasing concentration of the lime additive. Similar results were obtained with the strontium and barium additives; barium shortening the reduction time to 50 minues when present at the atom ratio of 1:3.3.

Melting of the Reduced Products

Acceleration of the roasting reactions in carbothermal reduction of the chromite with lime additions will cut the furnacing time, thus saving time and energy in the production of the ferroalloy. Roasting would normally be followed by smelting of the reduced products to separate the ferroalloy button from the slag phase. Here again addition of lime is beneficial in lowering the slag melting point and hencein enhancing phase segregation in the smelting step.

Panel A of figure 13 shows a cross section of a carbothermally reduced chromite sample at 0.1 torr after adding 17.5 percent (w/w) lime and soaking at 1,700° C for 20 minutes at atmospheric pressure. A button of Fe-Cr alloy had formed at the lower left side of the crucible. However, many droplets of the metallic phase were trapped in the slag phase indicating that complete

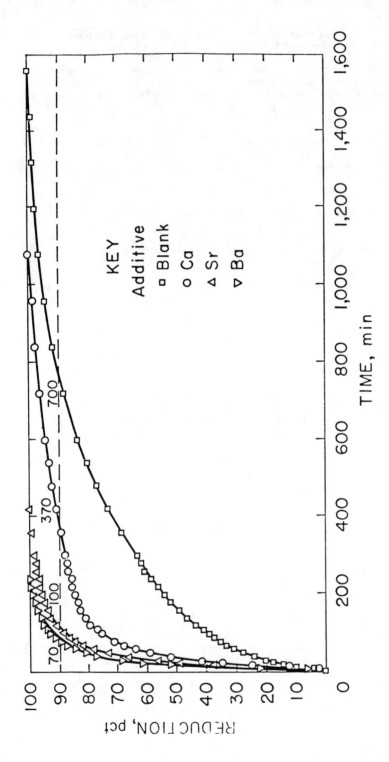

FIGURE 11. Effect of alkaline earth oxide additions on the carbothermal reduction of chromite.

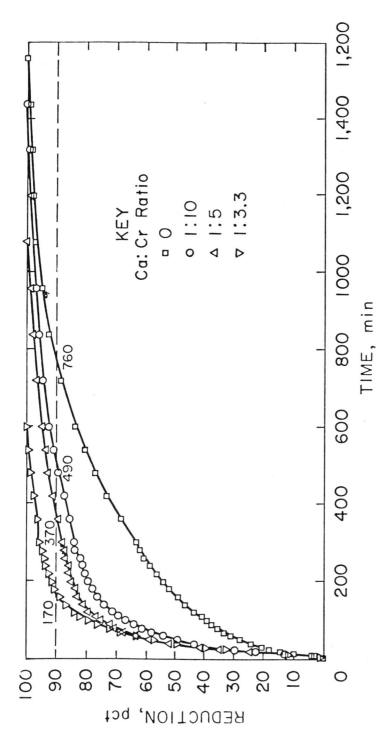

FIGURE 12. Effect of concentration of the calcium promoter on chromite reduction.

FIGURE 13. Photographs of cross sections of reduced product-additive mixtures after melting at 1,700° C. A, CaO addition; B, CaO and SiO$_2$ addition.

segregation of metallics from slag did not occur under these conditions.

Panel B of figure 13 is similarly prepared like panel A except for the addition of 16.1 percent SiO_2 with the lime before soaking at 1,700° C. Here a larger button of Fe-Cr alloy formed at the bottom of the crucible with better separation of the metallics from the slag.

A method has been devised (6) in which the lime and silica are added to the hot in situ reduced chromite after releasing the vacuum and raising the charge temperature from 1,300 to 1,700° C. This would result in further saving in the thermal energy required to heat the charge from room temperature to the smelting temperature. Instead of adding 17.5 percent lime and 16.1 percent SiO_2 for the smelting step, one can pre-mix part of the lime (13 percent) before the roasting step to accelerate its completion at reduced pressure, and then introduce the balance of 4.5 percent lime and 16.1 percent SiO_2 needed for the smelting step at 1,700° C. This enhanced productivity is, therefore, achieved at no cost for the lime accelerator since its addition is subsequently required to complete the process of ferroalloy production.

Summary and Conclusions

The reduction of chromite wsa found to proceed via the carbide intermediaries $(Fe,Cr)_7C_3$ and $(Fe,Cr)_{23}C_6$. Evidence of carbide intermediates analogous to Cr_3C_2 or Fe_3C was not found. Reduced products containing less than 2 weight-percent carbon and as low as 0.01 weight-percent carbon were obtained by allowing the reduction to proceed to various stages of completion. The extent of reduction increased with increased temperature and decreased pressure; however, operating conditions were limited due to the onset of appreciable chromium vaporization. The fastest reaction was obtained with pellets made with minus 400-mesh concentrate and minus 400-mesh reductant at 1,300° C under 1 torr pressure. Foundry coke, carbon black, and anthracite were found to be superior to graphite in the carbothermal reduction of chromite and are ranked in that order with respect to the speed of reduction. Buttons of ferrochrome alloys were made by melting a mixture of reduced products and appropriate amounts of CaO and SiO_2 at 1,700° C for 20 minutes. Production of ferrochromium alloys by vacuum reduction and simple melting is technologically feasible. Whether it is feasible from an energy or economic standpoint requires a larger scale demonstration of the process.

Production of low-carbon stainless steels by first producing a ferrochrome alloy with less than 2 percent carbon by low-pressure reduction and melting, and then decarburizing this alloy with the iron charge in an AOD should result in an energy savings equal to that required for decarburization of the iron charge and HCFeCr to the carbon content of the iron charge and ferrochrome alloy produced by the low-pressure reduction-melting method. Conservation of argon, chromium, ferrosilicon, and refractories are obvious advantages.

LCFeCr produced by the low-pressure reduction-melting method would eliminate the need for a submerged arc smelter and AOD since LCFeCr can be added with the iron charge directly to a melting furnace to make stainless steel. Capital savings would then be the difference in capital expenditure for a submerged arc furnace versus a vacuum furnace and melting furnace. The disadvantages of the AOD are eliminated and chromium is conserved.

References

1. Boericke, F. S. Equilibria in the Reduction of Chromic Oxide by Carbon and Their Relation to the Decarburization of Chromium and Ferrochrome. BuMines RI 3747, 1944, 34 pp.

2. Downing, J. H. Fundamental Reactions in Submerged-Arc Furnaces. Proc. Elec. Furnace Conf., 1963, Trans. Met. Soc., AIME, v. 21, 1963, pp. 288-296.

3. Industrial Heating, Enhancement of the AOD Reactor Concept, v. 42, No. 6, 1975, pp. 11-13.

4. Eramus, H. de W. (assigned to Union Carbide and Carbon Corp., New York). Purification of Ferrochromium. U.S. Pat, 2,473,019, June 14, 1949.

5. _____. Purification of Ferrochromium. U.S. Pat, 2,473,020, June 14, 1949.

6. Pahlman, J. E. and S. E. Khalafalla. (assigned to the U.S. Department of the Interior). Production of Ferrochromium Alloys. U.S. Pat. 4,306,905, December 22, 1981.

7. Robiette, A. G. E. Electric Smelting Processes, John Wiley & Sons, New York, 1973, pp. 150-176.

8. Samarin, A. M., and A. A. Vertman. Production of Chromium and Carbon Free Ferrochrome by the Method of Vacuum Sintering. Tr. Inst. Met., No. 1, 1957, pp. 60-66.

9. Spendelow, H. R., Jr., and H. de W. Eramus. (assigned to Union Carbide and Carbon Corp., New York). Production of Low Carbon Ferrochromium. U.S. Pat. 2,473,021, June 14, 1949.

10. Urquhart, R. The Production of High-Carbon Ferrochromium in a Submerged Arc Furnace. Mater. Sci. and Eng., v. 4, No. 4, 1972, pp. 48-65.

CONTINUOUS CASTING OF STEEL

Professor R. D. Pehlke

Department of Materials and Metallurgical Engineering
University of Michigan
Ann Arbor, MI 48109

Abstract

The history and present status of continuous casting of steel are reviewed, including a general description of current processes. Plant layout, process development and elements of machine component design are outlined, and key steps in integration into a steel production system are detailed.

Recent productivity advances and improvements in quality are described. Process developments are reviewed, and some future avenues for this technology are considered.

Introduction

Continuous casting is entering a new era of development, not only in view of its increasing application in the production of steel, but also in its own evolution as a process and its interaction with other processes in steel manufacture. Continuous casting output has shown an accelerating growth curve as illustrated in Figure 1 (1). More than 40% of current world steel production is continuously cast, and in 1983 the continuous casting ratio in Japan exceeded 80%. In the following paragraphs, the advantages of the process are outlined along with its development and current challenges for improvement.

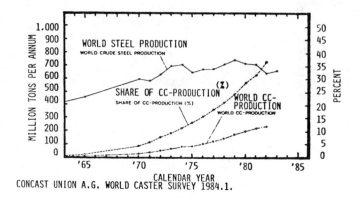

Figure 1 - Worldwide Production of Crude Steel and Share of Continuous Casting[1]

General Description of the Process

The purpose of continuous casting is to bypass conventional ingot casting and to cast to a form that is directly rollable on finishing mills. The use of this process should result in improvement in yield, in surface condition, and in internal quality of product.

Continuous casting involves the following sequence of operations:

1. Delivery of liquid metal to the casting strand.
2. Flow of metal through a distributor into the casting mold.
3. Formation of the cast section in a water-cooled mold.
4. Withdrawal of the casting from the mold.
5. Further heat removal from the casting by water spraying beyond the mold.
6. Cutting and removal of the cast bars or slabs.

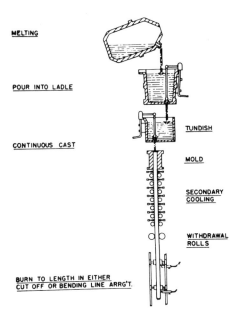

Figure 2 - Elements of a Continuous Casting Strand[2]

A schematic diagram showing the main components of a continuous casting machine is presented in Figure 2 (2). Molten steel in a ladle is delivered to a reservoir above the continuous casting machine called a tundish. The flow of steel from the tundish into one or more open-ended, water-cooled, copper molds is controlled by a stopper rod-nozzle or a slide gate valve arrangement. To initiate a cast, a starter or dummy bar is inserted into the mold and sealed so that the initial flow of steel is contained in the mold. After the mold has been filled to a desired height, the dummy bar is gradually withdrawn at the same rate as molten steel is added to the mold. The initial liquid steel freezes onto a suitable attachment of the dummy bar so that the cast strand can be withdrawn down through the machine. Solidification of a shell begins immediately at the surface of the copper mold. The length of the mold and the casting speed are such that the shell thickness is capable of retaining the molten metal core on exiting from the copper mold. To prevent sticking of the frozen shell to the copper mold, the mold is normally oscillated during the casting operation. The steel strand is mechanically supported by rolls below the mold where secondary cooling is achieved by spraying cooling water onto the strand surface to complete the solidification process. After the strand has fully solidified, it is sectioned into desired lengths by a cutoff apparatus. This final portion of the continuous casting machine also has provision for disengagement and storage of the dummy bar.

Several arrangements are now in commercial use for the continuous casting of steel. The types of continuous casting machines include vertical, vertical with bending, curved or S-strand with either straight or curved mold, curved strand with continuous bending and horizontal. Examples of the principal types of continuous casting machines currently producing slabs are presented in Figure 3 (3).

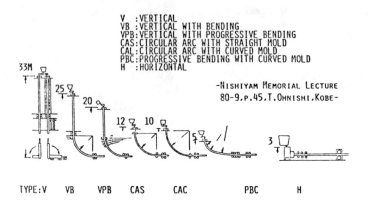

Figure 3 - Principal Types of Continuous Casting[3]

Most of the original continuous casting machines for steel were vertical machines. The vertical machines with bending, and also curved strand machines, although more complicated in their construction, were developed to minimize the height of the machine and allow installation in existing plants without modification of crane height. Four basic caster designs for slabs are shown in Figure 4 with an indication of the required installation height and the corresponding solidification distance or metallurgical length (ML) (4).

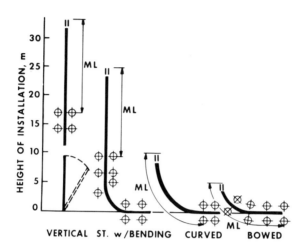

Figure 4 - Four Basic Caster Designs[4]

Historical Aspects of Continuous Casting

One of the earliest references to continuous casting is a patent granted to George Sellers in 1840 who had developed a machine for continuously casting lead pipe (5). There is some indication that this process had been underway prior to the Sellers' patent which was directed toward improvement of this continuous casting process. The first work on continuous casting of steel was by Sir Henry Bessemer who patented a process for "manufacture of continuous sheets of iron and steel" in 1846 and made plant trials on continuous casting of steel in the 1890's (6). Key developments are summarized in Table I (7).

Continuous casting had its start prior to the beginning of the 20th century but it was not until the mid-1930's in Germany that commercial production of continuously cast brass billets was introduced. Junghans, an active inventor of casting technology, provided many improvements in the process, in particular the introduction of the oscillating mold system to avoid sticking of the casting to the mold. Further development of the process for the casting of nonferrous metals continued including installation of processing units in North America. Mold lubrication in the form of oil or, more recently, low melting slag powders was introduced. Taper of the mold to compensate for metal shrinkage on solidification provided improved heat transfer.

Table I

Historical Survey of Continuous Casting
from, Continental Iron and Steel Trade Report
The Hague, August 27, 1970 (7)

Historical Survey in Brief

1840   The American G.E. Sellers was the first to gain a patent of a device for the continuous casting of lead tubings.

1843   Another patent concerning continuous casting of lead tubings was gained by J. Laing. Laing already mentioned the movement of the mandrel in order to prevent sticking of the cast material.

1846   H. Bessemer's basic process "Casting with rolls" was covered by an English patent, whereby his chief stress was given to the production of glass. The production of tin foil and other metals was also of some importance.

1857   H. Bessemer saw the necessity of preventing the sticking of the casting to its casting rolls.

1872   The Englishmen W. Wilkinson and E. Taylor were the first to document the idea of continuous casting with moving moulds.

1886   The basic principles of vertical casting of steel were covered by the American B. Atha. A machine based on his intermittent process is said to have been in operation on a production basis until 1910.

1889   The German M. Daelen designed a vertical type continuous casting machine similar to those still operating today.

1912   The Swede A.H. Pehrson recognized the advantages of a mould oscillation in direction of the strand with horizontal casting.

1915   When continuous casting still was far from industrial application, G. Mellan considered an automatic control for the feeding of metal into the mould depending on the metal level.

1921   C. W. van Ranst proposed the continuous relative motion between strand and mould.

1933   The father of modern continuous casting, S. Junghans, suggested a nonharmonic oscillation which would not influence the heat transfer between strand and mould.

1933   The first plant for industrial continuous casting of brass, according with the vertical casting method utilizing an open ended mould, was built in Germany by S. Junghans. A monthly production of up to 1700 tons of brass was achieved.

1935   A plant for continuous production of brass plates with casting rolls was in operation until 1937 at Scovill Manufacturing Co.,U.S.A.

1936   Vereinigte Leichtmetallwerke (Germany) started a semi-continuous casting machine for aluminum alloys.

1938   A semi-horizontal machine for casting steel with moving moulds after the Goldobin principle was built in Russia.

At the end of the 30's   S. Junghans gave certain non-German license-rights to I. Rossi, New York. Affter World War II this co-operation led to the creation of CONCAST AG, Switzerland, and on the other hand, Junghans laid the foundation for the activities of two German companies in the field of continuous casting, namely Mannesmaann and Demag, from which the present company DST (Demag Stranggiess-Technik GmbH) emerged - which at present holds the second place on the list of important suppliers of continuous casting equipment in the world.

1946   The first pilot plants for continuously casting steel were built at Babcock and Wilcox (U.S.A.), Low Moor (Great Britian), and Steel Tube Works. Amagasaki (Japan), - The next were the following three: Eisenwerk Breitenfeld (Austria), BISRA (Great Britain), and Allegheny-Ludlum (U.S.A.).

1951   The first big installation on a semi-continuous basis in the USSR was set up at Krasny Oktobri. Slabs of stainless steel up to 800 x 180 mm (32" x 7.2") were cast.

1952   The first production plant in the West for small steel billets was put into operation at Barrow (Breat Britain). Carbon steel was cast into billets of 2" to 4" (50 to 100 mm) square, and also some stainless steel was cast.

1952   A German patent by O. Schaeber describing the casting of a bent instead of a straight vertical strand was published.

1954   On the American continent Atlas Steels (Canada) set to work an installation for stainless steel slabs. This installation was planned by I. Rossi's Metalcast Company. In the same year I. Rossi created together with H. Tanner a company in Europe, Concast AG of Zurich (Switzerland).

1956   At Barrow (Great Britain), vertical cutting of the strand was replaced by horizontal cutting. For this purpose the strand was bent after having passed the withdrawal rolls. By this method the relatively large height of the vertical type casters could be reduced by approximately 20%.

1956   E. Schneckenburger and C. Kung (Switzerland) filed the "Concast model S" - patent, whereby the strand is formed in a bent mould, and subsequently guided and cooled in a circular strand guide.

1961   At Dillinger Steelworks (Germany) the first vertical type large slab machine with bending of the strand to horizontal discharge was started up.

1963   The first curved mould machine for billets became operational at Von Moos'sche Eisenwerke (Switzerland). The obvious advantages of this type of plant were looked upon by the steel industry as revolutionary for the industrial application of continuous casting.

1963   McLouth (U.S.A.): First continuous casting plant in which rimmed steel for use in car manufacture was cast. Similar statements were later made by U.S. Steel.

1963	I.M.D. Halliday presented at the invitation of the U.N.O., Centre for Industrial Development, in Prague a paper describing the curved mould casting technique, he also gave a world survey, totalling 61 machines in operation and 44 under construction, together 105 continuous casting machines.
1964	Shelton Iron and Steel (Great Britian): First new steelworks to turn out its entire production by continuous casting, consisting of 4 machines with 11 strands for medium to very large bloom sizes, and operating in connection with Kaldo converters.
1964	The first Concast S type curved mould machine for large slabs was started up at Dillinger Steelworks (Germany). The height of this type of machine was less than 50% of the corresponding height of a vertical type of machine. In the same year Mannesmann (Germany) presented their bow-type slab caster.
1964	I.M.D. Halliday reviewed "Continuous casting of steel in the USSR", in a publication of the Organization for Economic Co-operation and Development (OECD).
1966	After having neglected the changes of the newly developed bow and curved mould type machines since 1963, on March 8, 1966, Isvestia called the bow type caster "a useful machine", and announced at the same time that those machines are also built by UralMash in the USSR.
1968	The United Nation's Economic Commission for Europe published the hitherto most comprehensive report "Economic Aspects of Continuous Casting of Steel".
1968	The first 4-strand low head caster for large slabs in the West was started up at National Steel's (USA) Weirton Division. Subsequently, National pioneered successfully in casting slabs for tinplate applications.

Work on the continuous casting of steel in the United States was started in the 1930's by Allegheny-Ludlum at Brackenridege, PA, using the Junghans oscillating mold. However, it was not until the late 1940's and early 1950's that continuous casting was applied with some success at the Allegheny-Ludlum pilot plant in Watervliet, New York. By 1960, there were two commercial production machines in operation including the small slab casting unit at Atlas Steel in Welland, Ontario. It has been only in the past 20 years that rapid growth in use of this production process has come about which now accounts for a substantial fraction of total steel production.

Plant Layout

The design and layout of a steelmaking facility will often focus initially on continuous casting. The optimum plant layout will markedly vary from one installation to another. One major factor in the configuration is whether or not the steelmaking complex is being constructed on a greenfield site or whether the continuous casting facility is being added to an existing works as a "shoehorn" addition. Many of the major integrated steel works in Japan were constructed as greenfield sites during the period 1960 - 1975. Most of the mini-mills constructed throughout the world, and in particular in the United States, were also built as greenfield sites.

In building a greenfield site, the plant layout should incorporate two major features: a smooth and well organized arrangement for material handling and flow in the steelmaking system and the capacity for future expansion. Generally, these plants are designed for 100% continuous casting, and ingot facilities are not included at all. One excellent example is found in the layout of the Oita works, one of the finest steelmaking facilities in the world. A plant layout of the Oita works is shown in Figure 5 (8). Nearly all of the recently built mini-mills in-

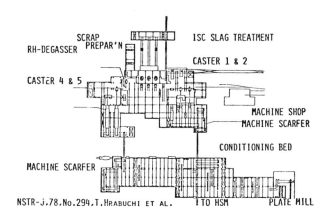

Figure 5 - Typical Greenfield Steelmaking Complex NSC-Oita Works[8]

corporate one or more electric furnaces and provide for 100% continuous casting of billets. Chaparral Steel at Midlothian, Texas is an excellent example. A profile of a proposed large scale electric furnace-billet casting plant is shown in Figure 6 (10.).

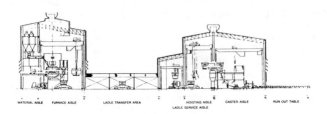

Figure 6 - Proposed meltshop with continuous feed of DR iron and producing 1.2 million tonnes billets annually[10]

A twin strand slab caster was shoehorned into the No. 4 BOF shop at Inland Steel. The addition of this caster substantially increased the output of this facility where ingot casting represented the limiting rate for production. The arrangement of this installation showing the caster and ingot facilities is presented in Figure 7 (11).

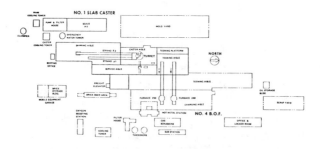

Figure 7 - Plant Layout, Inland Steel Company

An important characteristic of the plant layout, and in particular of the material handling facilities, is the concept that the continuous casting machine cannot wait. This design and operating concept has had a dramatic impact on steelmaking operations which have now become a slave to the continuous casting facilities. In this fashion, marked improvements in productivity have been developed for continuous casting as outlined below.

Marked increases in energy costs, as well as the desire for higher productivity, has led to the development of the "hot connection". Substantial savings in energy can be achieved by directly charging the hot continuously cast slab or billet to the re-heating furnaces of the rolling mill. Some installations have included direct in-line hot rolling of the cast product.

Process Development and Machine Design

A number of methods for distributing liquid steel to the mold or several molds of the casting machine have been investigated. Use of a tundish with appropriate flow control has been found to be a superior method for the production of quality steel. Considerable effort has been directed toward improvement of refractories and development of methods for preventing nozzle blockage. Geometrical arrangements, including the use of dams and weirs, in tundishes have provided suitable fluid flow characteristics to maximize the separation of non-metallic inclusions with an improvement in quality. Another factor in separation of non-metallic inclusions in the tundish is tundish size. Figure 8 indicates the trend in tundish size for major slab casting installations over the past 20 years (12).

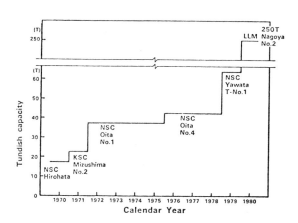

Figure 8 - Transition of the Largest Capacity of Tundish for 2 Strand Slab Caster[12]

Reoxidation of the molten steel is to be avoided. The use of refractory shrouds between the ladle and the tundish and the tundish and the mold have been adopted for slab casting as illustrated in Figure 9 (13).

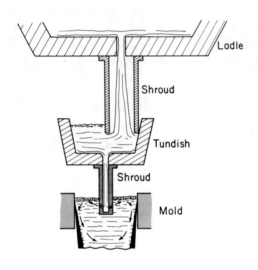

Figure 9 - Pouring Shrouds - Ladle to Tundish and Tundish to Mold[13]

One of the difficulties in continuous casting of small sections has been the protection of the pouring stream from mold to tundish because of the inapplicability of a pouring tube and the mechanical difficulty caused by the oscillation of the mold relative to the fixed tundish. In one instance, this has been overcome by the use of a flexible bellows, as illustrated in Figure 10, and successfully applied to the continuous casting of special product quality steel bars(13). Recently, further development has been made in the use of ceramic shrouds to protect pouring streams on billet machines.

Protection of the surface of the molten steel pool in the continuous casting tundish has been achieved through the use of suitable synthetic slags, gas blanketing and the use of a refractory cover to seal the tundish. Metal level in the tundish, as well as fluid flow pattern, is important in avoiding the ingestion of the slag layer into the metal stream flowing downward into the mold.

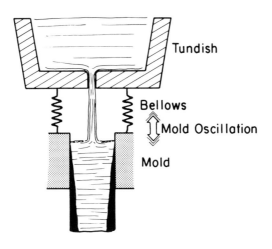

Figure 10 - Bellows between Tundish and Mold for Casting Billets[13]

The handling of the tundish is now accomplished either with two tundish cars or a double turret which provides for changing the tundish, often while continuing the casting process which allows extended sequence casting. The changing of the tundish not only makes a replacement at an appropriate time, but also allows for a change in steel composition with a minimal length of transition in the cast product.

The flow through, water cooled copper mold is the key element of the casting machine. Special attention has been given to problems associated with the design and material requirements for molds. A number of different designs have been used including thin-walled, tube-type molds, solid molds and molds made from plate. The use of plate molds was found to provide excellent mold life and to avoid the necessity for manufacture of molds from solid copper blocks in costs. Steel and brass, as well as copper, have been used for molds, but the outstanding material is nearly pure copper with small additions of elements that promote precipitation hardening or raise the recrystallization temperature, since both effects apparently provide longer mold life. Mold coatings are being applied to extend service life. Chromium is generally used, often with an intermediate layer of nickel for improved coherency.

The most suitable length for a continuous casting mold has been found to be between 20 and 30 inches, a range which seems to hold true regardless of section size. This surprising result can be explained by the fact that with smaller sections and higher casting rates, higher heat removal rates are achieved. Also, a thinner skin can be permitted on exiting from the mold for smaller sections than for larger sections. At higher casting rates the use of an increased taper in the mold is necessary to maintain high heat removal rates, particularly for the narrow faces of slab molds.

Oscillating molds, as used by Junghans, have been adopted almost universally, although fixed molds can be employed successfully with efficient lubricating systems. The oscillation is usually sinusoidal, a motion which can be achieved easily with simple mechanical arrangements. A fairly short stroke and a high frequency are used to provide "negative strip" in which the mean velocity of the mold movement is greater than the speed of withdrawl of the casting strand. Oscillation frequencies are being increased from 50-60 cps up to 250-300 cps with the benefit of shallower oscillation marks, less cracking and reduced conditioning requirements.

Molten metal in the slab mold is normally covered with a layer of special powder which protects the metal from reoxidation and absorbs inclusions. The powder has a low melting point and flows over the liquid steel to provide mold lubrication. Rape seed oil has been used typically to prevent sticking to the mold in billet casting. Metal flow rates are matched with slab casting speeds using a stopper rod in the tundish or a slide gate metering nozzle just above the shroud to control the delivery rate. Billets are normally cast with fixed metering nozzles and the strand speed is adjusted to any changes in steel flow rate.

Support of the thin steel shell exiting the mold is required, particularly for slab casters. Several systems have been employed, as illustrated in Figure 11, all of which provide intensified direct water cooling (14).

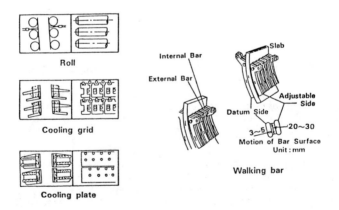

Figure 11 - Various Shell Supporting Systems[14]

Water spraying, i.e. secondary cooling, is critical to the process in that maximum cooling should be accomplished, but overcooling is to be avoided. The amount of heat removed by water sprays depends on the volume of the water, its temperature, and in particular the method of delivery, including spray pressure. This latter variable of pressure is important in that the spray should be sufficiently intense to penetrate the blanket of steam on the surface of the solidifying strand. The thermal conductivity of steel is relatively low; and consequently, as the surface temperature decreases and the shell thickness increases, cooling water has a lesser influence on the solidification characteristics. The interrelationship between support rolls, particularly for the casting of wide slabs, and the spray water and its delivery characteristics is particularly important.

Mixed spraying of air and water develops a unique secondary cooling system with a uniform water droplet size as contracted with conventional sprays. Air-mist cooling is illustrated along with the conventional water spray in Figure 12 as applied to slab casting (15). Although more costly, the air-mist spray system has a wide variation in spray pattern and intensity which offers excellent spray cooling control. Decreases in transverse and longitudinal cracking with air-mist cooling have been reported (16,17).

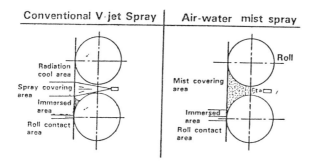

Figure 12 - Comparison of Spray Systems[15]

The straightening operation on curved strand casters has required special design and operating control. In general, temperatures at or above 900-1050°C have been recommended to avoid conditions where certain grades of steel have limited deformation capability and are susceptible to cracking. Multi- and four point straighteners have reduced imposed strains, and compression casting systems have reduced tensile stresses. Uniform temperature of the strand, including corners which tend to cool more quickly, is required.

The dummy bar which is used to stopper the mold for initiation of a cast has been fed from above and below in various installations. Some arrangements are shown schematically in Figure 13 (18). The top feeding arrangement which "chases" the last cast out of the machine offers a productivity advantage.

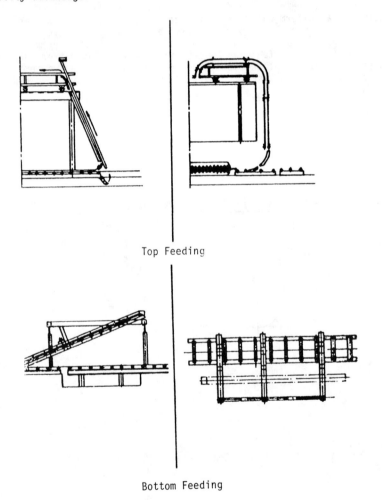

Top Feeding

Bottom Feeding

Figure 13 - Dummy Bar Arrangements[18]

Productivity Improvements

While continual increases in casting speeds over the years have led to improvements in casting machine productivity, the most dramatic factor has been sequence casting, i.e. continuous-continuous casting. Perfection of this development has required extraordinary achievement in design of ladle and tundish handling, and in design and maintenance considerations for long-term operation with processing times that extend for several days. A summary of sequence casting records in Japan has been presented by Harabuchi (19) and reported in Table II. In comparison, the January, 1983 casting record at Great Lakes Steel in the Detroit District involved a large slab caster (9½ by 99 inches) which cast 402 ladles and 91,480 tons of steel continuously. In addition to the bulk steel handling requirements, the ability to change nozzles, shrouds and tundishes at frequent intervals is also required. This string of casts at Great Lakes involved 35 tundish changes and 177 shroud changes over a period of 13 2/3 days.

While it has been shown that sequence casting extending beyond five or six heats does not dramatically increase productivity, provided a reasonable turn-around time and interactive scheduling with the steelmaking facilities exist, the capability for long-term sequence casting represents the opportunity for increased productivity with high quality at an essentially steady state operation of the caster.

An important characteristic of a casting operation with regard to productivity is the percentage of total clock time that steel is being processed in the machine. These percentages in high productivity casters often exceed 90% for highly utilized casting operations. Turn-around time for the dummy bar to re-start a cast is one of the factors in producing steel-in-mold results, but under idealized conditions scheduled maintenance will be a major factor and other items such as problems in steelmaking or difficulty on the strand will be minimized.

In the past and in many present installations, interruption of continuous casting production is required in order to make a width change. Several developments are now in operation which avoid this requirement. One such arrangement is that utilized at Great Lakes Steel where a very wide slab is cast and then slit into two or three optimum widths. Another approach is that taken at Oita Works of Nippon Steel Corporation where a sizing mill adjusts the slabs to various desired widths. Another, more versatile system involves the use of a variable width mold in which the taper and width are adjusted continuously in process. These techniques have permitted adoption of sequential casting with a minimum inventory based on width.

Another barrier to overcome in increasing productivity of continuous casting has been accomodation of ladle-to-ladle composition changes with adaption of sequence casting. Under ideal circumstances, where a plant produces a narrow range of compositions often with overlapping compositional requirements, compositional changes can be slowly stepped through each grade to accomodate the desired sequence of heats. However, when substantial variations in composition must be accomodated, physical barriers in the form of steel plates have been employed to provide physical isolation of the grade changes in the strand. In this fashion, the transition zone can be minimized and compositional changes accomodated without substantial losses in yield or downsngrading in quality.

TABLE II

RECORDS OF SEQUENCE CASTING IN JAPAN [19]

Compiled by T. Haribuchi, Nippon Steel Corporation
from the Iron and Steel Federation of Japan, 1984

	COMPANY	CC PLANT	HEATS/CAST	T/CAST	TIME OF ACHIEVEMENT
SLAB	Nippon Kokan	Keihin	270	24,775	Sep., 1974
	Nippon Kokan	Fukuyamaa	244	59,796	Aug., 1981
	Kawasaki Steel Co.	Mizushima No. 5	204	56,100	Nov., 1980
	NSC	Yamata No. 2	186	33,219	July, 1974
BLOOM	Sumitomo Metal Ind.	Wakayama No. 3	1,129	176,124	July, 1982
	Daido Steel Co.	Chita	320	21,000	Feb., 1984
	Sumitomo Metal Ind.	Kokura	281	19,413	Apr., 1982
	NSC	Kamishi	202	19,643	Jan., 1979
BILLET	Osaka Steel Co.	Okajima	214	19,200	July, 1978
	Kokko Steel Co.	Main Plant	174	5,569	Aug., 1978
	Funabashi Steel Co.	Main Plant	99	6,089	Oct., 1975

Quality Improvements

The operation of a continuous casting strand to produce good quality steel on a routine basis, uniformly and reliably, can be the most valuable asset of the process. Development work continues in an effort to improve the control and production reliability, particularly with regard to internal and surface quality in terms of freedom from inclusions and cracking.

Internal quality in terms of cracking is less important for those products which have large reduction ratios. Radial cracks, center looseness or centerline cracking, minor amounts of gas evolution, and other internal defects are not deleterious in heavily rolled products. However, in the manufacture of heavy plates, these conditions can represent serious product defects.

For aluminum killed steel, subsurface inclusions are usually in the form of aluminum oxide. In some cases surface scarfing can be effective in removing these inclusions which could provide a surface defect in a final rolled product. Unfortunately, this surface conditioning results in substantial yield losses. It has been reported (4) that use of multi-port shrouded nozzles can produce a fluid flow action in slab molds, which brings inclusions in contact with the molten mold powder covering to flux and dissolve oxide particles and provide a clean sub-surface zone for which no scarfing is required.

As noted above, the control of fluid flow and promotion of separation in ladle, tundish and mold can provide a substantial reduction in large inclusions.

Cracking, depending on steel grade, can be the result of intensive cooling or deformation in the casting strand. Surface cracking can be minimized by proper control of lubrication and cooling within the mold, and cooling and alignment in the upper spray zone. Other critical factors in cracking are control of spray cooling to avoid surface temperature rebound resulting in midway cracks or non-uniformity of cooling across or along the strand. Control of temperature distribution at the straightener has also been noted as being very important.

Future Developments

The emphasis on development of further improvements in continuous casting will be focused on control systems and automation for the process. The objectives of these systems will be to ensure high quality and high productivity. They will include monitoring of metal quality and will ensure that all aspects of the process are under proper control. In addition, various direct operating parameters, such as mold and slag or flux levels in ladle, tundish and mold, will be monitored directly. Sensors will be developed to control directly the process for proper cooling in the mold, first zone and throughout secondary cooling systems. Hot inspection techniques will be developed which will provide a direct measure of strand cast product quality as it leaves the casting machine and is directed to subsequent hot rolling processing.

Further development of the continuous casting process also will be directed toward horizontal continuous casting, particularly for larger sections. Horizontal casting systems have been explored for many years, but only in the recent past have they been successfully applied to steels. Oldsmobile Division of General Motors Corporation started casting 95 mm diameter steel bars in a horizontal mold in 1969 (20,21). Casts of up to 24 hours without interruption were made which is believed to be the longest sustained continuous casting operation for steel at that time.

Voest Alpine is developing a horizontal continuous casting machine for steels, and the Nippon Kokan (NKK) Steel Company of Japan has been producing horizontally cast steel sections up to 210 mm square (22). These systems depend on (a) a refractory nozzle or "break ring" at the entrance end of a stationary water cooled metallic mold, and (b) an intermittent motion for translating the bar, each forward stroke being followed by a descrete rest or dwell. A cross section of the tundish and mold arrangement of the NKK machine is shown in Figure 14. Casting of larger sections is being explored throughout the world, as are continuous casting processes for high speed casting of thin sections. As the technologies for cleaner steels evolve, these processes will come closer to commercial adoption.

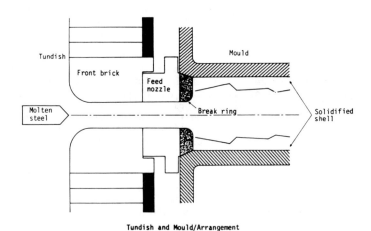

Tundish and Mould/Arrangement

Figure 14 - Tundish and Mold/Arrangement[23]

Several development groups throughout the world are now working intensively on direct casting of strip steel.

Summary

The dramatic growth and application of continuous casting reflect the primary advantages of substantially higher yield, a uniform product of excellent quality and higher productivity. An accelerating adoption of this technology is underway throughout the world. New concepts in design and operation are evolving, many of which will markedly influence steelmaking and subsequent finishing processes.

References

1. Concast A.G., World Caster Survey 1984, p. 1.

2. R. Clark, Continuous Casting of Steel, Institute for Iron and Steel Studies, New Jersey, 1970, p. 7.

3. T. Ohnishi, Nishiyama Memorial Lecture 80-9, p. 45.

4. R.W. Joseph and N.T. Mills, "A Look Inside Strand Cast Slabs", AISI, New York, May 1975.

5. G. Sellers, U.S. Patent No. 1908, (1840).

6. Sir Henry Bessemer, Excerpts from "On the Manufacture of Continuous Sheets of Malleable Iron and Steel Direct from Fluid Metal", Iron and Steel Institute Meeting of October 6, 1891 (J. Metals, Nov. 1965, pp. 1189-1191).

7. Continental Iron and Steel Trade Report, The Hague, August 27, 1970.

8. T. Haribuchi, et al., Nippon Steel technical report, Jan. 1978, No. 294.

9. R. Lincoln, "Melt Shop Retrofit at Chaparal /Goal 450,000 Tons Prime Billets", Electric Furnace Conference Proceedings-AIME, 36, 1978, pp. 168-171.

10. F. M. Wheeler and A.G.W. Lamont, "Current Trends in Electric Meltshop Design", Electric Furnace Conference Proceedings-AIME, 36, 1978, pp. 139-147.

11. C. R. Jackson and L. R. Schell, "Fifteen Years of Looking-One Year of Operating, The Start-up and First Year's Operation of Inland's No. 1 Slab Caster", Open Hearth Conference Proceedings-AIME, 57, 1974, pp. 55-66.

12. T. Haribuchi, Nippon Steel Corporation, Tokyo, private communication, May 1984.

13. R.D. Pehlke, "Reoxidation of Liquid Steel", Radex Rundschau, Heft 1/2, (Jan. 1981) pp. 349-367.

14. C.R. Jackson, Inland Steel Company, East Chicago, Indiana, private communication, May 1983..

15. Nippon Steel, Oita; Steelmaking Committee of ISIJ, No. 77.

16. M. Tokuda, et al., Iron and Steel Institute of Japan, 83, p. 919.

17. Y. Kitano, et al., Iron and Steel Institute of Japan, 84, p. 179.

18. K. Schwaha, Voest-Alpine, Linz, Austria, private communication, May, 1984.

19. T. Haribuchi, Nippon Steel Corporation, Tokyo, private communication, (from the Iron and Steel Federation of Japan, May 1984).

20. F.J. Webberre, R.G. Williams and R. McNitt, "Steel Scrap Reclamation Using Horizontal Strand Casting", GM Research Publication GMR-11, October 20, 1971.

21. W.G. Patton, "GM Casts In-Plant Scrap Into In-Plant Steel", Iron Age, (December 9, 1971) pp. 53-55.

22. F. G. Rammerstorfer, et al., "Model Investigations on Horizontal Continuous Casting", Paper No. 19, Voest-Alpine Continuous Casting Conference, 1984.

23. Private communication, U. Miyashita, NKK Nippon Kokan, Tokyo, Japan, June 1981.

Acknowledgement

This brief review of continuous casting technology has been an outcome in part of a continuing engineering education program offered at the University of Michigan each year since 1976. The contributions by C. R. Jackson and N. T. Mills of Inland Steel Company, R. Lincoln of Chaparral Steel Company, K. Schwaha of Voest Alpine A.G. and T. Haribuchi of Nippon Steel Corporation in developing and organizing material on recent developments in continuous casting of steel are gratefully acknowledged. Some highlights of their presentations are included in this manuscript.

Characterization and Utilization of Iron-Bearing Steel

Plant Waste Materials

Donald R. Fosnacht

Inland Steel Company
Research Laboratory
East Chicago, Indiana 46312

Summary

An investigation was initiated to obtain the scope of the problems associated with recycling various waste oxide materials generated at Inland Steel and to identify and test techniques that could possibly aid in increasing the amount of waste oxide materials recycled to primary operations. Based on the results of this work, several new techniques were developed and tested on a plant or laboratory scale. These include the use of cold-bond agglomeration methods for producing waste oxide agglomerates for use as a cooling agent or scrap substitute in steel-making operations, and the use of various techniques for beneficiating blast furnace flue dust and soaking pit rubble so that the upgraded materials can be utilized in sinter plant or ironmaking operations. The physical and chemical characteristics of the waste oxide materials and the results of the laboratory and plant testing are discussed in this paper.

Introduction

In the production of finished steel products, various waste oxide materials are generated (e.g., blast furnace flue dust and sludges, swarf, turnings, steelmaking dusts, soaking pit rubble, and mill scale). These materials present environmental and resource recovery problems. The amount of waste oxide material generated is dependent on both the quality of the materials used and the operating conditions employed in a given steel mill operation. At Inland Steel over 1 million Mg of waste oxide materials are generated annually. Future environmental, space, and economic considerations will necessitate new handling and treatment methods for these materials so that the contained iron content of the materials can be better utilized and the disposal problems associated with the materials lessened. At present, however, problems associated with the materials make direct recycling to blast furnace or steelmaking operations quite difficult.

Problems Associated With Direct Recycling of Waste Materials

Several problems currently restrict the amount of steel plant ferrous fines that can be directly recycled back to primary steel plant operations.[1] Some materials (e.g., blast furnace flue dust and sludges, and steelmaking dusts) contain tramp elements (e.g., Zn, Pb, S, P, Na, K) which are both undesirable and difficult to remove. Many materials are so fine (less than 0.01 mm) that they present difficulties in handling and assimilation into normal operations. Transfer of these fine materials from dust collection equipment to appropriate transportation media can lead to serious clean-up problems as dry dusts can easily become airborne and lead to emission problems. Wet-collected dusts are often very difficult to handle and transport, because they have poor sedimentation properties and are very difficult to dewater to satisfactory levels. Thus, "handling" and transportation of both dry and wet fines is difficult and often very expensive. Materials such as mill scale contain oil which can vaporize and lead to pollution control equipment operating difficulties and potential pollution control problems. Finally, the chemistry of some of the materials is quite variable and this can lead to difficulties in maintaining stable operation for facilities utilizing the materials.

Methods Used to Recycle Materials

Many methods have been proposed to make the waste oxide materials more amenable to recycling[1]. Cold bond agglomeration processes (e.g., Cobo, Pelletech, Grangcold, NCP, etc.), which will allow compaction of the materials into agglomerates that can be easily handled and subsequently used in various primary operations, have been proposed. Another technique used at some steel companies is to agglomerate fine steelmaking dusts in the form of uncured pellets (green balls) and use these materials in steelmaking. However, with these techniques the chemistry and variability of the dusts and other waste materials are not altered. Problems associated with using the materials containing high amounts of tramp impurities remain. In addition, potential pollution problems caused by oil-containing materials are not alleviated. High temperature agglomeration methods have also been proposed (i.e., sintering or pelletization with induration) and many steel plants today use these techniques to recycle large quantities of waste oxide materials.[2]

Other methods have been proposed or are being used to treat waste oxide materials. Various deoiling techniques which generally involve either solvent washing or thermal incineration are used to process plant mill scale (e.g., Luria, Hoesch, Oneida, Colerapa, Pedco, or Ransohoff Processes).[3] Some materials, such as blast furnace flue dust are beneficiated using hydrocyclone processing in order to remove zinc and allow recycling of the low zinc residue. Various high temperature reduction processes have been developed to both agglomerate the waste oxide materials and also remove many of the deleterious impurities. These processes generally produce a metallized product for use in either iron-making or steelmaking operations. Some processes of this type include: Sumitomo-SDR & SPM, Kawasaki, SL/RN, Waelz, BSC/BCI, and Ferrotech.[1] Even newer processes by-pass the metallized iron product and either produce a hot metal or steel equivalent (e.g., Elred, Inred, Plasmadust, INMETCO, McDowell-Wellman).[1] In general, the high temperature processes have not gained wide acceptance in the United States for processing the materials because of high capital and operating costs. The INMETCO Process, however, is being used to process high value stainless steel waste oxide materials [1].

Based on the problems generally associated with waste oxide materials and the known processes available to process the materials, four options appeared possible to increase the amount that could be recycled at Inland:

I) Limited recycling through sinter plant operations coupled with disposal in sanctioned landfills;

II) Physical beneficiation (if required) of the dust materials and increased usage of the upgraded materials in primary operations using either cold-bond or sintering techniques, with any unusable material being discarded into sanctioned landfills or sold for out-of-plant treatment;

III) Beneficiation of the materials using both chemical and physical methods (e.g., high temperature reduction followed by magnetic separation and screening) and increased usage of the upgraded materials in primary operations; and,

IV) Treatment of all waste oxide materials through a central high temperature reduction plant that would produce a metallized product, hot metal, or steel for use in either iron- or steel-making operations.

Based on the above considerations, it became very apparent that a thorough understanding of the raw material characteristics was necessary in order to determine what treatment techniques or processes could be used to treat Inland's materials.

Characteristics of Inland's Waste Oxide Materials

Experimental Methods Employed

Samples of mill scale, blast furnace flue dusts, swarf, soaking pit rubble, and steelmaking dusts were investigated by various techniques. Chemical analyses were made using standard methods, and the results reported on an elemental basis. X-ray diffraction analyses were performed using a Debye-Scherrer camera and chromium radiation. These techniques were supplemented by scanning electron microscopy, electron microprobe analysis and optical microscopy. The bulk densities of the dry-collected flue dusts and mill scale were determined by weighing the amount of material necessary to fill a 0.028 m^3 container; hand tamping was used to simulate settling. "Theoretical" densities were determined using a water pycnometer.

Particle size distribution analyses were made using various sizing techniques. Standard sieving methods were employed for mill scale, dry-collected blast furnace flue dust, swarf, and soaking pit rubble. A Cyclosizer Sub-Sieve Analyzer was used to examine the wet-collected dust materials. The steelmaking dusts were also evaluated using a Model T-AII Coulter Counter. Relative magnetic susceptibility measurements were made by means of a Satmagan saturation magnetization analyzer. "Percentage magnetic material" values were obtained as direct instrumental readings and are equivalent to that which would be obtained for samples containing various amounts of iron as magnetite dispersed in a magnetically inert material. Other testing procedures, such as magnetic separation, hydrocycloning, froth flotation, and briquetting, were also employed.

Mill Scale

This material accounts for approximately one third of the waste oxide material generated each year by Inland. It results from surface oxidation of ingots, slabs, and other steel materials in rolling and finishing operations. The surface oxide products fracture off the steel materials and are collected in scale pits. The scale is periodically cleaned from these pits and screened into fine and coarse products. The coarse (+6.35 mm) materials are used directly as blast furnace feed material. The undersize material is deoiled using a thermal incineration method and added to the sinter plant blend. Only the characteristics of this

Table I. Average Chemical Characteristics of Fine Mill Scale, Soaking Pit Rubble, and Swarf (%)

	Mill Scale	Soaking Pit Rubble	Swarf
Fe	73.7	57.0	93.2
Fe (Met)	1.2	0.4	64.9
Fe^{++}	48.5	10.5	23.2
Fe^{+++}	22.1	ND	5.8
Al	0.7	1.0	ND
Si	0.4	0.9	0.1
Ca	1.4	2.4	ND
Mg	0.1	3.4	ND
Mn	0.5	0.6	1.0
P	<0.1	<0.1	ND
S	<0.1	ND	0.2
Na	<0.1	<0.1	ND
K	<0.1	<0.1	ND
Oil	0.5	ND	<0.1
C	ND	ND	2.9

ND=Not Determined.

undersize material will be examined. The fine scale is still relatively coarse when compared to either the blast furnace dusts or the steelmaking dusts. It is dense; has a high iron content (e.g., 73.7% Fe); and is low in tramp impurities. It does contain a significant quantity of oil which necessitates a deoiling operation prior to its addition to the sinter

Table II. Physical Characteristics of Fine Mill Scale

Bulk Density = 2130 kg/m^3
Theoretical Density = 4300 kg/m^3
% Magnetic Material = 48.0

Size (mm)	2.38	1.41	1.00	0.59	0.297	0.210	0.149	0.074	0.044	-0.044
% Retained	7.8	8.4	11.8	4.6	36.2	14.2	9.2	1.5	5.8	0.5

plant blend. The material consists of metallic iron, various iron oxides, and gangue material. The predominant iron oxide phases are wustite and magnetite. Minor amounts of hematite are also present. Since the scale consists largely of metallic iron and partially oxidized iron oxides, a significant quantity of heat is released when the material is sintered due to oxidation of these phases to hematite. A significant portion of the material is magnetic due to its iron and magnetite contents.

Soaking Pit Rubble

This material is a by-product of the ingot heating furnaces. As with mill scale, oxidation of the ingot surface occurs and during processing the oxide skins break off and fall to the bottom of the soaking pit. The residence time and thermal conditions in the pit are such that these oxide materials fuse together and partially react with the refractory materials in the furnace. The pits are cleaned of this material when they periodically dug out for reline. During the course of material removal, the fused oxides are often contaminated by brickwork and other refuse (see Figure 1). The iron-bearing components in the rubble are a relatively

Figure 1 - Soaking Pit Rubble as Viewed at Collection Pit Under Soaking Pit Battery

pure mixture of wustite, magnetite, and some hematite. Magnetic separation of the material shows that both magnetic and non-magnetic iron oxides are present. The chemical analysis of the rubble is shown in Table I. As can be seen, the amount of tramp impurities such as alkali and phosphorus is quite low. Compared to other waste oxide materials, the particle size is quite coarse with over 74% of the material greater than 7.94 mm. Overall, the soaking pit rubble materials are a relatively coarse, high iron, largely tramp-free materials which are contaminated by refractories.

Table III. Size Analysis of Soaking Pit Rubble

Size (mm)	254	76.2	7.94	-7.94
% Retained	6.4	28.5	38.9	26.4

Swarf

This material is generated as a by-product of surface defect removal operations from billets and other steel products. The swarf is obtained as a high iron content, dry, largely oil-free, turning type material. It contains greater than 93% total iron and over 63% of this iron occurs in metallic form. The major deleterious impurity found in the material is sulfur, and for the samples tested the average sulfur content was 0.18%. X-ray diffraction analysis indicates that the predominant phases present in the material are metallic iron, wustite, and magnetite. Based on the average chemical analysis and the assumption that the amount of hematite present is negligible, it appears that there is roughly one-third as much wustite and one-eighth as much magnetite as metallic iron. This indicates that the material occurs largely in a highly reduced state. The screen analysis indicates that 90% of the material is greater than 0.297 mm and less than 2.38 mm.

Table IV. Size Analysis of Swarf Material

Size (mm)	6.35	4.76	2.38	1.68	1.41	1.19	0.59	0.42	0.297	0.21
% Retained	0.6	0.4	6.6	10.5	7.3	7.7	30.8	18.2	11.1	3.8

Size (mm)	0.149	-0.149
% Retained	1.4	1.6

Blast Furnace Flue Dusts

Dry-Collected Dusts. Two types of dry-collected dusts are generated at Inland: material from our large diameter No. 7 Blast Furnace and slightly different material from our older, smaller furnaces. The older furnaces produce a dust material slightly richer in iron and slightly lower in carbon than the materials produced at No. 7 Blast Furnace. Both materials are multi-component physical mixtures of degraded blast furnace burden materials (e.g., coke, pellets, mill scale, BOF slag, ore, sinter, and limestone) with a variable chemistry. The diversity of the mixture is reflected in the x-ray diffraction results which show that the significant phases present are hematite, magnetite, graphite, calcium carbonate, wustite, and silica. The bulk density of the material depends on the chemical composition (i.e., more carbon components results in lower bulk

Table V. Chemical Analysis of Blast Furnace Flue Dusts (%)

	Dry Collected		Wet-Collected	
	Older Furnaces	No. 7 BF	Older Furnaces	No. 7 BF
Fe_{++}	30.6	13.9	25.6	21.6
Fe	12.8	1.6	5.5	ND
C	31.3	55.8	42.3	49.3
Si	3.0	2.3	3.0	2.8
Al	0.9	0.7	1.1	1.3
Ca	5.1	4.2	3.6	1.7
Mg	1.2	1.0	1.1	0.5
Mn	0.6	ND	0.3	ND
Cr	<0.1	ND	ND	ND
S	0.5	0.6	0.4	ND
Zn	0.1	0.1	0.4	1.2
Pb	0.1	ND	0.1	ND
Na	0.1	0.2	0.1	ND
K	0.3	0.1	0.2	ND
P	0.1	<0.1	0.1	ND

ND=Not Determined.

density), but on average is about half that of mill scale. The chemical variability is quite high for the dust materials produced on either type of furnace (e.g., iron and carbon contents can vary by ±10%). The size distribution of particles in the dust materials is broad and ranges from a top size of approximately 2.4 mm to particle sizes less than 0.014 mm. All the components of the dust materials can be found over the whole particle size range, but some segregation according to particle size can be seen. The carbon components concentrate in the coarser size fractions (over 70% of the total carbon is found in the particle fractions greater than 0.149 mm). The iron components are more evenly distributed throughout the particle size range and only 50% of the iron components are greater than 0.149 mm. Most of the zinc contained in the dust is

Table VI. Physical Properties of Dry Collected Blast Furnace Flue Dust from the Older Blast Furnaces

Bulk Density (kg/m^3) = 980
Theoretical Density (kg/m^3) = 2700

Size (mm)	2.38	1.68	1.19	0.84	0.59	0.42	0.297	0.21
% Retained	3.4	1.6	2.1	3.6	7.7	19.4	12.4	19.6
Size (mm)	0.149	0.074	0.047	0.036	0.026	0.018	0.014	-0.014
% Retained	16.5	0.2	7.3	1.9	1.2	0.9	0.9	1.3

Comment: Size distribution for No. 7 Blast Furnace is comparable.

found in the smaller size fractions (over 62% of the total zinc is found in the size fraction less than 0.149 mm). The results shown in Table VII are based on samples of dust from the older furnaces. The dust from No. 7 BF behaves similarly.

Table VII. Segregation of Iron, Carbon, and Zinc According to Size for Dry-Collected Blast Furnace Flue Dust from the Older Furnaces

Screen Size (mm)	Chemical Analysis of Fraction (%)		
	Fe	C	Zn
0.59	11.0	63.8	<0.1
0.297	16.6	50.0	0.1
0.210	27.2	33.7	0.1
0.149	28.8	31.5	0.1
0.074	32.8	28.5	0.1
0.044	27.0	21.4	0.2
-0.044	57.8	10.8	0.4

Wet-Collected Dusts. The wet-collected dust materials are similar to the dry-collected dusts. Similar iron oxide, carbon, and gangue components are found in each. The materials contain some magnetic iron oxides, as indicated by x-ray diffraction and Satmagan testing. The dust is generally smaller in size consist and contains (on average) more carbon than the dry dust. Over 65% of the material is less than 0.07 mm and 31% less than 0.025 mm. These materials are concentrated in Dorr thickeners

Table VIII. Size Analysis of Wet-Collected Blast Furnace Flue Dust

No. 7 Blast

Size (mm)	0.149	0.105	0.074	0.053	0.044	0.037	0.025	-0.025
% Retained	17.2	14.5	12.1	9.6	6.2	3.8	5.4	31.0

Older Furnaces

Size (mm)	0.149	0.105	0.074	0.053	0.044	0.037	-0.037
% Retained	20.1	14.5	17.4	8.4	4.6	8.6	26.4

to an average water content of 58%. The water content is quite variable and can range from as low as 45% to greater than 70%. The material must be dewatered prior to its use. Like the dry dust, the chemistry of the material is variable (similar to that for the dry-collected dust). Segregation of some chemical species according to particle size is found for these dust materials. The zinc materials concentrate in the lower particle size range (less than 0.025 mm). The majority of the carbon components are found in the particle size range greater than 0.053 mm.

Table IX. Segregation of Iron, Carbon, and Zinc According to Size for the Wet-Collected Blast Furnace Flue Dust from No. 7 Blast Furnace

Screen Size (mm)	Chemical Analysis of Fraction (%)		
	Fe	C	Zn
0.149	10.2	84.0	0.1
0.105	11.0	76.9	0.1
0.074	29.7	69.2	0.1
0.053	28.1	65.8	0.1
0.044	26.2	51.2	0.1
0.037	35.2	36.6	0.1
0.025	44.6	25.4	0.1
-0.025	41.4	14.1	1.5

Steelmaking Dusts

Dry-Collected Dusts. At Inland, electric furnace and open hearth dusts are collected in the dry state using a baghouse and electrostatic precipitator, respectively. These dusts are similar in mineralogy,

Table X. Average Chemical Analysis of Steelmaking Dusts (%)

	Open Hearth	Electric Furnace	BOF-OG	BOF-OH
Fe_{++}	52.0	32.8	57.4	52.8
Fe_{+++}	1.7	1.1	36.2	16.5
Fe	ND	ND	ND	35.8
Si	0.2	1.1	0.9	0.8
Al	0.1	0.3	0.1	0.2
Ca	1.0	6.6	4.2	6.4
Mg	0.4	1.6	0.7	1.6
Mn	0.5	3.5	1.3	1.1
S	1.4	0.5	0.2	0.3
Zn	10.6	10.3	3.2	4.1
Pb	1.2	2.1	1.0	0.4
Na	0.5	0.9	0.1	0.1
K	0.7	1.0	0.1	0.1
P	0.1	0.1	0.1	0.1

magnetic properties, and size consistency, but have somewhat different chemistries. X-ray diffraction analyses show that the predominant phases in the materials are magnetite, hematite, and zincite. The materials are very fine as over 97% of the material is less than 0.064 mm. The open

Table XI. Size Analysis of Steelmaking Dusts*

Flue Dust	% of Total Dust Greater Than Particle Diameter (mm)							
	0.003	0.004	0.005	0.006	0.008	0.016	0.032	0.064
Open Hearth	100.0	78.8	65.0	54.0	44.4	21.0	9.9	1.6
Electric Furnace	77.8	66.3	59.1	53.1	47.8	34.5	19.8	2.2
BOF-OG	80.3	68.7	60.0	52.1	45.3	28.8	13.4	2.5
BOF-OH	82.6	70.1	60.5	53.5	48.4	33.2	16.5	2.9

*Coulter counter used for this analysis.

hearth dust is richer in iron than the electric furnace dust (e.g., 52% versus 33%) and contains less flux materials. The zinc concentrations for these materials depend on the amount of zinc-bearing scrap charged into the furnaces. When high levels are employed, the zinc content of the dusts can exceed 20%. Conversely, when little zinc-bearing scrap is charged, zinc contents below 5% generally result.

Wet-Collected Dusts. In both of Inland's BOF operations, Venturi scrubbers are used to collect the fine dusts generated during refining. Two different hood systems are employed at our furnaces. No. 4 BOF has the traditional open hood (OH) system which allows some air to mix with the off-gas. The newer No. 2 BOF has a closed hood off-gas system (OG) which greatly reduces air infiltration into the off-gas system. As a con-

3) The material from the first belt discharged onto a second belt and then over a magnetic head pulley.

4) The magnetic materials from the two separations were then combined and screened to produce three iron oxide products: (a) +76.2 mm, (b) -76.2 mm +9.53 mm, and (c) -9.53 mm.

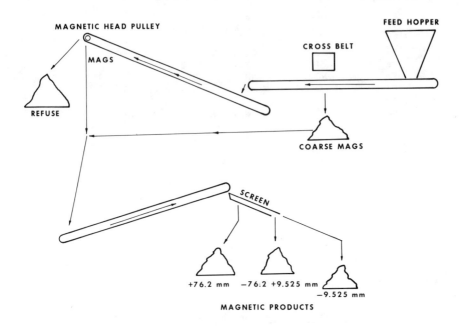

Figure 2 - Flowsheet Evaluated for Beneficiating Soaking Pit Rubble

Visual examination of the product piles indicated that roughly 80% of the original material was obtained as iron oxide products. Approximately 65% of the product was +9.53 mm with 50% greater than 76.2 mm. The iron oxide materials obtained were clean and refractory-free. Chemical analysis of the products indicated that the coarse materials contained greater than 65% iron and were free of tramp impurities. The fine material (-9.53 mm) contained in excess of 40% iron and was similar in chemistry to plant sinter. Based on the results of the plant test, elements of the proposed flowsheet were incorporated into the general processing scheme for treating plant mill scale, and, today, soaking pit rubble is being processed. The coarse materials are used as blast furnace feed and the fines are added to the sinter blend.

sequence, the off-gas from the OG system has a higher reduction potential. Because of this, the iron oxide forms in the No. 2 BOF dust are in a more reduced state. The main iron form in this material is wustite; in contrast, the No. 4 BOF material contains mostly a mixed iron spinel. The chemistry of the dusts is similar. Both materials contain over 52% iron and 3-4% zinc and have similar size consistencies as well. The size distributions are comparable to that found for the dry-collected steelmaking dusts.

Techniques Developed to Treat the Waste Oxide Materials

Current Recycling Methods

Inland currently uses its sinter plant as the main means to recycle the waste oxide, slag, and coke breeze materials generated in the plant. Since several of the waste materials are highly variable in chemistry, a materials blending and bedding program was developed to form a well-mixed blend of materials as feedstock to the sintering process. This was done to allow production of more consistent quality sinter and to allow better control of the sintering process itself. At the current time, thermally deoiled mill scale, raw mill scale, fine slag, and pellet fines are fed in a rateable fashion into a pug mill mixer and the mix is then bedded for future systematic reclamation and use at the sinter plant.[2] In addition, a separate bedding and blending facility is used to treat blast furnace flue dust (both dry- and wet-collected material) in order to reduce its chemical variability. This is especially important for the flue dust as it supplies part of the fuel requirement for the sintering process and high chemical variability causes uneven firing conditions during sintering. This in turn can drastically impact both plant productivity and sinter product quality.

Based on the above and the characterization work previously described, several alternative flowsheets were developed to process some of the waste oxide materials. These flowsheets were then tested at various scales. The results of the research work and the rationale behind the flowsheet development are described below.

Recovery of the Iron Units in Soaking Pit Rubble

Although the intrinsic value of the iron units contained in the soaking pit rubble had been recognized for some time, a method of recovering these values had not been developed and, as a consequence, the rubble was generally used as landfill. The primary problems associated with recycling the material, as noted previously, are its contamination with spent refractory materials and plant refuse. In addition, the iron oxide materials in the rubble have a wide size distribution and proper sizing of the materials is required to use the iron oxide materials in primary operations. Based on the physical and chemical characteristics of the material, it was felt that a simple flowsheet (Figure 2) incorporating screening and magnetic separation techniques should allow effective reclamation of the iron oxide values from the rubble. This flowsheet was tested in the plant and it involves the following steps:

1) Material was fed by front end loader to a feed bin which discharged the material onto a flat belt.

2) The material on the belt passed under a cross belt magnetic separator which removed some of the coarse magnetic iron oxide materials.

Table XII. Chemical Analysis of Products (%)

	+76.2 mm	-76.2 +9.53 mm	-9.53 mm
Fe	67.0	65.6	41.2
Ca	1.1	1.9	10.7
Mg	0.5	0.8	2.7
Si	1.5	1.6	4.9
Mn	0.7	0.7	1.6
Pb	0.02	0.02	0.02
P	0.03	0.02	0.02
S	0.03	0.03	0.09

Use of Waste Oxides Briquettes in Steelmaking Operations

In order to increase the amounts of fine steelmaking dusts recycled in the plant and to allow recycling of mill scale without prior deoiling or oxidation, a process was developed to agglomerate the waste oxides and use the agglomerates in steelmaking operations as an ore substitute, refining aid, or as a final coolant.[4-5] For this process, various waste oxide materials (e.g., mill scale, steelmaking dusts, swarf, blast furnace flue dust, etc.) are blended and then briquetted using a hydrated lime/molasses or other appropriate binder. Agglomeration tests have been conducted on laboratory, pilot, and commercial scales. The test results indicate that strong, stable briquettes can be produced without using high temperature or special curing methods. A plant test was conducted to determine the feasibility of using the briquettes in Inland's various steelmaking operations. During the plant test over 700 Mg of briquettes were produced at a commercial briquetting facility and used in three different steelmaking facilities at Inland.

For these tests, a waste oxide blend consisting of 50% open hearth dust, 30% oily (-6.35 mm) mill scale, and 20% swarf was used. Approximately 5% molasses and 2% hydrated lime were added to the basic blend as binder materials. The resulting briquettes contained 67.9% Fe, 0.4% S, 2.1% lime, and approximately 1.1% Zn. The results of the test program indicate that the briquettes can be used with little difficulty under BOF or BOH steelmaking conditions, and that a satisfactory practice can be developed for using the briquettes under electric furnace conditions. In most instances, handling did not result in excessive degradation of the briquettes. Some handling problems can occur if the briquettes are stored in an area where they are exposed to rain and subsequent freeze/thaw cycles. These problems can be greatly reduced by ensuring that the briquettes are stored under cover or directly delivered to shop bins for immediate use.

At the BOF and open hearth shops, the briquettes appear to be a viable substitute for iron ore pellets and hard ore, and a partial substitute for coolant scrap. At the electric furnace, the briquettes were found to make the bath more reactive and increase the kinetics of carbon removal from the steel. Because of the limited duration of the test, no attempt was made to optimize the steelmaking practices or the briquette composition. Based on other laboratory and pilot scale testing, it is important to note that the composition of the briquettes can be adjusted to provide a range of chemical and physical properties, such as iron content, carbon content, density, and thermochemical effect, as required by the individual steelmaking shop.

Table XIII. Proportions of Various Materials Used During Briquette Trials

Waste Oxide Blend: 50% Open Hearth Dust
                    30% -6.35 mm Oily Mill Scale
                    20% Swarf

Briquette Make-up: 93% Waste Oxide Blend
                    5% Molasses
                    2% Hydrated Lime

Table XIV. Chemical Analysis of Briquettes (%)

Fe	67.9
Fe(met)	17.7
Fe^{++}	21.9
Mn	0.57
Ca	1.43
Mg	0.17
Pb	0.14
Zn	1.12
S	0.40

Beneficiation of Blast Furnace Flue Dust

Several factors limit the amount of flue dust that can be utilized in sintering operations. First, the variability of the material itself is a problem, as mentioned previously. Second, is the heat balance of the strand. A proper heat balance must be maintained in order to assure efficient sinter plant operation. Since mill scale, blast furnace flue dust, and coke breeze are the main fuel sources at Inland's sintering operation, when one of the components is increased in the sinter blend it becomes necessary to curtail the use of others in order to maintain a proper thermal balance on the strand. This factor can cause a limitation on the amount of flue dust that can be recycled. Third, the flue dusts have significant levels of tramp elements. The amount of dust that can be used will be dependent on the zinc and alkali loading considerations for the furnaces where the sinter is to be employed. Finally, both the dry- and wet-collected dusts are relatively fine in particle size. Sinter plant efficiency can be detrimentally affected when the amount of fines used becomes too large because permeability is adversely affected. This can lead to decreased sinter quality and sinter strand productivity. Currently, Inland blends and beds the flue dust in order to reduce its chemical variability. The other factors noted are not impacted by our current handling techniques. Research was undertaken to determine if the carbon and iron units could be separated using various physical beneficiation methods. The results of this work have been previously reported [6] and the details will not be given here. The results of the test work indicate that carbon-rich concentrates containing greater than 70% C and iron-rich concentrates containing greater than 45% iron and less than 5% carbon can be produced using a combination of screening, low and high intensity magnetic separation, and flotation techniques. The iron concentrates can be used in sintering operations as is or after they have been micropelletized to increase their overall particle size. The use of size separation techniques allows removal of a significant amount of the contained zinc. The carbon concentrates can be used as a fuel in the plant.

Concluding Remarks

Based on the results of the research noted and the efforts of our operating departments, Inland has made steady progress in increasing the amount of waste oxide materials that are recycled on a routine basis. Most of the mill scale, fine soaking pit iron oxides, swarf, and blast furnace flue dusts are recycled through the sinter plant at this time. The development of the cold-bonded briquettes for use in steelmaking should allow increased recycling of steelmaking dust in the future by this route if future tests are promising and economic operating practices for using the briquettes are developed. Further work to develop more techniques which will allow even more recycling of the waste oxides is still underway.

Acknowledgment

The author wishes to recognize Mr. J. P. Gindl for his contributions in conducting the bulk of the experimental work during the characterization studies and during flowsheet development. The cooperation of the personnel from Inland's various primary operating departments was greatly appreciated during the plant trials. Finally, the Inland Steel Company is gratefully acknowledged for granting permission for publication of this work.

References

[1] D. R. Fosnacht, "Recycling of Ferrous Steel Plant Fines--State of the Art," Iron & Steelmaker, April, 1981, pp. 22-26.

[2] J. A. Ricketts and H. C. Boehme, "Sintering Without Ore Fines," Iron & Steelmaker, October, 1983, pp 17-23.

[3] "Hydrocarbon Removal from Ironbearing Wastes," Collaborative Technology Report No. 1, American Iron and Steel Institute, 1984.

[4] D. R. Fosnacht, Patent Pending, U. S. Application 314,107.

[5] D. R. Fosnacht, Canadian Patent 1,154,595, Oct. 4, 1983.

[6] D. R. Fosnacht, "Beneficiation of Dry- and Wet-Collected Blast Furnace Flue Dusts for Potential Use in Steel Plant Primary Operations", Transactions-ISS, I, 1982, pp. 21-28.

Author Index

Altman, R., 97
Archer, G., 3

Benn, F. W., 251
Bhappu, R. R., 397
Borowiec, K., 79
Brimacombe, J. K., 327
Bustos, A. A., 327

Chan, B., 117
Chang, Y. A., 41
Chaubal, P. C., 63

Davenport, W. G., 397
Davey, T. R. A., 23

Foot, D. G., 251
Forsen, O., 353
Fosnacht, D. R., 479

Goel, R. P., 97
Guzman, S. S., 49

Hager, J. P., 231, 277
Hettula, E., 353
Howie, B. S., 147
Huiatt, J. L., 251
Hsieh, K., 41

Jha, M. C., 179
Jones, W. D., 379
Jorgensen, D., 327

Kammel, R., 133
Kellogg, H. H., 3
Khalafalla, S. E., 433
Kim, Y. H., 3

Landau, U., 133
Li, T., 277
Lilius, K., 353
Liu, Q. G., 387
Loutfy, R. O., 263

Makinen, J. K., 289
Mateer, M. W., 165
May, W. A., 231
Meyer, G. A., 179
Moore, J. J., 417
Munroe, N. D. H., 289
Murawa, M. M., 417

Nagano, T., 311
Nakamura, T., 117
Natalie, C. A., 277

O'Keefe, T. J., 165
Opie, W. P., 379

Pahlman, J. E., 433
Parameswaran, K., 97
Pehlke, R. D., 457

Rajcevic, H. P., 379
Reid, K. J., 417
Richards, G. G., 327
Rosenqvist, T., 79
Rupert, M. C., 231

Sabacky, B. J., 179
Sharma, R. A., 147
Sohn, H. Y., 63
Stapurewicz, T., 3
Stavropoulos, G., 97
Szesny, B., 133

Themelis, N. J., 289
Tiwari, B. L., 147
Toguri, J. M., 117

Verdonik, D., 3

Willis, G. M., 23
Winand, R., 209
Worrell, U. L., 387

Young, S., 263

Subject Index

Activation energy of
  chalcopyrite oxidation, 63
Activity
  determination of, 25
  of iron in slags, 117
  in pseudobinary sulfide melts, 32
Additives
  in zinc electrolysis, 165
  organic in tin electrolysis, 133
Agglomeration
  steel plant wastes, 490
Air injection
  in copper converting, 327
Aluminum
  electrolysis, 264
  scrap, magnesium from, 147
Antimony in
  copper anodes, 353
  zinc electrolyte, 169
Arsenic in copper anodes, 353

Beryl
  flotation of, 259
Beta-naphthol
  in zinc electrolysis, 172
Binary systems
  Bi-Sb, 8
  $CaCl_2$-$MgCl_2$, 151
  Cu-Bi, 10
  Cu-Sb, 5
  Li_2SO_4-Ag_2SO_4, 389
Bismuth
  removal from copper, 365
  in copper anodes, 353
Blast furnace
  slag viscosity, 97
  smelting copper, 376
  steel plant dusts, 484

Carbothermic reduction
  of chromite, 422
  of tacomite, 422
  promoters for, 451
Chalcopyrite
  oxidation kinetics and
    thermodynamics of, 63
Chloride hydrometallurgy, 209
Chromia - carbon reactions, 437
Chromite reduction
  in plasma, 422
  in vacuum furnaces, 441

Cobalt
  electrodeposition, 211
  in zinc electrolyte, 167
Continuous casting
  machine design, 467
  plant layout, 465
  process, 467
  productivity, 473
  steel, 457
Copper
  anode impurities, 353
  bismuth removal from, 353
  chloride leaching, 218
  converting, 327
  electrolysis, 226, 368
  electrowinning, 217
  pyrometallurgical purification reduction
    with Fe alloy, 122
  solidification, 358
  transfer in Fayalite slags, 123
Copper converting, 327
  high pressure air in injection, 332
  reaction kinetics, 335
  tuyere blockage, 332
Copper smelting
  continuous of copper sulfide, 311
  secondary, 379
Copper sulfide concentrates
  continuous smelting of, 289
  fluid-bed roasting of, 311
Cryolite bath
  effect of LiF on, 266

Electrochemical sensors, 387
Electrolyses of
  cobalt, 211
  copper, 216, 368
Electrolytic extraction of
  magnesium from aluminum scrap, 147
  copper in chloride solutions, 216
  cobalt in chloride solutions, 211
  manganese in chloride solutions, 212
Electrowinning
  copper, 217
EMF study
  Fe-Ni-S-O system, 41
Emission control in copper smelting, 380
Entrainment in fluid-beds, 278

Fayalite slags
  copper and iron transfer in, 117
  oxygen pressure effect, 117
  iron activity in, 117
Ferrochrome production, 433
Fish Creek deposit, 251
Flash smelting
  mathematical model, 300
  Outokumpu process, 289
  rate phenomena, 300
Flotation of
  beryl, 259
  fluorite, 251
  muscovite, 257
  silicate, 258
Fluidized-bed reactor
  nickel chloride pyrohydrolysis, 179
  copper sulfide roasting, 65, 277
Fluid bed roasting
  elutriation, 278
  entrainment, 278
  pyrohydrolyser, 188
  residence times, 277
  sulfide roasting, 277
Fluorite recovery
  Fish Creek deposit, 251
  flotation, 254
Fluospar, 252
Free energy of formation
  magnesium titanates, 87
  tin chloride, 49

Glue in zinc electrolysis, 170

Hall cell electrolytes, 263
Horizontal continuous casting, 457
Hydrometallurgy, Chlorides, 209

Ilmenite smelting, 79
Inland Steel Company, 479
Iron in fayalite slags, 117
Iron bearing steel plant wastes, 479

Kinetics of
  chalcopyrite oxidation, 63
  copper converting, 327

Lithium fluoride
  effect in cryolite baths, 266

Magnesium
  from aluminum alloy scrap, 147
  impurity in titaniferrous slags, 79

Magnetic analyzer, 102
Magnetite
  concentration, 97
Manganese
  electrodeposition, 224
Margules equations, 4
Mitsubishi smelting process, 312
Morphology of
  cobalt deposits, 216
  copper anodes, 371
  zinc deposits, 169
Muscovite
  flotation of, 257

Naoshima smelter, 277
Nickel
  chloride refining process, 180
Nickel chloride
  pyrohydrolysis, 179
  refining, 180

Organic additives
  for tin electrolysis, 135
  for zinc electrolysis, 169
Outokumpu smelting process, 290
Oxidation of chalcopyrite, 63

Perrin Process, 436
Plasma reduction of minerals, 417
Potentiometric studies, 173
Pyrohydrolysis of nickel chloride, 179

Quartz
  depression of, 255
Quarternary system
  Cu-Ni-O-Sb, 375
  Fe-Ni-S-O, 41

Recycling steel plant wastes, 480
Reductants for Ferrochrome, 433
Residence times
  during fluid bed roasting, 277
Roasting, fluid-bed
  copper sulfide concentrates, 65, 277
  residence time, 277
Rotary kiln studies, 235

Scanning electron microscope
  for zinc deposits, 169
  in copper refining, 355
Secondary copper smelting, 379
  control of hydrocarbons in, 380
  control of solid emissions, 381

Silver
    extraction, 231
    chloride vaporization, 231
    chloride vapor pressure, 237
Silver chloride
    vapor complex, 231
    vapor pressure, 237
    volatilization, 235
    thermodynamics, 237
Simplex process, 436
Slags
    cleaning for metal recovery, 123, 383
    Fayalite, 117
    ionic nature of, 23
    metal solubilities in, 23
    oxidation states in, 102
    thermodynamics of, 25
    titaniferrous, 79
    viscosity measurements of, 97
Smelting
    continuous of copper sulfide, 311
Solidification
    copper anodes with As and Sb, 353
Solvent extraction
    in nickel recovery,
Steel plant wastes utilization, 479
Steelmaking
    continuous casting, 457
Sulfatization of impurities
    in titaniferrous slags, 79
Sulfide roasting
    fluid bed, 277
Sulfur dioxide detection, 387
Sulfur trioxide detection, 387

Taconite, carbothermic reduction, 422
Ternary systems
    Cu - Bi - Sb, 12
    Cu - Ni - O, 355
    KCl - NaCl - $CuCo_2$, 151
    TiO_2 - Fe_2O_3 - MgO, 81
Thermochemical properties of
    Cu - Sb - Bi system, 3
Thermodynamics of
    metal, matte and slag solution, 23
    chalcopyrite oxidation, 63
Tin
    electrolytic production of, 133
    formation of anodic films, 139
    in copper smelting, 379
    organic additives in electrolysis of, 133

Tin chlorides
    disproportionation, 49
    free energy of formation, 49
    thermodynamic data on, 50
Titaniferrous slags
    magnesium in, 79
    sulfatization of impurities in, 79
Transpiration technique, 233

Vacuum furnace reactor
    for ferrochrome, 379
Vapor pressure
    silver chloride, 231
    transpiration technique, 233
Vaporizing
    silver chloride, 231
Viscosity
    blast furnace slags, 97
Volatility of
    silver chloride, 231
Voltammetry, cyclic
    in zinc electrolysis, 167

Zinc electrolysis
    deposit structure, 165
    effect of cobalt, 167
    effect of B-Naphthol, 172
    effect of glue, 169
    effect of antimony, 167
    using cyclic voltammetry, 167
Zinc sulfate electrolyte, 165